Marine Auxiliary Machinery

Seventh edition

H. D. McGeorge
C Eng, FIMarE, MRINA, MPhil

BUTTERWORTH
HEINEMANN

OXFORD   AUCKLAND   BOSTON   JOHANNESBURG   MELBOURNE   NEW DELHI

Butterworth-Heinemann
Linacre House, Jordan Hill, Oxford OX2 8DP
225 Wildwood Avenue, Woburn, MA 01801-2041
A division of Reed Educational and Professional Publishing Ltd

℟ A member of the Reed Elsevier plc group

First published 1952
Second edition 1955
Third edition 1963
Fourth edition 1968
Reprinted 1971, 1973
Fifth edition 1975

Reprinted 1976, 1979
Sixth edition 1983
Reprinted 1987
Seventh edition 1995
Paperback edition 1998
Reprinted 1999, 2000 (twice)

British Library Cataloguing in Publication Data
Marine Auxiliary Machinery – 7th rev. edn
  I. McGeorge, H. David
  0623.8

Library of Congress Cataloguing in Publication Data
McGeorge, H. D.
  Marine Auxiliary Machinery/H. D. McGeorge – 7th edn
  Includes bibliographical references and index
  1. Marine engines.   2. Marine machinery   I. Title
  VM765.M38 1995
  623.8'6—dc20   95–3360 CIP

ISBN 0 7506 4398 6

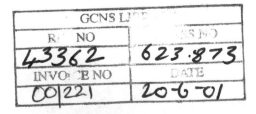
Typeset by Vision Typesetting, Manchester
Printed and bound in Great Britain by MPG Books Ltd, Bodmin, Cornwall

Contents

Preface

The preparation of the seventh edition of this established book on marine auxiliary machinery has necessitated the removal of some old material and the inclusion of new topics to make it relevant to the present day certificate of competency examinations. It is hoped that the line drawings, many of which were provided by Mr R. C. Dean, a former colleague in London, will be useful for the certificate of competency and other examinations. The majority of other illustrations and much of the basic text have been provided over the years by the various firms listed in the Acknowledgements. I am grateful to those firms who have supplied me with material added in this edition.

H. D. McGeorge

Acknowledgements

The author and publishers would like to acknowledge the cooperation of the following who have assisted in the preparation of the book by supplying information and illustrations.

Alfa-Laval Ltd.
APE-Allen Ltd.
ASEA.
Auto-Klean Strainers Ltd.
Bell & Howell Cons. Electrodynamics.
Blakeborough & Sons Ltd.
Blohm & Voss A.G.
Brown Bros & Co. Ltd.
B.S.R.A.
Bureau Veritas.
Caird & Rayner Ltd.
Caterpillar Traction Co.
Chubb Fire Security Ltd.
Clarke, Chapman Ltd.
Cockburn-Rockwell Ltd.
Crane Packing
W. Crockatt & Sons Ltd.
R. C. Dean
Deep Sea Seals Ltd.
The Distillers Co. Ltd (CO$_2$ Div.).
Donkin & Co. Ltd.
Fire Fighting Enterprises Ltd.
Fisher Control Valves Ltd.
G. & M. Firkins Ltd.
Foxboro-Yoxall Ltd.
G.E.C.-Elliott Control Valves Ltd.
Germannischer Lloyd.
Glacier Metal Ltd.
Hall Thermotank Ltd.
The Henri Kummerman Foundation
Howden Godfrey Ltd.
Hamworthy Engineering Ltd.
Harland & Wolff Ltd.
John Hastie & Co. Ltd.
Hattersley Newman Hender Ltd.
Hawthorn Leslie (Engineers) Ltd.
Hindle Cockburns Ltd.
James Howden & Co. Ltd.
F. A. Hughes & Co. Ltd.
W. C. Holmes & Co. Ltd.
Howaldtswerke-Deutche Werft A.G.
Hydraulics & Pneumatics Ltd.

IMI-Bailey Valves Ltd.
IMO Industri.
International Maritime Organisation.
KaMeWa.
Richard Klinger Ltd.
Kockums (Sweden).
K.D.G. Instruments Ltd.
Lister Blackstone Mirrlees Marine Ltd.
Lloyds Register of Shipping.
Mather & Platt Ltd.
Metering Pumps Ltd.
Michell Bearings Ltd.
Nash Engineering (G.B.) Ltd.
Navire Cargo Gear Int. AB.
Norwinch.
Peabody Ltd.
Penwalt Ltd.
Peter Brotherhood Ltd.
Petters Ltd.
Phillips Electrical Ltd.
Thos. Reid & Sons (Paisley) Ltd.
Ross-Turnbull Ltd.
Royles Ltd.
Ruston Paxman Diesels Ltd.
Simplex-Turbulo Marine Ltd.
Serck Heat Exchangers Ltd.
Spirax-Sarco Ltd.
Sofrance.
Sperry Marine Systems Ltd.
Stella-Meta Filters Ltd.
Stone Manganese Marine Ltd.
Stothert & Pitt Ltd.
Svanehoj, Denmark.
Taylor Servomax.
United Filters & Engineering Ltd.
Vickers Ltd.
Vokes Ltd.
Vosper Ltd.
The Walter Kidde Co. Ltd.
Weir Pumps Ltd.
Welin Davit & Engineering Ltd.
Wilson-Elsan Ltd.
Worthington-Simpson Ltd.

1

Main propulsion services and heat exchangers

The heat produced by running machinery, must be removed to ensure the satisfactory functioning of the equipment. Cooling is achieved primarily through circulation of water, oil and air but the abundant supply of sea water is normally reserved for use as an indirect coolant because the dissolved salts have a great potential for depositing scale and assisting in the setting up of galvanic corrosion cells. Pollution of coastal areas by industrial and other wastes has added to the problems of using sea water as a coolant.

Circulating systems for motorships

The usual arrangement for motorships (Figure 1.1) has been to have sea-water circulation of coolers for lubricating oil, piston cooling, jacket water, charge air, turbo-charger oil (if there are sleeve type bearings) and fuel valve cooling, plus direct sea-water cooling for air compressors and evaporators. The supply for other auxiliaries and equipment may be derived from the main sea-water system also.

There may be two sea-water circulating pumps installed as main and stand-by units, or there may be a single sea-water circulating pump with a stand-by pump which is used for other duties. The latter may be a ballast pump fitted with a primer and air separator. Ship side valves, can be arranged with high and low suctions or fitted to water boxes. High suctions are intended for shallow water to reduce the intake of sediment. Low suctions are used at sea, to reduce the risk of drawing in air and losing suction when the ship is rolling. A water box should be constructed with a minimum distance of 330 mm between the valve and the top, for accumulation of any air which is then removed by a vent. A compressed air or steam connection is provided for clearing any weed.

Ship side valve bodies for the sea-water inlet must be of steel or other ductile metal. Alternative materials are bronze, spheroidal graphite cast iron, meehanite or another high-quality cast iron. Ordinary grey cast iron has proved to be unreliable and likely to fail should there be shock from an impact or other cause. Permissible cast irons must be to specification and obtained from an approved manufacturer.

Bronze has good resistance to corrosion but is expensive and therefore tends to be used for smaller ship side valves. Steel is cheaper, but prone to corrosion. It may be cast or fabricated. Unprotected steel valve casings and pipes will, in

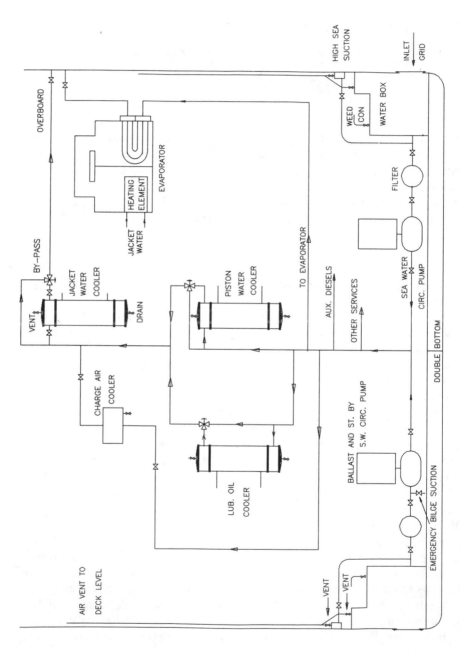

Figure 1.1 *Conventional sea-water circulation system*

the presence of sea water and bronze seats, valve lids and spindles, waste due to galvanic corrosion. However, the presence of corroding iron or steel confers benefits in sea-water systems. The metal acts as a sacrificial anode and additionally delivers iron ions which are carried through and give protection to other parts of system where they deposit.

The fresh-water circuit comprising jacket water circulating pumps, fresh-water coolers, cylinder jackets, cylinder heads, exhaust valves (if fitted), turbo-blowers and a branch to an evaporator, is under positive head, and therefore in a closed system with a header tank. It is normal for there to be a blanked connection between the sea-water system and engine jacket water circuit, for use in an emergency. If the engine pistons are fresh-water cooled, the circuit may be in parallel with the jacket circuit but it is more likely to be separate. Main and stand-by piston cooling water circulating pumps are mounted directly on the drain tank so that with flooded suctions no primer is required. The piston cooling system embraces a separate cooler, the inlet manifold, telescopic pipes, pistons, outlet manifold, drain tank and pumps.

The engine system temperatures are kept as high as practicable. The system shown has salt-water bypass valves on oil and water coolers for temperature control. These are valves controlled by thermo-pneumatic devices. It is usual to make provision for warming the fresh circulating water before the main engines are started, either by steam or by circulating from the auxiliary jacket water cooling circuit.

The auxiliary sea-water cooling circuit for generator diesel prime movers may have its own sea inlet and pumps for circulation, with a cross connection from the main sea-water circulation system. Air compressors together with the inter- and after-coolers may be supplied with sea-water cooling in parallel with the main system or alternatively, there may be crankshaft-driven pumps. Charge air coolers are sea-water circulated.

The jacket water system for generator diesel prime movers is similar to that for the main engines, usually with a separate header tank. Pumps for the services are duplicated or cross connected.

Sea-water pipes for circulation of cooling water, together with those for bilge and ballast systems, are prone to internal wastage from corrosion and erosion. External corrosion is also a problem in the tank top area. Steel pipes additionally suffer from rusting.

Control of temperature in heat exchangers

The three basic methods for controlling the temperature of the hot fluid in a heat exchanger when the cooling medium is sea-water, are:

1   to bypass a proportion or all of the hot fluid flow,
2   to bypass or limit the sea-water flow;
3   to control sea-water temperature by spilling part of the sea-water discharge back into the pump suction.

The last of these methods could be used in conjunction with one of the other two and it was resorted to when sea water was used for direct cooling of diesel engines. It enabled the sea water to be passed through jackets at a temperature warmer than that of the sea. Very cold sea water would cause severe thermal stress. The temperature of sea water for direct cooling was kept to between 40° and 49°C, the upper limit being necessary to limit scale formation.

Automatic control equipment for the system shown above, is based on using a control valve to bypass the sea water at the outlet side of the heat exchanger. This ensures that the heat exchanger is always full of sea water and is particularly important if the heat exchanger is mounted high in the sea-water system and especially if it is above the water line. Pneumatically operated valves may be fitted for temperature control, through bypassing the sea water.

The flow of hot fluid through a heat exchanger may be controlled by a similar bypass or by a control valve of the Walton wax-operated type, directly actuated by a temperature sensor.

Shell and tube coolers

Shell and tube heat exchangers for engine cooling water and lubricating oil cooling (Figure 1.2) have traditionally been circulated with sea water. The sea water is in contact with the inside of the tubes, tube plates and water boxes. A two-pass flow is shown in the diagram but straight flow is common in small coolers. The oil or water being cooled is in contact with the outside of the tubes and the shell of the cooler. Baffles direct the liquid across the tubes as it flows through the cooler. The baffles also support the tubes and form with them a structure which is referred to as the tube stack. The usual method of securing the tubes in the tube plates is to roll-expand them.

Tubes of aluminium brass (76% copper; 22% zinc; 2% aluminium) are

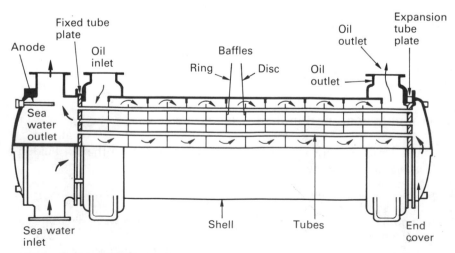

Figure 1.2   *Tube type cooler*

commonly employed and the successful use of this material has apparently depended on the presence of a protective film of iron ions, formed along the tube length, by corrosion of iron in the system. Unprotected iron in water boxes and in parts of the pipe system, while itself corroding, does assist in prolonging tube life. This factor is well known (Cotton and Scholes, 1972) but has been made apparent when iron and steel in pipe systems have been replaced by non-ferrous metals or shielded by a protective coating. The remedy in non-ferrous systems, has been to supply iron ions from other sources. Thus, soft iron sacrificial anodes have been fitted in water boxes, iron sections have been inserted in pipe systems and iron has been introduced into the sea water, in the form of ferrous sulphate. The latter treatment consists of dosing the sea water to a strength of 1 ppm for an hour per day for a few weeks and subsequently dosing again before entering and after leaving port for a short period.

Electrical continuity in the sea-water circulating pipework is important where sacrificial anodes are installed. Metal connectors are fitted across flanges and cooler sections where there are rubber joints and 'O' rings, which otherwise insulate the various parts of the system.

Premature tube failure can be the result of pollution in coastal waters or extreme turbulence due to excessive sea-water flow rates. To avoid the impingement attack, care must be taken with the water velocity through tubes. For aluminium-brass, the upper limit is about 2.5 m/s. Although it is advisable to design to a lower velocity than this — to allow for poor flow control — it is equally bad practice to have sea-water speeds of less than 1 /sec. A more than minimum flow is vital to produce moderate turbulence which is essential to the heat exchange process and to reduce silting and settlement in the tubes.

Naval brass tube plates are used with aluminium-brass tubes. The tube stacks are made up to have a fixed tube plate at one end and a tube plate at the other end (Figure 1.3) which is free to move when the tubes expand or contract. The tube stack is constructed with baffles of the disc and ring, single or double segmental types. The fixed end tube plate is sandwiched between the shell and water box, with jointing material. Synthetic rubber 'O' rings for the sliding tube plate permit free expansion. The practice of removing the tube stack and replacing it after rotation radially through 180 degrees, is facilitated by the

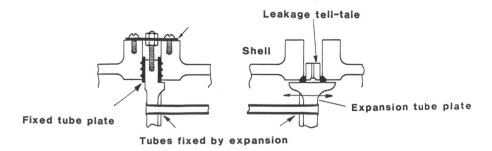

Figure 1.3 *Detail of cooler expansion arrangement*

type of cooler described. This may prolong cooler life by reversing the flow so that tube entrances, which are prone to impingement damage, become outlets.

Cooler end covers and water boxes are commonly of cast iron or fabricated from mild steel. Unprotected cast iron in contact with sea water, suffers from graphitization, a form of corrosion in which the iron is removed and only the soft black graphite remains.

The shell is in contact with the liquid being cooled which may be oil, distilled or fresh water with corrosion inhibiting chemicals. It may be of cast iron or fabricated from steel. Manufacturers recommend that coolers be arranged vertically. Where horizontal installation is necessary, the sea water should enter at the bottom and leave at the top. Air in the cooler system will encourage corrosion and air locks will reduce the cooling area and cause overheating. Vent cocks should be fitted for purging air and cocks or a plug are required at the bottom, for draining.

Clearance is required at the cooler fixed end for removal of the tube stack.

Plate type heat exchangers

The obvious feature of plate type heat exchangers, is that they are easily opened for cleaning. The major advantage over tube type coolers, is that their higher efficiency is reflected in a smaller size for the same cooling capacity.

They are made up from an assembly of identical metal pressings (Figure 1.4a) with horizontal or chevron pattern corrugations; each with a nitrile rubber joint. The plates, which are supported beneath and located at the top by parallel metal bars, are held together against an end plate by clamping bolts. Four branch pipes on the end plates, align with ports in the plates through which two fluids pass. Seals around the ports are so arranged that one fluid flows in alternate passages between plates and the second fluid in the intervening passages, usually in opposite directions.

The plate corrugations promote turbulence (Figure 1.4b) in the flow of both fluids and so encourage efficient heat transfer. Turbulence as opposed to smooth flow causes more of the liquid passing between the plates to come into contact with them. It also breaks up the boundary layer of liquid which tends to adhere to the metal and act as a heat barrier when flow is slow. The corrugations make the plates stiff so permitting the use of thin material. They additionally increase plate area. Both of these factors also contribute to heat exchange efficiency.

Excess turbulence, which can result in erosion of the plate material, is avoided by using moderate flow rates. However, the surfaces of plates which are exposed to sea water are liable to corrosion/erosion and suitable materials must be selected. Titanium plates although expensive, have the best resistance to corrosion/erosion. Stainless steel has also been used and other materials such as aluminium-brass. The latter may not be ideal for vessels which operate in and out of ports with polluted waters.

The nitrile rubber seals are bonded to the plates with a suitable adhesive. Removal is facilitated with the use of liquid nitrogen which freezes, makes

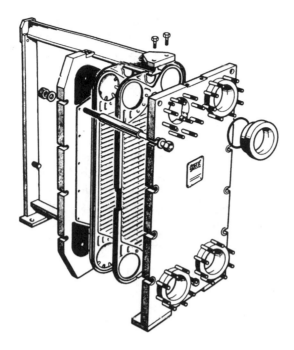

Figure 1.4a *Plate type heat exchanger*

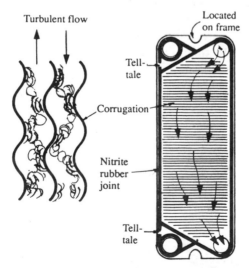

Figure 1.4b *Turbulence produced by plate corrugations*

brittle and causes contraction of the rubber seal which is then easily broken away. Other methods of seal removal result in plate damage.

Nitrile rubber is suitable for temperatures of up to about 110°C. At higher temperatures the rubber hardens and loses its elasticity. The joints are squeezed when the plates are assembled and clamping bolts are tightened after cleaning.

Overtightening can cause damage to the plates, as can an incorrect tightening procedure. A torque spanner can be used as directed when clamping bolts are tightened; cooler stack dimensions can also be checked.

Titanium

The corrosion resistance of titanium has made it a valuable material for use in sea-water systems whether for static or fast flow conditions. The metal is light weight (density 4.5 kg/m³) and has good strength. It has a tolerance to fast liquid flow which is better than that of cupro-nickel. It is also resistant to sulphide pollution in sea water. While titanium has great corrosion resistance because it is more noble than other metals used in marine systems, it does tend to set up galvanic cells with them. The less noble metals will suffer wastage unless the possibility is reduced by careful choice of compatible materials, coating of the titanium, insulation or the use of cathodic protection.

Charge air coolers

The charge air coolers fitted to reduce the temperature of air after the turbo-charger and before entry to the diesel engine cylinder, are provided with fins on the heat transfer surfaces to compensate for the relatively poor heat transfer properties of air. Solid drawn tubes with a semi-flattened cross section, have been favoured (Figure 1.5a). These are threaded through the thin copper fin plates and bonded to them with solder for maximum heat transfer. Tube ends are fixed into the tube plates (Figure 1.5b) by being expanded and soldered.

Cooling of the air results in precipitation of moisture which is removed by

Solid drawn tubes

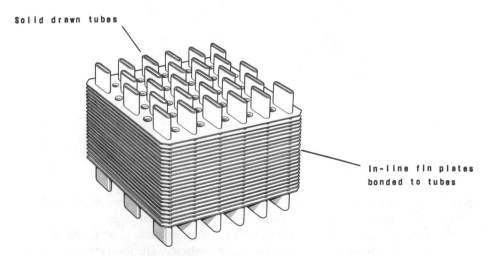

In-line fin plates
bonded to tubes

Figure 1.5a   *Detail of charge air cooler tube arrangement*

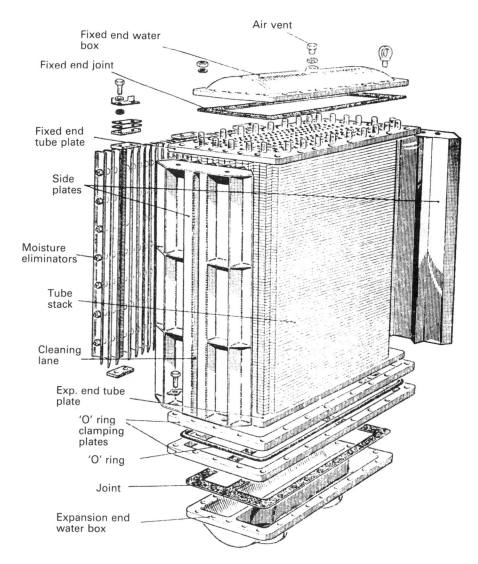

Figure 1.5b *Charge air cooler (courtesy Serck Heat Exchangers Ltd)*

water eliminators fitted at the air outlet side. A change of direction is used in some charge air coolers to assist water removal. Condensate is removed by a drain connection beneath the moisture eliminators.

Maintenance of heat exchangers

The only attention that heat exchangers should require is to ensure that the heat transfer surfaces remain substantially clean and flow passages generally clear of obstruction. Indication that fouling has occurred, is given by a progressive increase in the temperature difference between the two fluids, and change of pressure.

Fouling on the sea-water side is the most usual cause of deterioration in performance. The method of cleaning the sea-water side surfaces depends on the type of deposit and heat exchanger. Soft deposits may be removed by brushing. Chemical cleaning by immersion or *in situ*, is recommended for stubborn deposits. With shell and tube heat exchangers the removal of the end covers or, in the case of the smaller heat exchangers, the headers themselves, will provide access to the tubes. Obstructions, dirt and scale can then be removed, using the tools provided by the heat exchanger manufacturer. Flushing through with fresh water is recommended before a heat exchanger is returned to service. In oil coolers or heaters, progressive fouling may take place on the outside of the tubes. Manufacturers may recommend a chemical flushing to remove this *in situ*, without dismantling the heat exchanger.

Plate heat exchangers are cleaned by unclamping the stack of plates and exposing the surfaces. Plate surfaces are carefully washed using a brush or dealt with as recommended by the manufacturer to avoid damage. If the plate seals require replacement they may be removed with the method described in the section on plate coolers. Prising seals from their bonding, e.g. with sharp tools, causes plate damage.

Corrosion by sea water may occasionally cause perforation of heat transfer surfaces with resultant leakage of one fluid into the other. Normally the sea water is maintained at a lower pressure than the jacket water and other liquids that it cools, to reduce the risk of sea water entry to engine spaces. Leakage is not always detected initially if header or drain tanks are automatically topped up or manual top up is not reported. Substantial leaks become evident through rapid loss of lubricating oil or jacket water and operation of low level alarms.

The location of a leak in a shell and tube cooler is a simple procedure. The heat exchanger is first isolated from its systems and after draining the sea water and removing the end covers or headers to expose the tube plates and tube ends, an inspection is made for evidence of liquid flow or seepage from around tube ends or from perforations in the tubes. The location of small leaks is aided if the surfaces are clean and dry. The fixing arrangement for the tube stack should be checked before removing covers or headers to ensure that the liquid inside will not dislodge the stack. This precaution also underlines the need for isolation of a cooler from the systems.

To aid the detection of leaks in a large cooler such as a main condenser, in which it is difficult to get the tubes dry enough to witness any seepage, it is usual to add a special fluorescent dye to the shell side of the cooler. When an ultra-violet light is shone on to the tubes and tube plates leaks are made visible because the dye glows.

Plate heat exchanger leaks can be found by visual inspection of the plate surfaces or they are cleaned and sprayed with a fluorescent dye penetrant on one side. The other side is then viewed with the aid of an ultra-violet light to show up any defects.

Leaks in charge air coolers allow sea water to pass through to the engine cylinder. This can be a problem in four-stroke engines because there is a tendency for salt scale to form on air inlet valve spindles and this makes them stick. The charge air manifold drain is regularly checked for salt water. Location of the leak may be achieved by having a very low air pressure on the air side and inspecting the flooded sea-water side for air bubbles. Soapy water could be used as an alternative to having the sea-water side flooded.

If a ship is to be out of service for a long period, it is advisable to drain the sea-water side of heat exchangers then clean and flush through with fresh water, after which the heat exchanger should be left drained, if possible until the ship re-enters service.

Venting and draining

It is important that any heat exchanger through which sea water flows should run full. In vertically-mounted single-pass heat exchangers of the shell-and-tube or plate types, venting will be automatic if the sea-water flow is upwards. This is also the case with heat exchangers mounted in the horizontal attitude, with single- or multi-pass tube arrangements, provided that the sea-water inlet branch faces downwards and the outlet branch upwards. With these arrangements, the water will drain virtually completely out of the heat exchanger when the remainder of the system is drained.

With other arrangements, a vent cock fitted at the highest point in the heat exchanger should be opened when first introducing sea water into the heat exchanger and thereafter periodically to ensure that any air is purged and that the sea-water side is full. A drain plug should be provided at the lowest point.

Heat exchange theory

The rate of flow of heat through a heat exchanger tube or plate from the fluid at the higher temperature to the one at the lower (Figure 1.6) is related to the temperature difference between the two fluids, the ability of the material of the tube or plate to conduct and the area and thickness of the material.

If neither fluid is moving, the conductivity of the fluids has also to be taken into account and the fact that with static conditions as one fluid loses heat and the other gains, the temperature difference is reduced and this progressively slows down the rate of heat transfer.

With slow moving liquids at either side of a jacket cooler heat exchange surface, there is likely to be a constant temperature difference provided the hotter fluid is receiving heat from a steady source (as from a cylinder water jacket) and there is a continuous source for the cooler fluid (circulation from the sea). Laminar flow (Figure 1.7) occurs in slow moving liquids with the highest velocity in the centre of the liquid path and a gradually slower rate towards containing surfaces. A static boundary layer tends to form on containing surfaces and heat flow through such a layer relies on the ability of the layer to conduct. The faster moving layers also receive heat mainly by conductivity.

The temperature profile across an element of wall surface may be considered

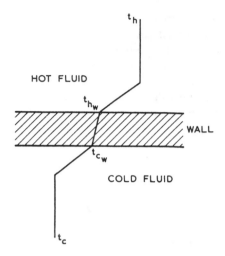

Figure 1.6   *Temperature gradient between fluids*

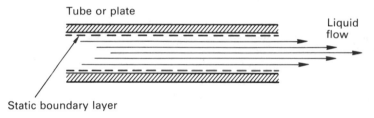

Figure 1.7   *Laminar flow*

as approximating to that depicted in Figure 1.6. The temperature of the hot fluid falls through its boundary layer from that of the bulk of the fluid (t_h) to (t_{hw}) that of the wall. There is a further drop through the wall from (t_{hw}) to (t_{cw}) and then through the boundary layer on the cold side from (t_{cw}) to (t_c) which is taken as the general temperature of the cold fluid.

Considering a rate of heat flow δQ through the element of wall surface area δA:

$$\delta Q = h_1 \, (t_h - t_{hw})\delta A = (k/y)(t_{hw} - t_{cw})\delta A = h_2(t_{cw} - t_c) \, \delta A$$

where:

h_1 = co-efficient of heat transfer on the hot fluid side;
h_2 = co-efficient of heat transfer on the cold fluid side;
k = thermal conductivity of the wall material;
y = thickness of the wall.

If the overall co-efficient of heat transfer between the hot and cold fluid is defined as:

$$U = \frac{\delta Q}{(t_h - t_c) \, \delta A}$$

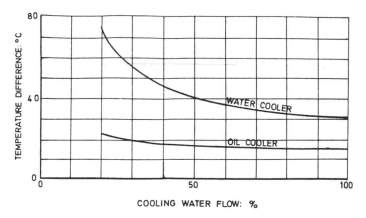

Figure 1.8 *Effect of variation in cooling water flow*

then

$$\frac{1}{U}=\frac{1}{h_1}+\frac{1}{h_2}+\frac{y}{k}$$

This is the basic equation governing the performance of a heat exchanger in which the heat transfer surface is completely clean. Additional terms may be added to the right hand side of the equation to represent the resistance to heat flow of films of dirt, scale, etc. The values of h_1 and h_2 are respectively determined by the fluids and flow conditions on the two sides of wall surface. Under normal operating conditions, water flowing over a surface gives a relatively high co-efficient of heat transfer, as does condensing steam, whereas oil provides a considerably lower value. Air is also a poor heat transfer fluid and it is quite usual to modify the effect of this by adding extended surface (fins) on the side of the wall in contact with the air.

In a practical heat exchanger, the thermal performance is described by the equation.

$$Q=U\,\theta\,A$$

where:

Q = rate of heat transfer;
θ = logarithmic mean of the temperature differences at the inlet and outlet of the heat exchanger: this is a maximum if the fluids flow in opposite directions (counterflow);
A = surface area of heat transfer wall.

It is sometimes important to appreciate the effect of variation of cooling water flow through a heat exchanger. The graph in Figure 1.8 illustrates two typical instances, one a jacket water cooler and the other a lubricating oil cooler (both sea-water cooled), in which the difference in temperature between the hot fluid and the sea-water is plotted against sea-water flow, assuming constant hot fluid flow and rate of heat transfer.

A dye can be used to demonstrate laminar flow in a liquid and also the effect of speeding up the flow (Figure 1.9) so that turbulence is produced. Turbulence is an agitation of the liquid caused by faster flow. If a dye is present when the flow rate is increased, the agitation is made evident in a random movement which rapidly disperses the colouring substance. Turbulence is beneficial in a heat exchanger, because it rotates particles of the liquids so that they tend to break up the boundary layer and remove heat by direct contact with the heat transfer surfaces. The price for the benefit of turbulence along a heat exchange surface is that at tube entrances, or the entry area between pairs of plates in plate type coolers, the turbulence is more extreme and damage from corrosion/erosion occurs. This type of attack is termed impingement.

A second advantage of turbulent flow, is that the scouring action tends to keep cooler surfaces clean.

Central cooling system

The corrosion and other problems associated with salt water circulation systems can be minimized by using it for cooling central coolers through which fresh water from a closed general cooling circuit is passed. The salt water passes through only one set of pumps, valves and filters and a short length of piping.

Figure 1.10 shows a complete central cooling system in which all components are cooled by fresh water. The three sections are (1) the sea-water circuit; (2) the high temperature circuit; and (3) the low temperature circuit.

The duty sea-water pump takes water from the suctions on either side of the machinery space and after passing through the cooler it is discharged straight overboard. The main and stand-by pumps would be of the double entry centrifugal type but, as an alternative, a scoop arrangement can be incorporated (Figure 1.11) with central cooling. A main circulating pump must have a direct bilge suction for emergency duty, with a diameter not less than two thirds that of the main sea-water inlet. In motor ships a direct suction on another pump of the same capacity is acceptable.

Materials for the reduced salt-water system for the central cooling arrangement will be of the high quality needed to limit corrosion/erosion problems.

Water in the high temperature circuit, is circulated through the main engine and auxiliary diesels by the pumps to the left of the engine in the sketch. At the outlet, the cooling water is taken to the fresh water distiller (evaporator) where the heat is used for the evaporation of sea water. From the outlet of the

Tube or plate

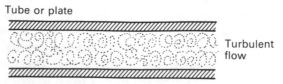

Turbulent flow

Figure 1.9   *Turbulent flow of a fluid*

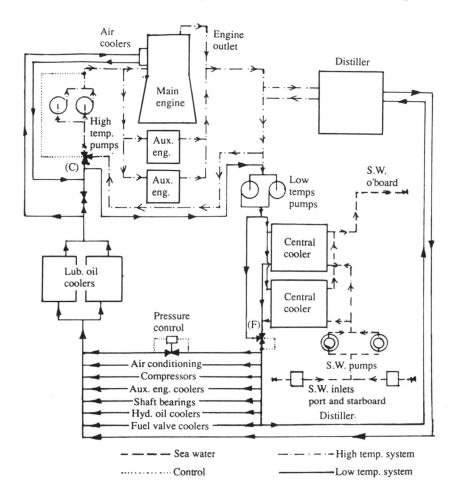

Figure 1.10 *Central cooling system*

evaporator, the cooling water is led back to the suction of the high temperature pump through a control valve (C) which is governed by engine inlet temperature. The control valve mixes the low and high temperature streams to produce the required inlet temperature, which is about 62°C. Engine outlet temperature may be about 70°C.

For the low temperature circuit, the heat of the water leaving the central coolers is regulated by the control valve (F). Components of the system are arranged in parallel or series groups as required. The pressure control valve works on a bypass. The temperature of the water after the cooler may be 35°C and at exit from the main engine lubricating oil coolers it is about 45°C.

The fresh water in the closed system is treated with chemicals to prevent corrosion of the pipework and coolers. With correct chemical treatment, corrosion is eliminated in the fresh water system, without the need for expensive materials.

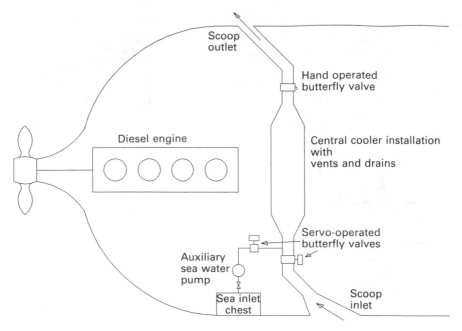

Figure 1.11   *Scoop arrangement for motorship central cooling system*

Scoop arrangement for a motor ship

A scoop (Figure 1.11) designed to supply sea-water circulation through the central coolers while the vessel is underway, may be installed instead of a conventional sea-water circulating pump. The scoop imposes some extra drag on the hull so that the power for sea-water circulation is supplied from the main propulsion instead of from the generators and electrical system. Economic advantages are claimed for a correctly designed scoop but the arrangement is viable only for a simple straight through flow as for central coolers or the large condenser of a steam ship. The electrically driven pump is used only for manoeuvring or slow speeds. It is of smaller capacity than would be required for an ordinary circulating pump.

Circulating systems for steamships

The main sea-water circulating system for a ship with main propulsion by steam turbine is similar to that of a motorship with a central cooling system. The difference is that the sea water passes through a large condenser and an oil cooler rather than central coolers and then to the overboard discharge. The main sea-water inlets, port and starboard, may be arranged as for the motorship example (Figure 1.1) with high and low suctions, orthodox double entry circulating pump (with emergency bilge suction) and a stand-by pump. Alternatively, the arrangement may be based on a scoop to supply the main condenser. Scoops have been preferred for fast, high-powered steamships, with circulation only through a single large condenser and sufficient speed to ensure that the scoop gives an adequate flow of water. Small axial flow circulating pumps (Figure 1.12) have been installed in conjunction with some scoop arrangements, with the idea that at speed, the pump impeller would idle and provide very little resistance to the scoop flow. The axial flow pump, intended for slow speed and manoeuvring, suffered from thrust problems when idling in a number of installations.

Closed feed system and feed heating

To ensure trouble-free operation of water-tube boilers the feed water must be of high quality with a minimal solid content and an absence of dissolved gases. Solids are deposited on the inside surfaces of steam generating tubes, as the water boils off, and the scale so formed causes overheating and failure. Dissolved gases tend to promote corrosion. Distilled water used as boiler feed has a great affinity for gases in the atmosphere. Chemical treatment is aimed at preventing corrosion problems but the boiler closed feed system must play its part in maintaining minimal contact between feed water and air at every stage and in promoting the dissociation and removal of any air or other gases. The feed system also assists efficient operation by condensing used steam and returning it to the boiler as feed at the highest temperature attainable economically. In practical terms, this means maximum recovery of the latent

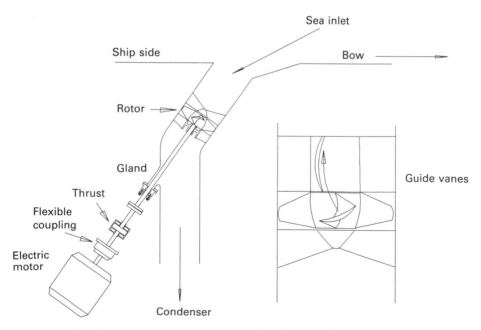

Figure 1.12　*Scoop system for steamship with axial flow pump*

heat in steam. Regenerative condensers and de-aerators are two major components in the complex feed systems which have been evolved.

The feed system for a tanker (Figure 1.13) shows a superheated steam supply from two water-tube boilers to the main propulsion turbine, the turbo-alternator and for cargo pumps. Each of these is served by a separate condenser and extraction pumps which return condensate to the closed feed system.

The condenser

A condenser is a vessel in which a vapour is deprived of its latent heat of vaporization and so is changed to its liquid state, usually by cooling at constant pressure. In surface condensers, steam enters at an upper level, passes over tubes in which cold sea water circulates, falls as water to the bottom and is removed by a pump (or flows to a feed tank).

The construction of condensers is similar to that of other tubular heat exchangers, with size variation extending up to the very large regenerative condensers for main propulsion steam turbines. Some smaller condensers may have U tubes for a two-pass flow and free expansion and contraction of tubes. The cooling water for straight tube condensers, circulates in one or two passes, entering at the bottom. With a scoop, there is one pass flow. A water box, of cast iron or steel, is fitted at each end (one end with U tubes) of the shell. Sandwiched between the flanges of the boxes and the shell are admiralty brass (70% Cu, 29% Zn, 1% Sn) tube plates. These are drilled and when soft-packing is used, counter bored and tapped.

Tubes may be of cupro-nickel (70% Cu, 30% Ni) or aluminium brass (76% Cu, 22% Zn, 2% Al) and of 16–20 mm outside diameter. Straight tubes can be expanded into the tube plates at both ends (Figure 1.3), expanded at the outlet end and fitted with soft packing at the other, or fitted with soft packing at both ends (Figure 1.14). An expansion allowance (Figure 1.15), provided where tubes are expanded into tube plates at both ends, may take the form of a shell expansion joint. Tubes are prevented from sagging by a number of mild steel tube support plates. A baffle plate at the entrance to the steam space, prevents damage from the direct impact of steam on the tubes.

Access doors are provided in the water box end covers of very large condensers for routine inspection and cleaning, with one or more manholes in the shell bottom for the same purpose.

Corrosion by galvanic action is inhibited by zinc or mild steel sacrificial anodes or alternatively, impressed current protection may be used. Dezincification of brasses may be prevented by additives, such as 0.04% arsenic, to the alloy.

Tube failure is likely to be caused by impingement, that is corrosion/erosion arising from entrained air in, or excessive speed of, circulating water. Failure could otherwise be from stress/corrosion cracking or dezincification of brass tubes. Defective tubes can be plugged temporarily.

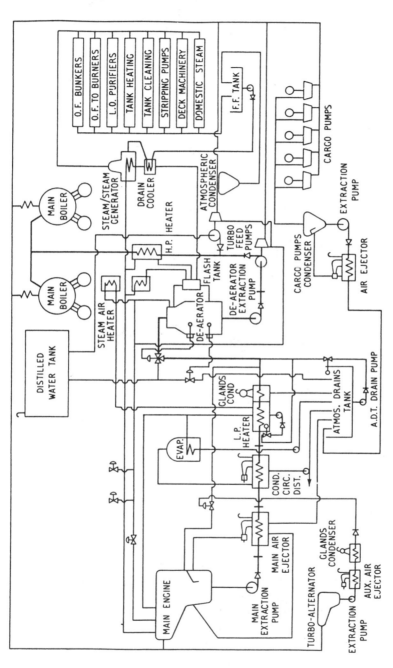

Figure 1.13 *Closed feed system for a tanker (Weir Pumps Ltd)*

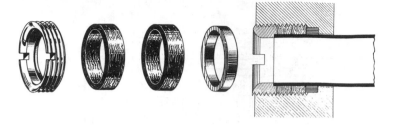

Figure 1.14   *Example of safe packing assembly for condenser tubes (courtesy of Crane Packing)*

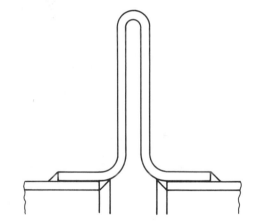

Figure 1.15   *Shell expansion joint*

The regenerative condenser

As it expands through a turbine, as much as possible of the available useful work is extracted from the steam by maintaining vacuum conditions in the condenser. Part of the function of the condenser is to condense the steam from the low pressure end of the turbine at as low a pressure as possible.

The effective operation of a condenser requires that the sea water is colder than the saturation temperature of the exhaust steam and this means that undercooling will occur. Any undercooling must be made good during the cycle which turns the feed water back to steam, and undercooling increases the temperature range through which the condensate, returning to the boiler, must be raised again before it boils off. To avoid this thermal loss, condensers are built with regenerative ability in that paths (Figure 1.16) are arranged between and below the tube banks for direct flow of part of the steam to the lower part of the condenser. This steam then flows up between the tubes and meets the condensate from the main part of the exhaust, dripping from the tubes. The undercooled condensate falls through this steam atmosphere and heat transfer occurs, resulting in negligible undercooling in the final condensate.

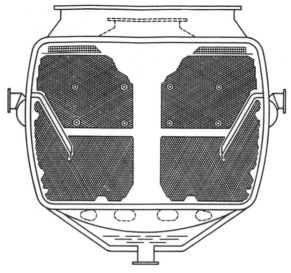

Figure 1.16 *Weir's regenerative condenser (courtesy G & J Weir Ltd, Glasgow)*

The condensate, dripping from the tubes, may be below the saturation temperature corresponding to the vacuum, by as much as 5°C, initially.

The de-aeration performance of a condenser is also related to undercooling in that the amount of gas, such as oxygen, that can remain in solution in a water droplet at below saturation temperature is dependent on the degree of undercooling.

Theoretically, if a water droplet is at the saturation temperature then no gas will remain in solution with it. One method of reducing the degree of undercooling when sea water temperature is low, is to recirculate a portion of the cooling water to enable the condenser to be worked at its design condition, whenever possible.

The feed system and feed heating

Non-condensable gases and some vapour are removed from the main condenser (Figure 1.13) by an air ejector, cooled by the main condensate and released in the ejector condenser. The condensed ejector steam passes with other clean drains (gland steam condenser, low pressure feed heater, evaporator) to a drains tank from which a pump draws, to discharge, with the mains condensate, to the de-aerator. It is common practice to reflux these drains, that is to return them to the main condenser in the form of a spray at a high level where, meeting the turbine exhaust, they are de-aerated before mixing with the main body of the condensate and being removed by the condensate extraction pump.

Heating steam for the steam/steam generator, the de-aerator and the low pressure feed heater are bled from the main turbines at appropriate stages, so that all of the latent heat is recovered. The feed pump exhaust is treated similarly.

The steam/steam generator, providing low pressure steam for services whose condensate may be contaminated, has its own separate feed system.

The centrifugal extraction pump, driven either by electric motor or steam turbine, draws from the condenser and delivers to a de-aerating heater through the heat exchangers already mentioned. Another extraction pump passes the de-aerated feed to a multi-stage centrifugal pump, also either electric motor or, more often, turbine driven. The pump delivers the feed to the boilers at a temperature approaching that of saturation through a high pressure feed heater, supplied with steam bled from the high pressure turbine. Make-up feed is produced by evaporation and distillation (sometimes double) at sub-atmospheric pressure, stored in a tank and introduced to the boilers from the main condenser with the refluxed drains or through the de-aerator. The feed pumps, feed piping and fittings are duplicated.

Steam-jet air ejector

A steam-jet ejector may be used to withdraw air and dissolved gases from the condenser. In each stage of the steam-jet ejector, high pressure steam is expanded in a convergent/divergent nozzle. The steam leaves the nozzle at a very high velocity in the order of 1220 m/s and a proportion of the kinetic energy in the steam jet transferred, by interchange of momentum, to the body of air which is entrained and passes along with the operating steam through a diffuser in which the kinetic energy of the combined stream is re-converted to pressure energy. The maximum pressure ratio that can be obtained with a single stage is roughly 5:1 and consequently it is necessary to use two or even three stages in series, to establish a vacuum in the order of 724 mm Hg, with reasonable steam consumption.

There are a variety of ejector designs in service which work on the same principle. Older units have heavy cast steel shells which serve as vapour condensers and also contain the diffusers. These are arranged vertically, the steam entering at the top (Figure 1.17). More recent designs have the diffusers arranged externally and the vapour condenser shell is somewhat lighter in construction. Horizontal and vertical arrangements can be found and some units are arranged as combined air ejectors and gland steam condensers.

Horizontal single element two stage air ejector

An air ejector which has been commonly used, is shown schematically in Figure 1.18. The unit comprises a stack of U-tubes contained in a fabricated mild steel condenser shell on which is mounted a single element two stage air ejector.

The condensate from the main or auxiliary condenser is used as the cooling medium, the condensate circulating through the tubes whilst the air and vapour passes through the shell. The high velocity operating steam emerging from the first stage ejector nozzle entrains the non-condensables and vapour from the

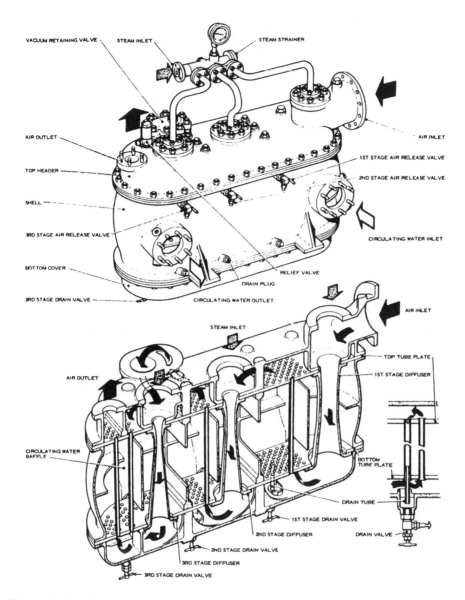

Figure 1.17 *Three stage air ejector with internal diffusers*

main condenser and the mixture discharges into the inter (or first stage) condenser.

Most of the steam and vapour is condensed when it comes into contact with the cool surface of the tubes, falls to the bottom of the shell and drains to the main or auxiliary condenser. The remaining air and water vapour are drawn into the second stage ejector and discharged to the after (or second stage) condenser. The condensate then passes to the steam drains tank and the non-condensables are discharged to the atmosphere through a vacuum retaining valve.

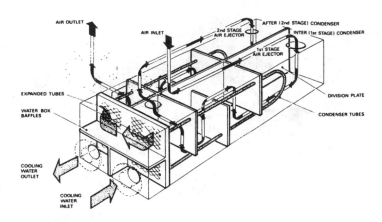

Figure 1.18   *Horizontal single element two stage air ejector*

The vacuum retaining valve is shown in Figure 1.19 which is fitted as a safety device to reduce the rate of loss of vacuum in the main condenser if the air ejector fails. It is mounted on a pocket built out from the second stage condenser, and consists essentially of a light stainless steel annular valve plate which covers ports in a gunmetal valve seat. When the pressure inside the after condenser exceeds atmospheric pressure the valve lifts and allows the gases to escape to atmosphere. A relief valve is fitted on the first stage condenser shell of the twin element unit.

The ejector stages, Figure 1.20, consist of monel metal nozzles in mild steel holders discharging into gunmetal diffusers. Expansion is allowed for by sliding feet at the inlet end.

Nash rotary liquid ring pumps

Nash rotary liquid ring pumps, in association with atmospheric air ejectors, may be used instead of diffuser-type steam ejectors and are arranged as shown in Figure 1.21. The pump, discharging to a separator, draws from the condenser

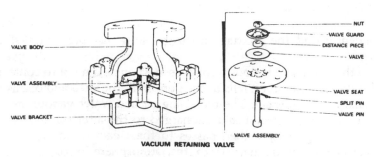

Figure 1.19   *Vacuum retaining valve*

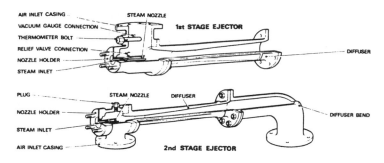

Figure 1.20 *First and second stage ejectors*

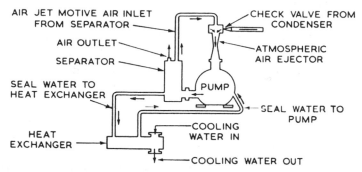

Figure 1.21 *Nash liquid pump ring*

through the atmospheric air ejector, creating a partial vacuum of about 600 mm Hg. At this stage the ejector, taking its operating air from the discharge separator, which is vented to atmosphere, comes into action, sonic velocity is attained and vacua in the order of 725 mm Hg maintained in the condenser. The liquid ring pump is sealed with fresh water recycled through a sea-water cooled heat exchanger.

The Nash pump, using recycled fresh water for sealing, may be also used to provide oil-free instrument air at a pressure of 7 bar (Figure 1.22).

Condensate extraction pump

Removal of condensate from a condenser imposes very difficult suction conditions on the pump. The available net positive suction head (NPSH) is minimal because the condenser is situated low in the ship permitting a static suction head of only 450–700 mm and the condensate is at, or near its vapour pressure. It is necessary therefore to ensure that the pump's required NPSH is correspondingly low and to this end suction passages and inlets are given ample area. The pumps used for the duty, are two stage units (Figure 1.23a) the first stage impeller being arranged as low as possible in the pump with an upward facing eye. This impeller feeds a second impeller via suitable passages in the pump casing.

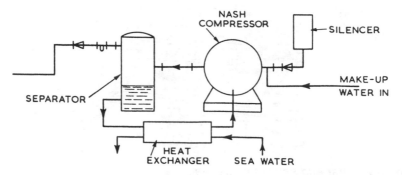

Figure 1.22   *Using recycled fresh water for sealing (Nash Engineering Co. (GB) Ltd)*

Where extraction pumps are fed from de-aerators or drain coolers, the pump suction level is maintained constant by a specially designed float control. In instances where the pump is drawing from the main condenser however, it is common practice to operate the pump on a free suction head. This means that the pump must operate with a variable NPSH at varying flow rates. Figure 1.23b shows the head/quantity (H/Q) curve for a two stage extraction pump operating under a variable NPSH. The system resistance curve is interposed on the diagram.

When operating at the specified maximum capacity, the flow rate corresponds to that at the intersection of the natural H/Q characteristic of the pump and the system resistance curve. As the level of condensate in the suction sump falls below the point where the available NPSH intersects with the minimum required NPSH the pump starts to cavitate and its output is regulated. The natural H/Q characteristic is modified as the discharge pressure is reduced to that required to overcome the resistance of the system at the reduced flow rate. When the suction level increases the flow rate increases. Therefore, inherent control of the flow rate is achieved without the use of a float controlled regulator. Such a system is usually referred to as cavitation control. The extraction pump is a cavitating pump.

Feed water heaters

Surface or direct contact feed heaters, play an important part in the recovery of latent heat from exhaust steam. Direct contact feed heaters are also known as de-aerators.

Surface feed heaters

These are shell and tube heat exchangers, made with materials and scantlings appropriate to their working temperatures and pressures. It can be shown that the minimum economic terminal difference (i.e. the temperature difference

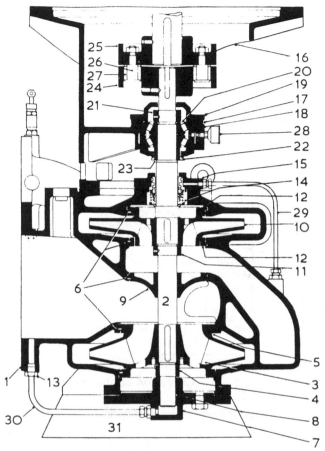

Figure 1.23 (a) *Weir two stage extraction pump*

1. Pump casing (split)
2. Pump spindle
3. Impeller (1st stage)
4. Impeller nut and tag washer (1st stage)
5. Casing ring (1st stage)
6. Dowel pins
7. Bottom cover
8. Bottom bush liner
9. Intermediate bush
10. Impeller (2nd stage)
11. Impeller nut and lock screw (2nd stage)
12. Casing ring (2nd stage)
13. Plug
14. Mechanical seal
15. Seal clamping plate
16. Motor stool
17. Thrust bearing housing
18. Thrust bearing cover
19. Thrust bearing end cover
20. Thrust bearing
21. Thrust nut and lock screw
22. 'V' ring
23. Distance piece
24. Flexible coupling (pump half)
25. Flexible coupling (motor half)
26. Coupling bolt and nut
27. Coupling pad
28. Grease lubricator
29. Water return pipe
30. Water supply pipe
31. Pump feet

between the heating fluid inlet and the heated fluid outlet) is about 6°C. If, however, the heating fluid (steam) is superheated by at least 110°C above saturation temperature and the heated fluid (feed water) leaves the heater at this saturation temperature or near to it, this terminal difference will be very small or zero. This can be achieved by using the exit section of the feed heater as a de-superheater.

By grouping a number of surface feed heaters in one module, an economy in

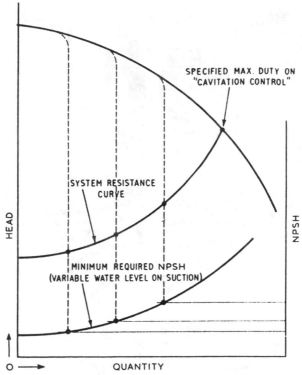

Figure 1.23 (b)   *Extraction pump cavitation control curves*

space and pipe connections can often be achieved and there is a movement in that direction in practice.

Many parts of the feed system are now installed as packaged units or modules. Figure 1.24 shows a package arrangement which combines no less than four different units, viz. steam jet air ejectors, glands steam condenser, the de-aerator vent condenser and the low pressure heater drains cooler. Such packaging obviously leads to compactness, centralization and reduction of branches to be connected by the shipbuilder. Other 'packaged' feed systems incorporate complete feed systems and Figure 1.25 shows an arrangement of a packaged feed system for a 67 000 tonne tanker.

De-aerators

Mention has been made of the need for clean, neutral boiler feed, free from dissolved gases and of the consequent use of efficient de-aerators. Figure 1.26 shows one of several which liberate the dissolved gases from the feed and provide a measure of feed heating simultaneously. This type of de-aerator has a great range of capacity and given a temperature rise of at least 20°C, an oxygen content of 0.2 cc/litre can be reduced to 0.005 cc/litre, when working between one-half full load and full load in a closed feed system.

Normally, the de-aerator is mounted directly on a storage tank, into which the de-aerated water falls, to be withdrawn through a bottom connection by a

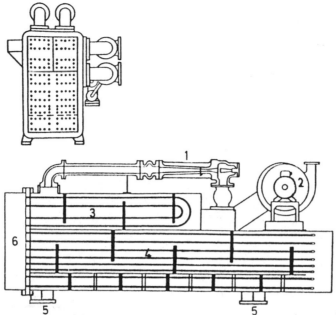

Figure 1.24 *This module combines four closed feed system components and is claimed to offer a 14% reduction in weight and a 19% reduction in volume. 1 Steam jet air ejector, 2 electrically driven vapour extractor, 3 gland steam condenser, 4 LP heater/drains cooler, 5 supporting feet, 6 water heater*

pump or by gravity. The tank usually has a capacity sufficient for 10 minutes' running supply of water but this is not necessarily the case.

The feed water enters the de-aerator head and so that its surface area may be increased to the maximum possible, it is divided into sprays of minute droplets by being forced through the spray nozzles into the shell; here it meets the heating steam and is brought rapidly to its saturation temperature. Most of the dissolved gases are released and with some vapour rise to the vapour release opening. The header may be divided and provided with two feed inlet connections, so that the efficiency of de-aeration may be maintained at low rates of flow, by reducing the number of nozzles in use.

Cascade trays

Three cascade trays are set one above the other in the lower part of the shell. The upper and lower of these trays have a raised lip on the outer periphery, have the central opening blanked and have a series of perforations arranged in rings towards the raised lip. The middle tray has a central opening with a raised lip and is perforated similarly. The falling spray collects on the upper tray and is again broken up as it passes through the perforations to the middle tray where the process is repeated, to be repeated again as it passes through the lower tray to the tank below. The combination of spray, heating and cascade ensures the liberation of all but a minute fraction of the gases in solution or suspension. The final water temperature depends upon the pressure of the controlled steam supply.

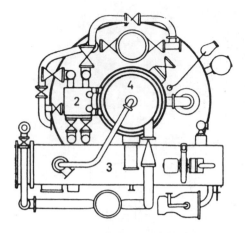

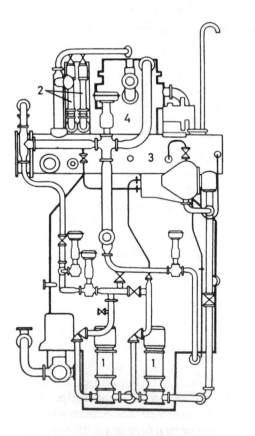
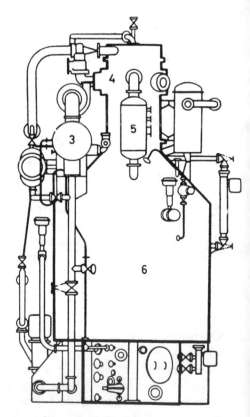

Figure 1.25 *Package feed system for tanker*

1. Extraction pumps
2. Steam jet air ejectors
3. L.P. heater and drain cooler

4. De-aerator
5. Flash chamber
6. De-aerated water storage tank

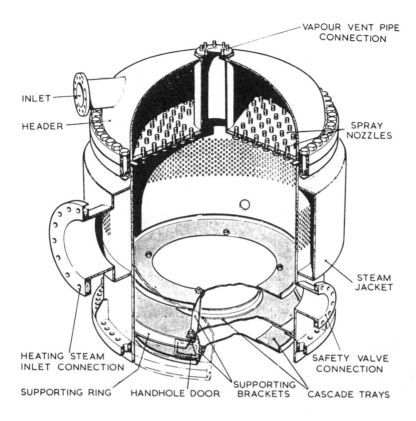

Figure 1.26 *Typical de-aerator (Weir Pumps Ltd)*

It will be apparent that de-aerators of this type must be installed at such a height in the engine room that the pressure head at the extraction or feed pump suction is greater than that corresponding to the water temperature.

The maintenance required is the control of corrosion, the cleaning of the nozzles, the renewal of those showing signs of erosion (which will seriously impair the efficiency), the overhaul of fittings and the maintenance of the safety valve.

Devaporizers

If the de-aerator cannot be vented to atmosphere or to a gland condenser satisfactorily, a devaporizer (Figure 1.27 is connected to the vapour outlet condensing the vapour vented with the non-condensable gases and cooling these gases before they are discharged. In the process the feed water is raised slightly in temperature. In design and construction devaporizers are similar to the other small heat exchangers working at moderate pressures. Leaving the devaporizer the feed water enters the de-aerator header.

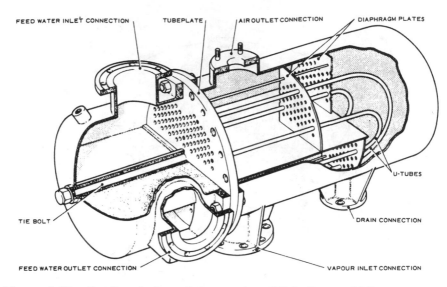

FEED WATER INLET CONNECTION   TUBEPLATE   AIR OUTLET CONNECTION   DIAPHRAGM PLATES

U-TUBES

TIE BOLT   DRAIN CONNECTION

FEED WATER OUTLET CONNECTION   VAPOUR INLET CONNECTION

Figure 1.27   *Sectional view of devaporizer (Weir Pumps Ltd)*

Weir multi stage turbo-feed pump

The water tube boiler, with its low storage capacity in relation to its steaming capability, demands a steady supply of pure feed water. The robust and reliable turbine driven centrifugal pump, is ideal for use as a water tube boiler feed pump. Additionally its uncontaminated exhaust can be used for feed heating and this improves cycle efficiency.

The successful development of water lubricated bearings and their use in the Weir turbo-feed pump, permitted the turbine and pump to be close-coupled in a very compact unit (Figure 1.28). During normal running, a multi-plate restriction orifice allows feed water from the first stage impeller discharge to flow through a two-way non-return valve and a strainer to the bearings. A relief valve is incorporated. A secondary supply of lubricating water is introduced through the two way non-return valve from an outside source such as the main condensate extraction pumps, to protect the bearings from damage during starting, stopping, or stand-by periods of duty. Figure 1.29 shows the arrangement of the lubricating water system.

The overspeed trip is triggered by a spring loaded unbalanced bolt mounted in the shaft between the two journals.

Pressure governor

A discharge-pressure operated governor (Figure 1.30) and a rising head/capacity curve from full load to no load gives inherent stability of operation. The main feature of the governor is that if the pump loses suction the steam ports are opened wide, allowing the pump to accelerate rapidly to the speed at which the

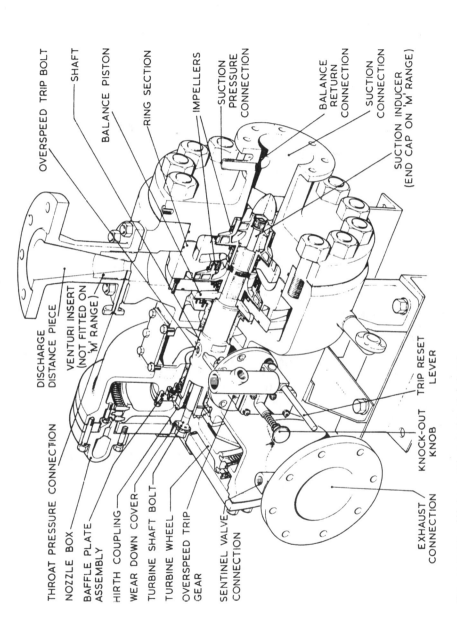

OVERSPEED TRIP BOLT

SHAFT

BALANCE PISTON

RING SECTION

IMPELLERS

SUCTION PRESSURE CONNECTION

BALANCE RETURN CONNECTION

SUCTION CONNECTION

SUCTION INDUCER (END CAP ON 'M' RANGE)

DISCHARGE DISTANCE PIECE

VENTURI INSERT (NOT FITTED ON 'M' RANGE)

THROAT PRESSURE CONNECTION

NOZZLE BOX

BAFFLE PLATE ASSEMBLY

HIRTH COUPLING

WEAR DOWN COVER

TURBINE SHAFT BOLT

TURBINE WHEEL

OVERSPEED TRIP GEAR

SENTINEL VALVE CONNECTION

EXHAUST CONNECTION

KNOCK-OUT KNOB

TRIP RESET LEVER

Figure 1.28 *Weir multi-stage turbo-feed pump*

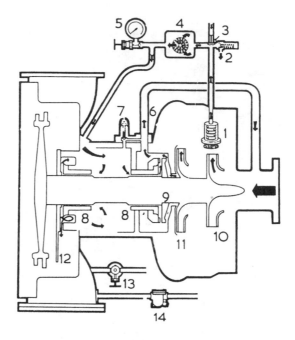

Figure 1.29   *Lubricating water system (Weir Pumps Ltd.)*

1. Multi-plate restriction orifice
2. Relief valve
3. Two-way non-return valve
4. Strainer
5. Pressure gauge
6. Balance chamber leakage
7. Leak-off control valve
8. Bearing
9. Balance piston
10. 1st stage impeller
11. 2nd stage impeller
12. Baffle plate
13. Drain valve
14. Drain trap

emergency trip acts. An adjusting screw, the collar of which bears against a platform in the casing, is threaded into the upper spring carrier and allows the compression of the spring to be altered by varying the distance between the upper and lower spring carriers.

The piston, fitted with an O-ring and a spiral back-up ring, slides in a close fitting liner. A flange on this liner locates in a recess in the governor casing cover and is sealed by an Armco iron joint ring.

When the pump is started, the throttle valve moves upwards due to the increasing discharge pressure under the piston until the desired pressure is reached. At this point, the upward force exerted on the governor spindle by the piston is equal to that being applied downwards by the spring, and the throttle valve is admitting the correct quantity of steam to the turbine to maintain the desired discharge pressure. With an increased demand on the pump the discharge pressure falls, allowing the pressure governor piston to move down under compulsion of the spring until the throttle valve opens to provide steam to satisfy the new demand. The reverse action takes place when the demand decreases.

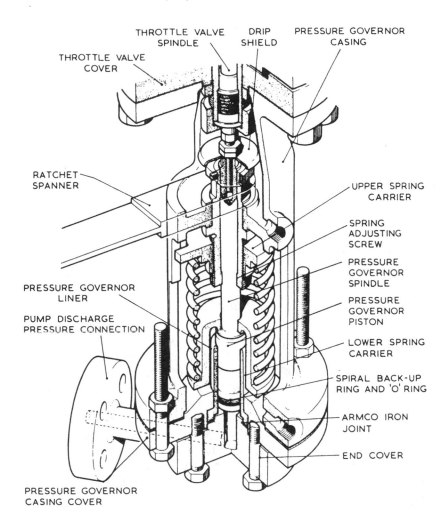

THROTTLE VALVE SPINDLE
THROTTLE VALVE COVER
DRIP SHIELD
PRESSURE GOVERNOR CASING
RATCHET SPANNER
UPPER SPRING CARRIER
SPRING ADJUSTING SCREW
PRESSURE GOVERNOR SPINDLE
PRESSURE GOVERNOR LINER
PUMP DISCHARGE PRESSURE CONNECTION
PRESSURE GOVERNOR PISTON
LOWER SPRING CARRIER
SPIRAL BACK-UP RING AND 'O' RING
ARMCO IRON JOINT
END COVER
PRESSURE GOVERNOR CASING COVER

Figure 1.30 *Weir discharge-pressure operated governor*

Safety (overspeed) trip

The 'Bolt' type overspeed trip (Figure 1.31) consists essentially of a spring-loaded stainless steel bolt which, due to its special design, is heavier at one end than the other. The rotary motion of the turbine shaft tends to move the bolt outwards, while the spring retains it in its normal position until the turbine speed reaches a pre-determined safety level. At this speed the centrifugal force exerted by the heavier end of the bolt overcomes the spring opposing it and the bolt moves outwards to strike the trip trigger. This in turn disengages the trip gear, allowing the steam stop valve to shut.

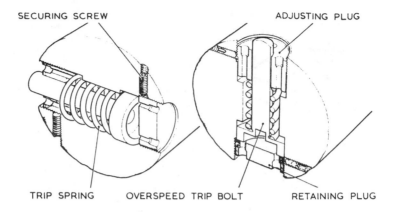

SECURING SCREW   ADJUSTING PLUG

TRIP SPRING   OVERSPEED TRIP BOLT   RETAINING PLUG

Figure 1.31   *Overspeed trip (bolt type)*

Turbine driven oil-lubricated pump

Prior to the introduction of the water lubricated turbo-feed pumps an oil-lubricated pump with a relatively long horizontal shaft was commonly used. In this particular pump the overspeed trip (Figure 1.32) was of the ring type. The pump was mounted on a taper at the end of the turbine shaft adjacent to the trip gear and was secured by a mild steel set bolt tapped into the end of the shaft. This set bolt was locked in place by a copper lock washer. The overspeed mechanism consisted of a case-hardened steel ring which, bored eccentrically, was weighted off-centre. The ring, however was spring loaded to maintain concentricity with the shaft until the speed of the turbine reached a predetermined safety limit. At that speed the centrifugal force exerted by the ring would overcome the force of the opposing spring and would then move outwards to strike the trip trigger.

Hydraulic balance mechanism

To control the axial movement of the rotating assembly, a balance piston (Figure 1.33) is arranged to counteract the effect of the thrust of the turbine and impellers. The arrangement keeps the rotating assembly in its correct position under all conditions of loading. Water at the approximate pressure of the pump discharge passes from the last stage of the pump between the impeller hub and the balance restriction bush C into the annular space B dropping in pressure as it does so. The pressure of water in the chamber B tends to push the balance piston towards the turbine end. When the thrust on the balance piston overcomes the turbine and the impeller thrust, the gap A between the piston and balance ring widens and allows water to escape. This in turn has the effect of lowering the pressure in chamber B allowing the rotating assembly to move back towards the pump end.

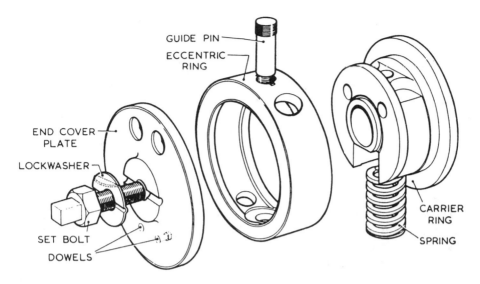

Figure 1.32 *Overspeed trip (ring type)*

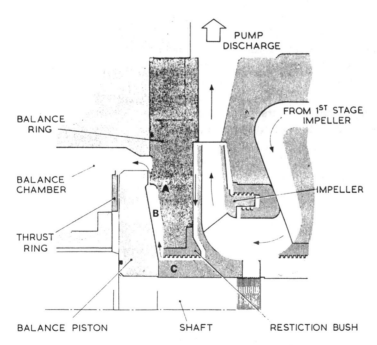

Figure 1.33 *Hydraulic balance*

Theoretically this cycle will be repeated with a smaller movement each time until the thrust on the balance piston exactly balances the other axial forces acting on the assembly. In practice the balancing of the forces is almost instantaneous and any axial movement of the shaft is negligible.

Weir electro-feeder

The Weir electro-feeder (Figure 1.34) is a multi-stage centrifugal pump mounted on a common baseplate with its electric motor. The number of stages may vary from two to fourteen depending upon the capacity of the pump and the required discharge pressure. The pump body consists of a number of ring sections fitted with diffusers and held in position between a suction and discharge casing by a ring of steel tie bolts.

The unit is supported on pads on the baseplate by two feet on each of the end casings, these feet being drilled to accommodate the holding down bolts. Tapered dowels are used to maintain the correct alignment, and the driving torque from the electric motor is transmitted through a flexible coupling.

The shaft assembly is supported on two ring lubricated white metal lined journal bearings bedded into plummer blocks, the lower sections of which form oil sumps. An internal hydraulic balancing arrangement similar to that found in the turbo-feed pump automatically maintains the shaft assembly in its correct axial position at all loads during running. To avoid excessive wear on this balancing arrangement when starting the pump, it is essential that the discharge pressure be built up quickly, and for this purpose, and to eliminate the possibility of reverse flow, the pump is fitted with a spring loaded non-return discharge valve.

Condensate cooled stuffing boxes packed with high quality packing, are used for shaft sealing and these can be additionally cooled by water-circulated cooling jackets in the suction casing and the balance chamber cover. A pressure-operated cut-out switch may be fitted which will automatically isolate the driving motor from its supply if the first stage discharge pressure falls to a predetermined value due to loss of suction pressure, cavitation or other reason.

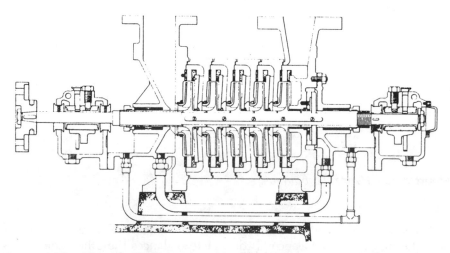

Figure 1.34   *Electrically-driven multi-stage feed pump (Weir Pumps Ltd)*

Further reading

Cotton, J. B. and Scholes, I.R. (1972) Titanium in marine engineering, *Trans I Mar E,* **84**, paper 16.

Conde, J. F. G. (1985) New materials for the marine and offshore industry, *Trans I Mar E,* **97**, paper 24.

Shone, E. C. and Grim, G. C. (1985) 25 Years experience with sea water cooled heat transfer equipment in shell fleets, *Trans I Mar E,* **98**, paper 11.

2

Machinery service systems and equipment

Service systems are necessary for the main machinery and for generators in addition to the circulating systems described in the previous chapter. The supply of compressed air for starting and control systems requires the provision of compressors and air receivers. Modern residual fuels need a handling system with settling tanks, centrifuges, heating, filtration, and sometimes with homogenization and blending equipment. Lubricating oil also benefits from being centrifuged as well as being filtered.

Air compressors and systems

Air at a pressure of 20 to 30 bar is required for starting main and auxiliary diesel engines in motorships and for the auxiliary diesels of steamships. Control air at a lower pressure is required for ships of both categories and whether derived from high pressure compressors through reducing valves or from special control air compressors, it must be clean, dry and oil free.

A starting air system for main diesels (Figure 2.1) normally has two air compressors and two reservoirs with sufficient capacity for twelve main engine starts (six if a non-reversible engine). The receivers must store sufficient air for the starts without the need for top up from the compressors.

Safety valves are normally fitted to the air receivers but in some installations the reservoirs are protected against overpressure by those of the compressors. There is a requirement that if the safety valves can be isolated from the reservoirs, the latter must have fusible plugs fitted to release the air in the event of fire. Reservoirs are designed, built and tested under similar regulations to those for boilers.

Explosions can and do occur in diesel engine starting air systems. Also air start valves and other parts are sometimes burned away without explosion. These problems have been caused by cylinder air start valves which have leaked or not closed after operation and have allowed access from the cylinder to the air start system of the flame from combustion. Carbon deposits from burning fuel and oily deposits from compressors are available as substances which may be ignited and produce an explosion in the air start system. If no explosion occurs, the flame from the cylinder and high temperature air from compression can cause carbon deposits in the system to burn. Careful maintenance of air start valves, distributors and other parts is vital as is regular cleaning of air start system components to remove deposits. The lubrication of

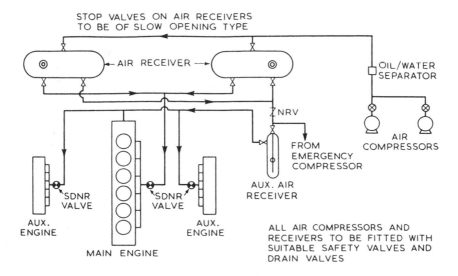

STOP VALVES ON AIR RECEIVERS
TO BE OF SLOW OPENING TYPE

AIR RECEIVER

OIL/WATER
SEPARATOR

NRV

FROM
EMERGENCY
COMPRESSOR

AIR
COMPRESSORS

AUX. AIR
RECEIVER

SDNR
VALVE

SDNR
VALVE

AUX.
ENGINE

AUX.
ENGINE

MAIN ENGINE

ALL AIR COMPRESSORS AND
RECEIVERS TO BE FITTED WITH
SUITABLE SAFETY VALVES AND
DRAIN VALVES

Figure 2.1 *Starting air system for main diesels*

components is limited as excess lubrication could cause the air start valves to be stuck by grease which has become hardened by the heat, and oil could accumulate in the pipes from this source. The draining of compressor coolers and air receivers is important. Drains on air start systems are also checked. Flame traps or bursting caps are fitted at each air start valve but it should be noted that protection of the latter type failed to prevent a serious explosion with the loss of seven lives in 1960.*

Starting air for auxiliary engines may be taken directly from the main engine air start receivers or from a small auxiliary receiver which can be kept at full pressure.

The low pressure control air system receiver is supplied ideally from a low pressure, oil free compressor. The supply may be obtained from the main air start reservoirs through reducing valves or pressure regulators, driers, oil traps and filters.

For steamships, the starting air arrangements for diesel auxiliary engines are similar to those in motorships. The low pressure air for instrumentation and remote control devices is likely to be supplied from a low pressure, oil free compressor.

Air compressors

A single stage compressor used to provide air at the high pressures required for diesel engine starting, would unfortunately generate compression temperatures of a level similar to those in a diesel. Such heat would be sufficient to ignite vaporized oil in the same way as in a compression-ignition engine. The heat produced in a single stage of compression would also be wasteful of energy.

*Merchant Shipping Act (1894).

This heat of compression adds energy and produces a resultant rise in pressure apart from that pressure rise expected from the action of the piston. However, when the air cools the pressure rise due to the heat generated is lost. Only the pressure from compression remains. The extra pressure due to heat, is of no use and actually demands greater power for the upward movement of the piston through the compression stroke.

Perfect cooling for the cylinder of a single stage compressor, with constant (isothermal) temperature during the process, would remove the problems, but is impossible to achieve. Multi-stage air compressor units with various cylinder configurations and piston shapes (Figure 2.2) are used in conjunction with intermediate and after cooling to provide the nearest possible approach to the ideal of isothermal compression.

Cycle of operation

On the compression stroke (Figure 2.3) for a theoretical single cylinder compressor, the pressure rises to slightly above discharge pressure. A spring-loaded non-return discharge valve opens and the compressed air passes through at approximately constant pressure. At the end of the stroke the differential pressure across the valve, aided by the valve spring, closes the discharge valve, trapping a small amount of high pressure air in the clearance space between the piston and the cylinder head. On the suction stroke the air in the clearance space expands, its pressure dropping until such time as a spring-loaded suction valve re-seats and another compression stroke begins.

Cooling

During compression much of the energy applied is converted into heat and any consequent rise in the air temperature will reduce the volumetric efficiency of

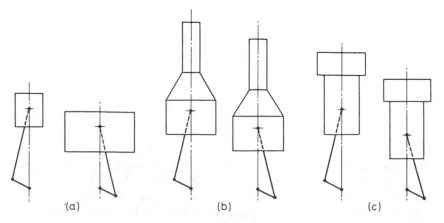

(a)   (b)   (c)

Figure 2.2   *Air compressor configurations*

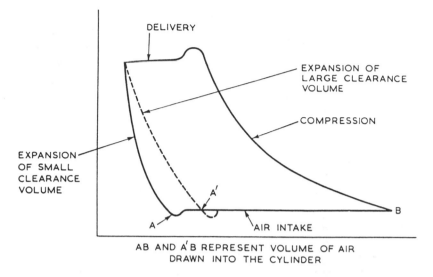

DELIVERY

EXPANSION OF LARGE CLEARANCE VOLUME

COMPRESSION

EXPANSION OF SMALL CLEARANCE VOLUME

A'

A

AIR INTAKE

B

AB AND A'B REPRESENT VOLUME OF AIR DRAWN INTO THE CYLINDER

Figure 2.3 *Compressor indicator diagram (courtesy Hamworthy Engineering Ltd)*

the cycle. To minimize the temperature rise, heat must be removed. Although some can be removed through the cylinder walls, the relatively small surface area and time available, severely limit the possible heat removal and as shown in Figure 2.4 a practical solution, is to compress in more than one stage and to cool the air between the stages.

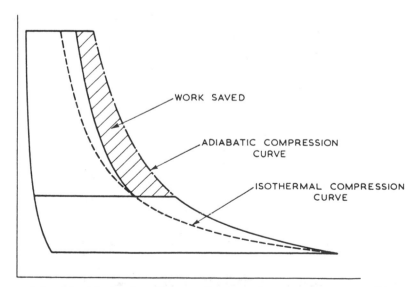

WORK SAVED

ADIABATIC COMPRESSION CURVE

ISOTHERMAL COMPRESSION CURVE

Figure 2.4 *Ideal indicator diagram for two stage compressor with intercooling (courtesy Hamworthy Engineering Ltd)*

For small compressors air may be used to cool the cylinders and intercoolers, the cylinder outer surfaces being extended by fins and the intercoolers usually being of the sectional finned-tube type over which a copious flow of air is blown by a fan mounted on the end of the crankshaft. In larger compressors used for main engine starting air it is more usual to use water-cooling for both cylinders and intercoolers.

Sea water is commonly used for this purpose with coolant being circulated from a pump driven by the compressor or it can be supplied from the main sea-water circulating system. Sea water causes deposits of scale in cooling passages. Fresh water from a central cooling systems serving compressors and other auxiliaries is preferable (see Chapter 1).

Two stage starting air compressor

The compressor illustrated in Figure 2.5 is a Hamworthy 2TM6 type which was designed for free air deliveries ranging from 183 m³ per hour at a discharge pressure of 14 bar to 367 m³ per hour at 42 bar.

The crankcase is a rigid casting which supports a spheroidal graphite cast

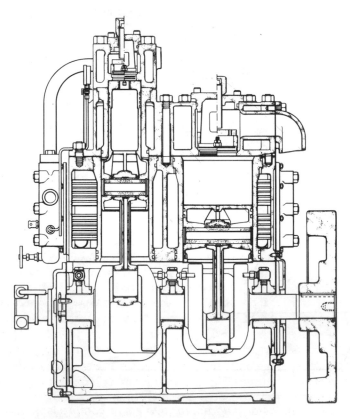

Figure 2.5   *Hamworthy 2TM6 air compressor*

iron crankshaft in three bearings. The crankshaft has integral balance weights and carries two identical forged steel connecting rods.

Both the first and second stage pistons are of aluminium alloy with cast iron compression rings. Scraper or oil control rings are fitted to return to the crankcase, most of the oil being splashed on to the cylinder walls from the bottom ends. The cylinder walls are lubricated by the splashed oil. The pistons have fully floating gudgeon pins; connecting rod top ends house phosphor bronze bushes. The bushes are an interference fit in the connecting rods and are so toleranced that the collapse of the bore when fitting is allowed for, to provide the correct running clearance. Steel backed white metal lined 'thin shell' main and crankpin bearings are used and all of the bearings are pressure lubricated by a chain driven gear pump.

Air suction and discharge valves are located in pockets in the cylinder heads. The valves are of the Hoerbiger type and are as shown in Figure 2.6. The moving discs of the valves have low inertia to permit rapid action. Ground landings are provided in the pockets on which the valve bodies seat. The bodies are held in place by set screws which pass through the valve box covers, capped nuts being fitted to the ends of the set screws. A combined air filter and silencer is fitted to the compressor air intake.

The intercooler is of the single pass type. The shell forms an integral part of the cylinder block casting, with the air passing through the tubes. The aftercooler is of the double pass U-tube type. Again the shell is integral with the cylinder block. Relief valves are fitted to the air outlets of each stage and are set to lift at 10% above normal stage pressure. The actual stage pressures vary according to the application. To protect the water side against over pressure in the event of a cooler tube failure, a spring loaded relief valve or bursting diaphragm, is fitted on the cylinder jacket.

Protection against overheating in the compressor discharge, is afforded by a fusible plug fitted on the aftercooler discharge head. Overheating sufficient to melt the alloy material of the plug can be the result of carbon build up around the discharge valve.

Operation and maintenance

Compressors must always be started in the unloaded condition otherwise pressures build up rapidly producing very high starting torques (Figure 2.7). During running there is an accumulation of oil carried over from the cylinders and water from moisture, precipitated in the coolers. The emulsion is collected in separators at cooler outlets and these must be drained off regularly, to reduce carry over. This is extremely important, first to prevent any large quantity of water and oil emulsion reaching a subsequent compression stage and causing damage to a further stage and secondly to reduce the amount carried over to the air receivers and starting air lines. Moisture in air receivers can give rise to corrosion and despite the proper operation of compressor cooler drains, a large amount tends to collect, particularly in humid conditions or wet engine rooms. It is good practice to check air reservoir drains regularly to assess the quantity

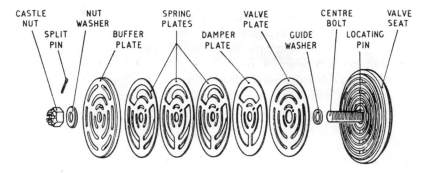

Figure 2.6   *Air compressor valve*

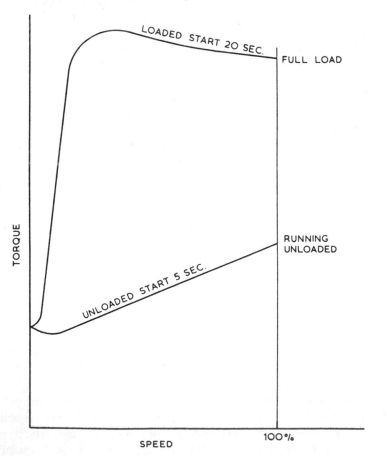

Figure 2.7   *Compressor torques*

of liquid present. In extreme conditions, drains may have to be used daily to
remove accumulated emulsion. This is very important if air for control systems
is derived from the main receivers, to prevent problems with the reducing
valve, moisture traps and filters. Moisture traps for the control air system also

require regular checking and possibly daily draining. A compressor is unloaded before being stopped by opening the first and second stage drains.

The maker's instructions normally recommend the use of a light oil for crankcase and cylinder lubrication. Diesel engine crankcase oils, which are likely to be a blend of light and heavy stocks, tend to produce deposits. The oil is inclined to emulsify and should be changed at frequent intervals. Excessive operation of automatically controlled compressors (usually due to a large number of leaks rather than actual use of air) usually means that valves require frequent cleaning and maintenance. Failure to keep valves in good condition, results in valve leakage and overheating with an associated accumulation of carbon on the valves and in valve pockets. The combination of carbon, excessive temperature and the high concentration of oxygen in compressed air can result in fire or explosion in the discharge pipe. The effect of fire in a compressor discharge, has been known to weaken the pipe causing a split through which flame, supported by the compressed air, has emerged.

Poor valve condition can be detected by observation of the stage pressures and is usually accompanied by excessive discharge temperature although the latter can also be a symptom of poor cooling.

Adequate attention must be given to the water cooling system. Overcooling can cause condensation on the cylinder walls, adversely affecting lubrication, while poor cooling, due probably to scale formation in a sea-water cooled machine, will result in a fall-off in volumetric efficiency and rapid valve deterioration. Inter- and after-coolers require periodic cleaning to remove scale (where they are sea-water cooled) and oily deposits from the air side. With respect to temperature limitation, classification societies require that the compressor should be so designed that the air discharge to the reservoir should not substantially exceed 93°C.

Automatic operation

Before the general introduction of control equipment, air compressors were stopped and started by engine room staff, as necessary, to maintain air receiver pressure. In port or at sea, this usually meant operating one compressor for about half an hour daily unless air was being used for the whistle (during fog), for work on deck or for other purposes. Whilst manoeuvring, the compressors would be started and stopped very frequently unless they were steam driven, when demand could be met by varying the speed. Some compressors were fitted with unloaders to hold the suction valve plates off their seats when receiver pressure reached the maximum and this gave a degree of automatic operation. To drain the coolers continuously during running, an automatic device was fitted to each cooler.

Compressors are now normally arranged for automatic stop and start as dictated by demand through pressure variation. Figure 2.8 shows a scheme for the automatic starting and stopping of two machines. Either machine can be selected as 'lead' machine. This will run preferentially during manoeuvring and at other times, automatically stopping and starting under the control of a

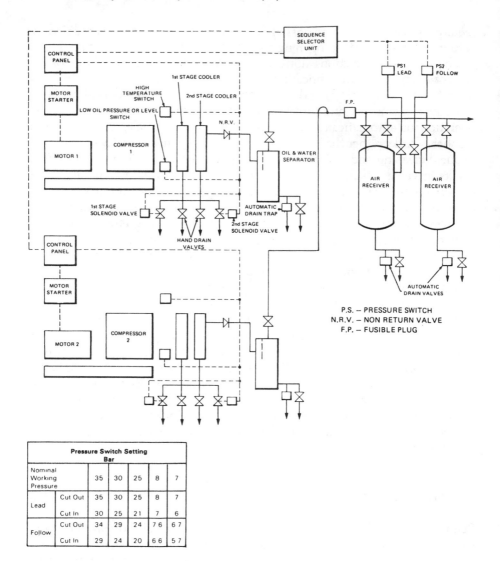

Figure 2.8   *Automatic operation of air compressors (Hamworthy Ltd)*

Pressure Switch Setting Bar						
Nominal Working Pressure		35	30	25	8	7
Lead	Cut Out	35	30	25	8	7
	Cut In	30	25	21	7	6
Follow	Cut Out	34	29	24	7 6	6 7
	Cut In	29	24	20	6 6	5 7

pressure switch on the air receiver. The 'follow' machine is arranged to back-up the 'lead' machine during manoeuvring, cutting in after the 'lead' machine when the receiver pressure falls below a pre-set value (see the table in Figure 2.8). When the pressure switch stops the compressor, the drain valves open automatically.

First and second stage cooler drains (Figure 2.8) may be operated by solenoids or other means. The drain valves are also opened briefly on an intermittently timed cycle thus providing automatic draining as well as unloading.

The cooler drain valves are normally open whilst the compressors are not in use to provide unloaded starting. Two timers are located in the control panel.

One delays the closure of the cooler drains to give the compressor time to run up to speed and to control the time for which drain valves remain open during periodic draining. The other timer controls the frequency of the periodic draining.

In certain applications involving the filtration of compressed air — particularly those associated with control equipment and the safety of human operators — it is imperative that an efficiency of virtually 100% be achieved in the removal of oil, both in the droplet stage and in its vapour form, together with moisture and other pipeline impurities.

A non-return valve in the discharge line of each compressor is necessary. It should be of the low inertia type. Automatic shut-off of the cooling water supply, where independent of the compressor, should also be arranged. This could be achieved with a valve, normally closed, which is opened by the first stage pressure acting on a diaphragm.

Control system air

Pneumatic control equipment is sensitive to contaminants which may be in the air. Viscous oil and water emulsions can cause moving parts to stick and produce general deterioration of diaphragms and other parts made of rubber. Water can cause rust build up which may also result in parts sticking or being damaged by rust particles. Metallic wear and other small particles can cause damage by abrasion. Any solids mixed with oil and water emulsions can conspire to block small orifices. Clean and dry control air is thus essential for the trouble free operation of systems.

Air leaving a conventional compressor usually contains oil carried over from the cylinder and water precipitated in the coolers. The two liquids combine to form an emulsion as witnessed when testing the drains. Dust and other small particles are carried through the compressor with the air, because the suction filters are necessarily fairly coarse. Usually compressor coolers are drained automatically. The receivers may be drained automatically or by engine room staff, sometimes as often as twice per day. The emulsion removed from compressor coolers and air receivers, has a viscosity which varies with machinery space conditions. Some emulsions are very viscous and they cause most problems with sticking.

When the source of control and instrument air is main air compressors and reservoirs, then special provision is necessary to ensure that air quality is high. The reducing valve which brings the main air pressure to the 7 or 8 bar required by the control air system, can be affected by emulsion carry over and can require frequent cleaning to stop it from sticking. Automatic drain traps may be fitted to the control air system, but many have traps which require daily draining by staff.

A moderate amount of free moisture in the control air could be removed by ceramic filters but to give the desired dryness factor an absorbent type drier may be considered necessary or a drier using refrigeration.

Control air from starting air at 15 to 40 bar

A three stage filtration system (Figure 2.9) employing a pre-filter, a carbon absorber, and an after filter, may be installed to deliver good quality air. The pre-filter contains a medium grade porous ceramic element and removes the gross atmospheric impurities from the air. The absorber is packed with activated carbon and provides a deep bed producing an evenly distributed flow for the removal of vapours (it can also remove flavours or taints). The after filter contains a fine grade porous ceramic element preventing the ingress of any migrating carbon particles to the pipeline, thereby assuring a pure supply.

Air from the main engine starting air receiver enters the filter assembly fully saturated, that is, carrying some free water and oil vapour. Virtually all of the free water and oil should be removed. Due to precipitation through the filter, more moisture will be removed from the air reducing the dew point by approximately 5°C. When the pressure reducing valve drops the pressure to 7 bar, the dew point falls sufficiently for the air to be suitable for immediate use in a control air system without the use of absorption driers. The reducing valve for this arrangement is also protected from the effects of emulsion in the air.

Oil-free and non-oil-free rotary compressors

Both these machines deliver wet air which must be dried as described above but the non-oil-free machine passes over some oil which must also be removed. This is usually effected by a pre-filter followed by a carbon absorber, which removes the oil, followed by an after filter (to remove the remaining free moisture). Either system can be found at sea.

Compressed air systems for steamships

A compressed air system is necessary to supply air for boiler soot-blower air motors, hose connections throughout the ship and possibly diesel generator starting. A general service air compressor would supply air at 8 bar but greater pressure (as for diesel ships) would be necessary for diesel starting. A general service air compressor can be shut down completely when air is not required but can be operated on either a stop/start or load/unload cycle, with the regulator controlling the pressure between 7 and 8.5 bar.

The instrument air system and general service air system should be separate and cross connected only in emergency. The instrument air is supplied typically by oil-free, water cooled compressors, arranged for discharge to the air reservoir through after coolers. Three compressors may be installed, to operate with two units running continuously on a load/unload cycle between 5.5 and 7 bar with the other unit ready on stand-by to start if the pressure in the reservoir falls below 5.3 bar. The air may be delivered to the control air system through two of three air dryers (a third drier in reserve) fitted with automatic drain traps. Air dryers based on the cooling of the air by refrigeration, have a

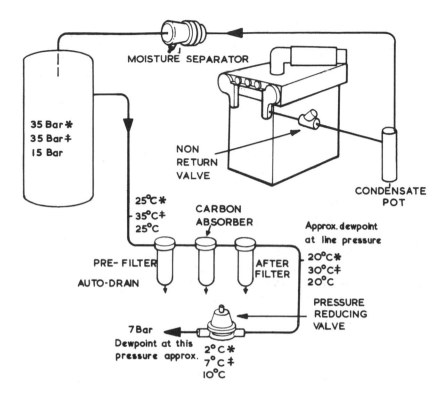

Figure 2.9 *Three stage filter for control air (Hamworthy Ltd)*

small sealed unit refrigeration compressor. In normal service they will reduce the dew point of the air to about −25°C and a high dew point alarm is fitted to warn of malfunction. The units are referred to as dehumidifiers; being rated typically at 170 m³/hr for an air consumption of about 150–160 m³/hr. Normally it is arranged that one of the units is able to carry the full load if the other units are not available. The filter element of each refrigifilter should be inspected and changed if necessary every six months.

Although the instrument air system is fitted with many individual drain traps and cocks, no moisture should be present anywhere in the system after the air dryers and if any is found the cause must be immediately investigated.

Control air consumption can be reduced at sea by shutting off the air supply to systems not in use such as tanker pumproom and cargo control room regulators and controllers.

Fuel handling and treatment

Fuels and lubricating oils are obtained from crude primarily by heating the crude oil, so that vapours are boiled off and then condensed at different temperatures. The constituents or fractions are collected separately in a distillation process. Crude oil contains gaseous fuels, gasoline (petrol), kerosene (paraffin), gas oils, distillate diesel fuels and lubricating oils which can be collected from the fractionating tower (Figure 2.10) where they condense out at the different levels maintained at appropriate temperatures. The crude is heated in a furnace, as shown at the left of the sketch.

The boiling process produces a residue which is very dense as the result of having lost the lighter parts. This high-density remainder has much the same hydrocarbon make up as the lighter fractions and is used as a fuel. Unfortunately the initial refining process not only concentrates the liquid but also the impurities.

Vacuum distillation, a second process, removes more of the lighter fractions, to leave an even heavier residue. As can be seen from Figure 2.10, the refinery can have additional conversion equipment. Vis-breaking or thermal cracking is one process using heat and pressure to split heavy molecules into lighter components giving a very dense residue. Catalytic cracking is a process that uses a powdered silica-alumina based catalyst with heating, to obtain lighter fractions, with, however, an increasingly heavy residue. In the latter process catalyst powder is continuously circulated through the reactor then to a regenerator where carbon picked up during the conversion reaction is burnt off. Unfortunately, some of the catalyst powder can remain in the residue which may be used for blending bunker fuel oils. The very abrasive silica-alumina catalytic fines have caused severe engine wear when not detected and removed by slow purification in the ship's fuel treatment system.

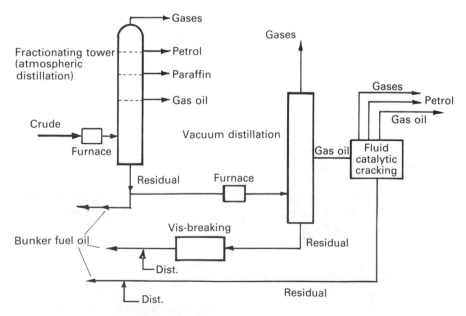

Figure 2.10   *Oil Refinery processes*

Fuel testing

Bunkers are classified as Gas Oil, Light and Marine Diesel Oil, Intermediate Fuel Oil and Marine or Bunker (C) Fuel Oil. The delivery note specifies the type of fuel, amount, viscosity, specific gravity, flash point and water content. Trouble frequently results from inferior fuels and there can be insuffient information to give warning. Fuel grading schemes and more detailed delivery notes are being used.

Some of the Classification Societies and specialist firms provide testing services and on-board testing equipment is available. A representative sample is needed to give an accurate test result and this is difficult to obtain unless a properly situated test cock is fitted in the bunker manifold where flow is turbulent. The sample is taken after flushing the test cock. Because of the variation in heavy fuel, small quantities are taken into the test container over the period of bunkering, to give a representative sample.

A full analysis can be given by the shore laboratory. On-board tests are limited to those which give reliable results and kits for specific gravity, viscosity, pour point, water content and compatibility are on the market. Flash point is found with a Pensky–Martin closed cup apparatus which has been carried on some ships.

Oil fuel transfer

The oil fuel system (see Figure 3.1 Chapter 3) provides the means for delivering fuel from the receiving stations at upper deck level, port and starboard, to double-bottom or deep bunker tanks. Sampling cocks are fitted at the deck connections to obtain a representative specimen for (a) shore analysis; (b) on board testing; and (c) retention on the ship. A filter on the downpipe removes large impurities.

Transfer

The fuel oil can be pumped from storage to settling tanks and also transferred if necessary, between forward and aft, port and starboard storage tanks, by means of heavy oil and diesel oil fuel transfer pumps. Transfer from settling to service tanks (Figure 2.11) in motorships, is via centrifuges. The latter are arranged as purifiers or clarifiers.

Fire risk

Fire is an ever-present hazard with liquid fuel because the vapour from it can form a flammable/explosive mixture with air. A hydrocarbon and air mixture containing between about 1% and 10% of hydrocarbon vapour, can be readily

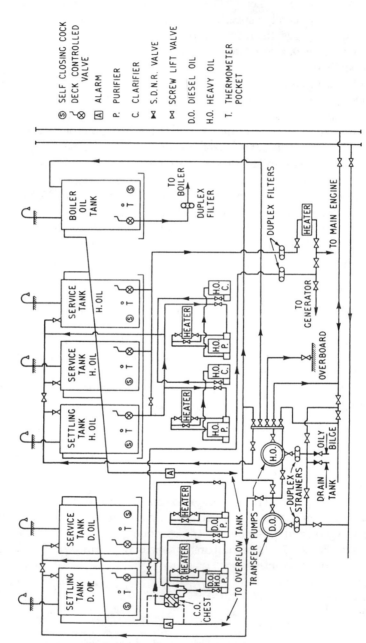

Figure 2.11 Settling and service tanks

ignited by a naked flame or spark. Combustion will also occur if the flammable mixture is in contact with a hot surface which is at or above the ignition temperature of the mixture. Ignition temperature for a hydrocarbon mixture may be about 400°C or less.

A crude oil will give off flammable gases such as methane, at ambient temperature, so that the space above it is very likely to contain a flammable mixture. Fuel oils are produced from crude in the refinery, when the gases such as methane are also extracted. Oil fuels are safer than the original crude oil.

Heat will cause the evolution of hydrocarbon vapour from fuel with the quantity being related to the volatility and temperature of the fuel. The hotter the fuel, the greater the amount of vapour accumulating in the space above it. The test for the closed flash point of a fuel, is based on the heating of a fuel sample in a closed container. The heat causes evolution of vapour which accumulates in the air space above the liquid and mixes with the air. A naked flame is dipped into the container at set temperature rise intervals. When the lower flammable limit (LFL) is reached (about 1% hydrocarbon vapour in air) ignition will occur and the mixture will burn with a brief blue flash. The temperature at which this occurs, is termed the closed flash point.

There are rules governing acceptable flash points and permissible temperatures for storage and handling of oil fuel. The closed flash point of fuels for general use should be not less than 60°C, but slightly higher figures are suggested by some authorities. Oil should not be heated to more than 51°C in storage and not more than 20°C beneath its known closed flash point. For purification and while in the service tank, then during delivery to the engine, temperatures are increased as necessary, to reduce viscosity. Settling tanks must have thermometers and the sounding arrangements must be proof against accidental egress of oil. Drain cocks must be self-closing and the outlet valves should be capable of being closed from safe positions outside the engine or boiler room. In passenger ships, this applies also to suction and levelling valves on deep tanks. Overflow pipes and relief valves not in closed circuit should discharge to an overflow tank having an alarm device, the discharge being visible. Tank air pipes must have 25% more area than their filling pipes and should have their outlets situated clear of fire risks. They should also be fitted with detachable wire gauze diaphragms. Provision should be made for stopping oil fuel transfer pumps from outside the machinery spaces.

From the filling station pipes descend to the oil fuel main(s). These will probably be two pipes, one for heavy oil and one for diesel fuel. The system extends forward and aft in the machinery spaces, possibly extending along the shaft tunnel and, in some ships, in a duct keel or pipe tunnel. The pipes connect to the fuel transfer pump(s) and to distribution valve (or cock) chests, from which pipes run to the fuel tanks. It was the practice to carry water ballast in empty fuel tanks and change over chests were arranged so that simultaneous connection to oil and ballast mains was not possible.

Transfer pumps draw from the oil main(s), from overflow and drain tanks and from the oily bilges — parts of the engine and boiler room bilges separated from the remainder by coamings — to which oil spillage is led. The pumps discharge to settling tanks, the oily water separator and the oil main(s). In passenger ships, it must be possible to transfer oil from any tank to any other tank without use of the ballast main, but this is possible with most systems. The heavy oil and diesel oil transfer pumps can usually by cross connected if necessary. Detail and arrangements will vary with the size, type and trade of the ship.

In steamships, the fuel is heated in the settling tanks by steam coils to assist water separation, and is then delivered to the burners through heaters and filters by the oil fuel pressure pumps.

In motorships, residual fuel is pumped from storage to one of, ideally, two settling tanks (Figure 2.11) of 24 hour capacity. Steam heating assists settling over 24 hours if possible, when any large quantities of water together with sludge, will gravitate to the bottom for removal via the drain or sludge cock. While the settling progresses in one tank, fuel from the other is being purified to the service tank which is not in use.

The purification of heavy fuel on many ships relies on one heater and a single self-sludging purifier. If problems with catalytic fines are likely or there are exceptional amounts of solid impurities, then two or even three centrifuges may be installed in series (Figure 2.12) or parallel (Figure 2.13) arrangement. With the series arrangement, the first machine acts as a purifier, removing any remaining water and most of the solids in suspension. The second machine, set up as a clarifier, removes the finer solids remaining. The parallel arrangement shown uses two machines set up as purifiers, each having a slow throughput to permit a longer dwell time for the fuel. The separators can have their own pumps but the separate pumps shown, allow flow to be controlled without restricting it. Closing in a valve to throttle flow, causes turbulence which can mix fuel and impurities more closely. Clean fuel is delivered to the service tank which is not in use. Fuel from the duty service tank passes to the engine booster pump and so to the fuel pumps and injectors, through further heaters. Diesel fuel is treated similarly but more simply, with a single stage of separation and no heating.

Sludge from centrifugal separators passes to a sludge tank from which it is removed by a pump capable of handling high viscosity matter.

It may be mentioned here, because it is not always understood, that fuel is heated for combustion, in order to bring it to a viscosity acceptable to the fuel injectors or burners.

Centrifuges

Liquids with a specific gravity or relative density difference can be separated in a settling tank by the effect of gravity and the process can be represented mathematically by

$$F_s = \frac{\pi}{6} D^3 \ (\rho_w - \rho_o) \ g. \tag{1}$$

Clearly in a standing vessel the acceleration cannot be altered to enhance the

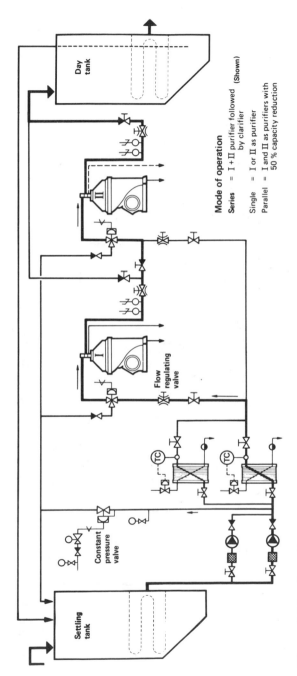

Figure 2.12 *Centrifuges arranged in series (courtesy Alpha-Laval)*

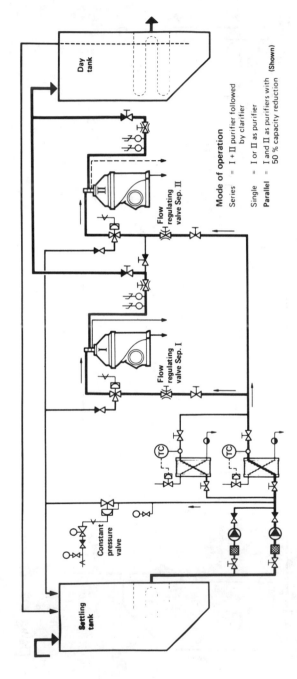

Figure 2.13 *Centrifuges arranged in parallel (courtesy Alpha-Laval)*

Within the figure:

Day tank

Settling tank

Constant pressure valve

Flow regulating valve Sep. II

Flow regulating valve Sep. I

Mode of operation

Series = I + II purifier followed
          by clarifier

Single = I or II as purifier

Parallel = I and II as purifiers with
          50 % capacity reduction **(Shown)**

separation force Fs, but by subjecting the operation to centrifugal force the above expression can be replaced by

$$F_s = \frac{\pi}{6} D^3 (\rho_w - \rho_o) \omega^2 r \tag{2}$$

where:

ω = angular velocity
r = effective radius.

Both the rotational speed and the effective radius are controllable within certain engineering limitations. Thus if a settling tank is replaced by a rotating cylinder the separating force and hence the speed of separation can be increased. This, effectively, is what happens in a centrifuge.

For many years marine centrifuges were designed for batch operation, that is the machines were run for a period during which solids accumulated in the bowl then the machine was stopped for manual cleaning. Batch centrifuging is still beneficial for the purification of lubricating oil because manual cleaning enables the operator to check the effectiveness of the operation. Modern centrifuges for fuel run continuously, being automatically sludged during operation. Most modern lubricating oil purifiers are also designed for continuous operation with an automatic self-sludging programme. Two distinct types of batch operated machines have been used. These are illustrated in Figure 2.14 and Table 2.1.

The obsolete tubular bowl machine was physically able to withstand higher angular velocities than the wide or disc bowl type, hence a higher centrifugal

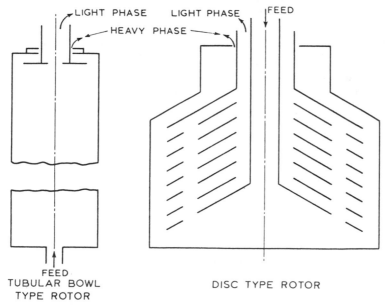

Figure 2.14 *Comparison of narrow and wide bowl types (Penwalt Ltd)*

Table 2.1   *Details of tubular and disc type centrifuges used for batch treatment of liquids*

	Maximum centrifugal force	Bowl dia. (mm)	Length dia. (of bowl)	Drive
Batch tubular Bowl type	13 000 – 20 000 × g	Up to 180	Up to 7	Bowl suspended from above
Batch disc bowl Type	5 000 – 8 000 × g	Up to, say, 600	Generally <1	Bowl supported from below

effect was available. The heavier phase (water) had only a short distance to travel before coming to the bowl wall where solids were deposited and the heavy phase liquid (water) was guided to the water discharge. However, the sludge retention volume and the liquid dwell time for a given throughput could only be increased by lengthening the bowl. This gave rise to bowl balancing and handling problems.

The wide bowl type was able retain more sludge before its performance was impaired and was much easier to clean. On the other hand settling characteristics in a wide bowl machine are relatively poor towards the bowl centre and the distance the water has to travel before reaching the wall is great.

To overcome these problems a stack of conical discs (Figure 2.15) spaced about 2–4 mm apart is arranged in the bowl. The liquid is fed into the bottom of the stack and flows through the spaces between adjacent plates. The plates then act as an extended settling surface, with the heavy impurities impinging on the under surfaces of the discs. As the particles impinge of the disc surfaces, they accumulate and eventually slide along the discs towards the periphery. At the disc stack periphery, water globules and solid particles continue to move out towards the bowl wall with the water being sandwiched between the solids and the oil, which orientates itself towards the bowl centre. The boundaries at which substances meet are known as interfaces.

The oil/water interface is very distinct and is known as the e-line. To gain the fullest advantage from the disc stack the e-line should be located outside of it. On the other hand if the e-line is located outside the water outlet baffle (top disc) discharge of oil in the water phase will take place.

Referring back to gravity separation in a settling tank, if the tank is partitioned as shown in Figure 2.16 continuous separation will take place. Since the arrangement is a very crude U-tube containing two liquids of different specific gravities, the height of the liquid in the two legs will have the relationship

$$\rho_l\,(e-l) \simeq \rho_h\,(e-h) \tag{3}$$

where:

ρ_l = density of oil
ρ_h = density of water.

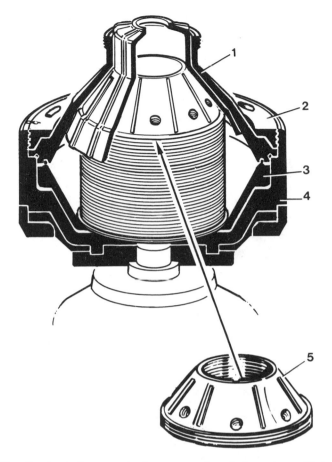

Figure 2.15 *Conical disc stack. Separator bowl: (courtesy Alpha-Laval)*

1. Bowl hood
2. Lock ring
3. Sliding bowl bottom

4. Bowl body
5. Disk stack

A very similar condition is found in the centrifuge (Figure 2.17) for which the equation is

$$\omega^2 \; \rho_l \; (e^2 - l^2) = \omega^2 \; \rho_h \; (e^2 - h^2) \tag{4}$$

or

$$\frac{\rho_h}{\rho_l} = \frac{e^2 - l^2}{e^2 - h^2} \tag{5}$$

The mechanical design of the centrifuge requires that the e-line is confined within certain strict limits. However variations in ρ_l will be found depending upon the port at which the vessel takes on bunkers. It is necessary therefore to provide means of varying h or l to compensate for the variation in specific gravity. It is usually the dimension h which is varied, and this is done by the use of dam rings (sometimes called gravity discs) of different diameters. Normally a

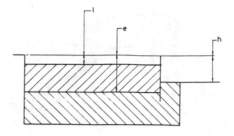

Figure 2.16   *Hydrostatic situation in gravity settling tank*

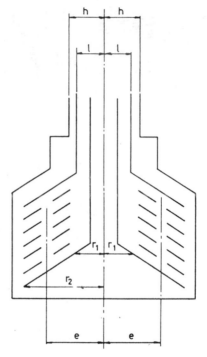

Figure 2.17   *Hydrostatic seal in a disc type centrifugal bow (Penwalt Ltd)*

table is provided in the instruction book for the machine, giving the disc diameter required for purifying oils of various specific gravities. Alternatively the disc diameter D_h may be calculated from the following formula which is derived from (5)

$$D_h = 2\sqrt{\left[l^2 \frac{\rho_l}{\rho_h} + e^2 \left(1 - \frac{\rho_l}{\rho_h}\right) \right]} \tag{6}$$

The dimension e can be taken as the mean radius of one thin conical plate and the heavy top conical plate (outlet baffle). If oil is discharging in the water outlet the gravity disc is too large.

It is important to realize that variation in oil temperature will cause a proportional variation in specific gravity (SG) and for this reason the oil temperature must remain constant. To some extent the feed rate will have an effect on the e-line (because it can alter the overheight of the liquid flowing over the lip of the gravity disc). Excessive feed should be avoided in any event since the quality of separation deteriorates with an increase in throughput. To prevent oil passing out of the water side on start up it is necessary to put water into the bowl until the water shows at the water discharge. The bowl of a purifier is, in effect, sealed by this water.

A clarifier is a centrifuge in which the dam ring or gravity disc is replaced by a blank disc with no aperture. The straight clarifier is intended to remove only solids, not water. Thus it has no requirement for sealing water when started.

Alfa-Laval intermittent discharge centrifuge

Figure 2.18 shows a centrifuge bowl capable of being programmed for periodic and regular dumping of the bowl contents to remove the sludge build-up. The sludge discharge takes place through a number of slots in the bowl wall. Between discharges these slots are closed by the sliding bowl bottom, which constitutes an inner, sliding bottom in the separating space. The sliding bowl bottom is forced upwards against a seal ring by the pressure of the operating liquid contained in the space below it. This exceeds the counteracting downward pressure from the process liquid, because the underside of the sliding bowl bottom has a larger pressure surface (radius R_1), than its upper side (radius R_2). Operating liquid is supplied on the underside of the bowl via a device known as the paring disc. This maintains a constant operating liquid annulus (radius R_3) under the bowl, as its pumping effect neutralizes the static pressure from the supply.

When the sludge is to be discharged, operating liquid is supplied through the outer, wider supply tube so that if flows over the lower edge of the paring chamber (radius R_4) and continues through a channel out to the upper side of a sliding ring, the operating slide. Between discharges, the operating slide is pressed upwards by coil springs. It is now forced downwards by the liquid pressure, thereby opening discharge valves from the space below the sliding bowl bottom so that the operating liquid in this space flows out (b).

When the pressure exerted by the operating liquid against the underside of the sliding bowl bottom diminishes, the latter is forced downwards and opens, so that the sludge is ejected from the bowl through the slots in the bowl wall. Any remaining liquid on the upper side of the operating slide drains through a nozzle g (c). This nozzle is always open but is so small that the outflow is negligible during the bowl opening sequence.

On completion of sludge discharge, the coil springs again force the operating slide upwards (d), thus shutting off the discharge valves from the space below the sliding bowl bottom. Operating liquid is supplied through the outer, wider tube, but only enough to flow to the space below the sliding bowl bottom and force the latter upwards so that the bowl is closed. (If too much

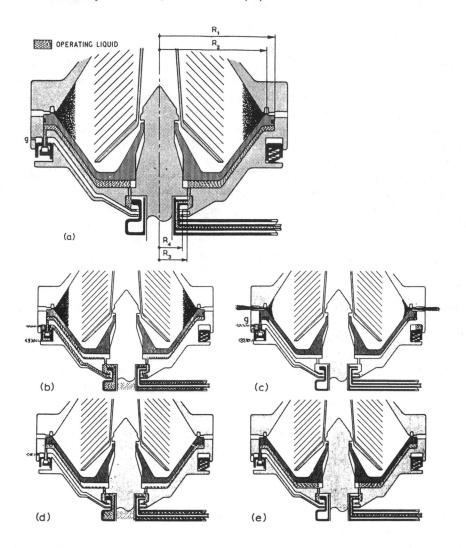

Figure 2.18   *Self-sludging centrifuge. Sequence of operation (Alpha-Laval)*

liquid is supplied, it will flow into the channel to the operating slide and the bowl will open again.)

The outer, wider inlet is now closed while the inner, narrower one is open (e). The paring disc counter-balances the static pressure from the operating liquid supply, and the bowl is ready to receive a further charge of oil. The situation is identical with that shown in the first illustration of the series but with the difference that the sludge discharge cycle is now accomplished.

Periodically the purifier bowl should be stripped and thoroughly cleaned. It is important to remember that this is a precision built piece of equipment, which has been carefully balanced and all parts should be treated with the utmost care.

High density fuel treatment (ALCAP System)

The density of a fuel tested at 15°C may approach, be equal to or greater than that of water. With high density fuels, the reduction in density differential between fuel and water can cause a problem with separation but not with the usual solid impurities. Heating of the fuel (Figure 2.19) will reduce the density and this may be sufficient in itself to obviate the problem of water separation. The change in density of water with temperature (dotted line) is not so pronounced, as can be seen from the graph, so that heating produces a differential. Some caution must be exercised in heating the fuel.

The Alfa-Laval design of centrifuge (Figure 2.20) intended for dealing with high density residual fuels, is a self-sludging machine which has a flow control disc that makes it virtually a clarifier. There are no gravity discs to be changed to make the machine suitable for fuels of different specific gravity/density. Heating is used to reduce the density (and viscosity) of the fuel so that water

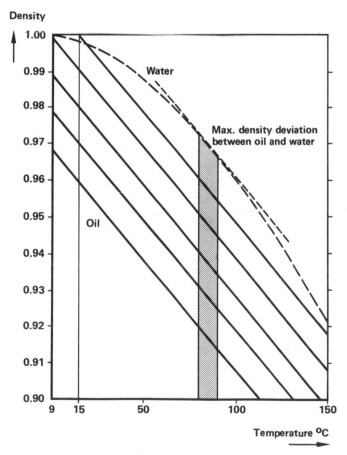

Figure 2.19 *Temperature/density graph (courtesy Alpha-Laval)*

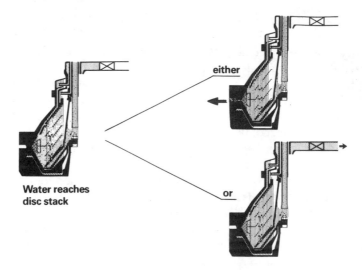

**Water reaches
disc stack**

Figure 2.20   *FOPX separator showing discharge of water (courtesy Alpha-
Laval)*

and sludge accumulate in the outer part of the bowl, as the result of the
centrifugal effect. As the interface moves inwards, but before reaching the disc
stack, water droplets flow through to reach a water sensing transducer (Figure
2.21). Via micro-processor circuitry, the transducer causes the bowl to
self-sludge or the water to be discharged through the water drain valve. The
system is said to be capable of handling fuels with densities as high as
1010 kg/m³ at 15°C.

Ancillary fuel equipment

The system which delivers residual fuel from the daily service tank to the diesel
or boiler, must bring it to the correct viscosity by heating. Filtration of the hot
fuel is important for diesels and a homogenizer may be installed.

Fuel heater

For burning heavy fuel oil in a boiler furnace, or a compression-ignition engine,
it is necessary to pre-heat it. This may be done in a shell and tube unit either
with plain tubes (Figure 2.22) or tubes with fins bonded to them (Figure 2.23)
and the oil flowing on the outside of the tubes. The heating medium is normally
steam but additional electric heaters are useful for start up from dead ship
condition.

The heating steam is used most effectively if it condenses during its passage
through the heater, and donates the large amount of latent heat as it reverts
from steam to water. A steam trap (see Chapter 4) is fitted at the steam outlet

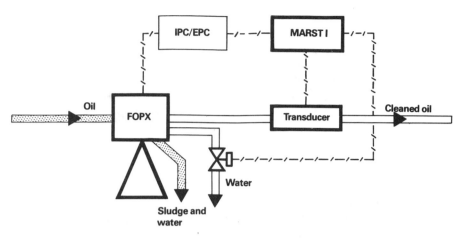

Figure 2.21 *ALCAP system showing water sensing transducer (courtesy Alpha-Laval)*

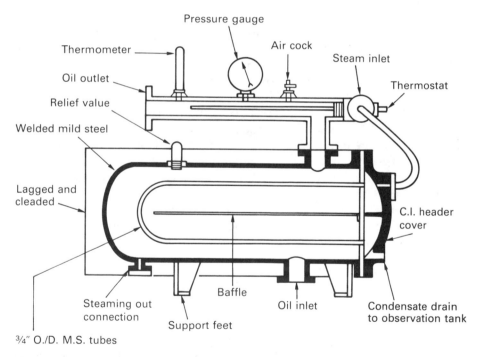

Figure 2.22 *Fuel heater with plain tubes*

from the heater, to make sure that only water returns to the observation and drain tank. If the trap fails, it is necessary to close in the steam return valve to achieve approximately the same effect (about a half or a quarter turn open).

Thermostat control may be employed for fuel heaters (Figure 2.22) with the setting based on a chart showing variation of viscosity with temperature. The charts may not of course be accurate for a particular fuel. Better results are

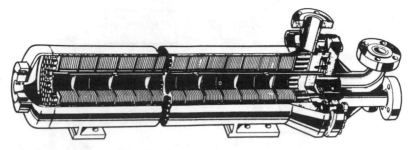

Figure 2.23   *Fuel heater with fins bonded to tubes*

achieved with viscosity controllers, which are used to control viscosity directly, through control of the steam supply.

Viscosity controllers (Viscotherm)

The basic principle of the Viscotherm viscosity monitoring device is shown in Figure 2.24. A continuous sample of the fuel is pumped at a constant rate through a fine capillary tube. As the flow through the tube is laminar, pressure drop across the tube is proportional to viscosity.

In this unit an electric motor drives the gear pump through a reduction gear, at a speed of 40 rpm. The pump is positioned in the chamber through which the fuel is passing from the heater to the fuel pumps or combustion equipment. Tapping points are provided to enable the pressure difference to be measured by means of a differential pressure gauge. The gauge is calibrated directly in terms of viscosity. Parts in contact with the fuel are of stainless steel for corrosion resistance.

A differential pressure transmitter (Figure 2.25) provides an analogue of viscosity to a pneumatic controller, which regulates the supply of fuel heating steam through a control valve.

Homogenizer

The homogenizer (Figure 2.26) provides an alternative solution to the problem of water in high density fuels. It can be used to emulsify a small percentage for injection into the engine with the fuel. This is in contradiction to the normal aim of removing all water, which in the free state, can cause gassing of fuel pumps, corrosion and other problems. However, experiments in fuel economy have led to the installation of homogenizers on some ships to deal with a deliberate mixture of up to 10% water in fuel. The homogenizer is fitted in the pipeline between service tank and engine so that the fuel is used immediately. It is suggested that the water in a high density fuel could be emulsified so that the fuel could be used in the engine, without problems. A homogenizer could not be used in place of a purifier for diesel fuel as it does not remove abrasives such

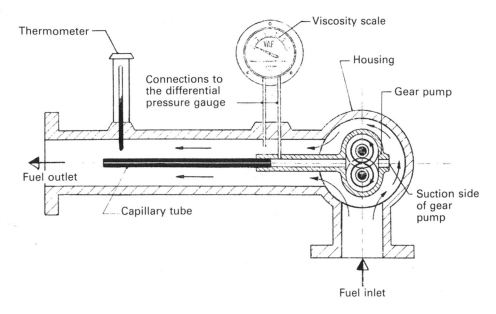

Figure 2.24 *Viscotherm viscosity monitoring device*

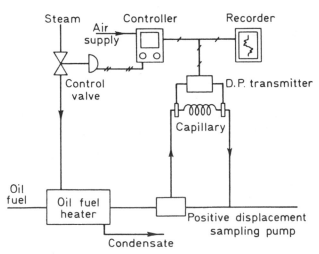

Figure 2.25 *Differential pressure transmitter*

as aluminium and silicon, other metallic compounds or ash-forming sodium which damages exhaust valves.

The three disc stacks in the rotating carrier of the Vickers type homogenizer are turned at about 1200 rev/min. Their freedom to move radially outwards means that the centrifugal effect throws them hard against the lining tyre of the homogenizer casing. Pressure and the rotating contact break down sludges and water trapped between the discs and tyre, and the general stirring action aids mixing.

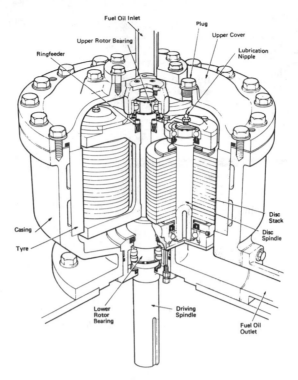

Figure 2.26   *Homogenizer (Vickers type)*

Package boiler combustion system

The elementary automatic combustion system based on a two flame burner
(Figure 2.27) is used for many auxiliary boilers. The burner is drawn oversize to
show detail. Various different control systems are employed for the
arrangement.

The burner has a spring loaded piston valve which closes off the passage to
the atomizing nozzle when fuel is supplied to the burner at low pressure. If the
fuel pressure is increased the piston valve will be opened so that fuel passes
through the atomizer. The system can supply the atomizer with fuel at three
different pressures.

The solenoid valves are two-way, in that the fuel entering can be delivered
through either of two outlets. The spill valves are spring loaded. When either
one is in circuit, it provides the only return path for the fuel to the suction side
of the fuel pressure pump. The pressure in the circuit will be forced, therefore to
build up to the setting of the spill valve.

A gear pump with a relief arrangement to prevent excessive pressure, is used
to supply fuel to the burner. Fuel pressure is varied by the operation of the
system and may range up to 40 bar.

Combustion air is supplied by a constant speed fan, and a damper
arrangement is used to change the setting.

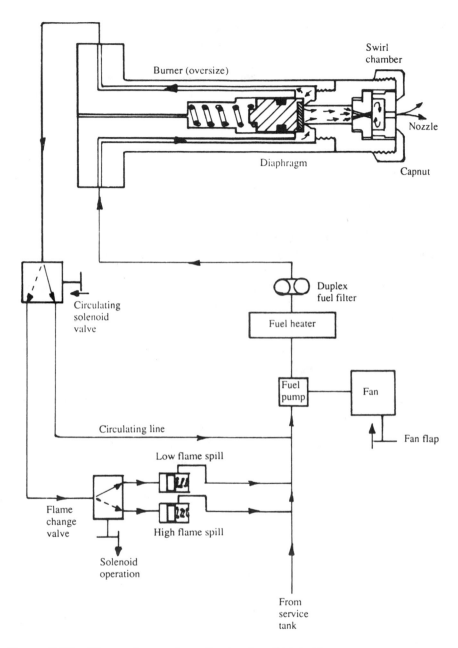

Figure 2.27 *Elementary automatic combustion system*

System operation

Control of the setup may be through various combinations of electrical,
electronic or mechanical systems. An electrical control scheme is employed in
this description. Electrical circuits are arranged so that when the boiler is
switched on (assuming water level and other factors are correct) the system will

(1) heat up and circulate the fuel; (2) purge the combustion space of unburnt gas; and (3) ignite the flame and, by controlling it, maintain the required steam pressure.

When the boiler is started, current is supplied first to the fuel heater. The electric heating elements are thermostatically controlled and when oil in the heater reaches the required atomizing temperature, another thermostat switches in the fan and oil circulating pump. Air from the fan purges the combustion spaces for a set time, which must be sufficient to clear any unburnt gases completely. If not removed an air/gas explosive mixture may be present, so that flame ignition could result in a dangerous blowback. The oil circulates from the pump and heater through the system via the oil circulating valve. This ensures that the oil flows through the burner until it is hot and thin enough to atomize.

When the oil circulating solenoid is operated, the fuel no longer returns to the suction side of the pump but is delivered to the low flame spill through the oil change valve. With the ignition arc 'on', oil pressure builds up sufficiently to push open the piston valve in the burner. The atomized fuel is ignited and once the flame is established, control of the oil change valve and fan damper depends on steam pressure. With low steam pressure, the oil control valve is actuated to deliver the fuel to the high flame spill. Pressure increases until this spill opens and the higher pressure forces a greater quantity of fuel through the burner. When steam pressure rises, the fuel is switched back to the low flame spill. The fan damper is operated at the same time to adjust the air delivery to the high or low flame requirement. The solenoid or pulling motor for the operation of the high/low flame is controlled by a pressure switch acted on by boiler steam pressure.

Boilers with automatic combustion systems have the usual safety valves, gauge glasses and other devices fitted for protection with additional special arrangements for unattended operation.

The flame is monitored by a photo-cell and abnormal loss of flame or ignition failure, results in shut down of the combustion system and operation of an alarm. Sometimes trouble with combustion will have the same effect if the protective glass over the photo-cell becomes smoke blackened.

Water level is maintained by a float-controlled feed pump. The float chamber is external to the boiler and connected by pipes to the steam and water spaces. There is a drain at the bottom of the float chamber. A similar float switch is fitted to activate an alarm and shut-down in the event of low water level (and high water level on some installations). Because float chambers and gauge glasses are at the water level, they can become choked by solids which tend to form a surface scum on the water. Gauge glasses must be regularly checked by blowing the steam and water cocks through the drain. When float chambers are tested, caution is needed to avoid damage to the float. Frequent scumming and freshening will remove the solids which are precipitated in the boiler water by the chemical treatment.

The boiler pressure will stay within the working range if the pressure switch is set to match output. If a fault develops or steam demand drops, then high steam pressure will cause the burner to cut out and the fuel will circulate as for warming through.

Incorrect air quantity due to a fault with the damper would cause poor combustion. Air delivery should therefore be carefully monitored.

Many package boilers burn a light fuel and heating is not required. Where a heater is in use, deviation from the correct temperature will cause the burner to be shut off.

The automatic combustion system is checked periodically and when the boiler is first started up. The flame failure photo-cell may be masked, so to test its operation or some means − such as starting the boiler with the circulating solenoid cut out − may be used to check flame failure shut down. Cut outs for protection against low water level, excess steam pressure, loss of air and change of fuel temperature are also checked. Test procedures vary with different boilers. At shut down the air purge should operate; the fan being set to continue running for a limited time.

Fuel blender for auxiliary diesels

Conventionally, the lower cost residual fuels are used for large slow speed diesel main engines and generators are operated on the lighter more expensive distillate fuel. The addition of a small amount of diesel oil to heavy fuel considerably reduces its viscosity and if heating is used to further bring the viscosity down then the blend can be used in generators with resultant savings.

The in-line blender shown in Figure 2.28 takes fuels from heavy oil and light diesel tanks, mixes them and supplies the mix directly to auxiliary diesels. Returning oil is accepted back in the blender circulating line. It is not directed back to a tank where there would be the danger of the two fuels settling out.

Fuel is circulated around the closed loop of the system by the circulating pump against the back pressure of the p.s. (pressure sustaining) valve. Thus there is supply pressure for the engine before the valve and a low enough

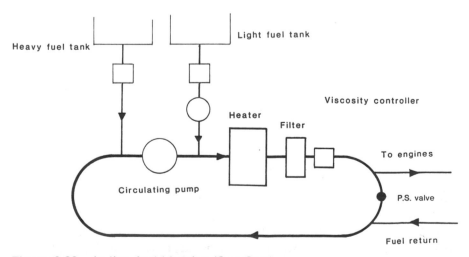

Figure 2.28 *In-line fuel blender (Sea-Star)*

pressure after it, to allow returning oil back into the loop. Sufficient light diesel is injected into the loop by the metering pump for light load running. As increased load demands more fuel, this is drawn in from the heavy oil tank by a drop in loop pressure on the suction side of the circulating pump. The extra fuel made necessary as the load increases is supplied from the residual fuel tank. At full load the ratio may be 30% diesel with 70% heavy fuel.

A viscotherm monitors viscosity and controls it through the heater. The hot filter removes particles down to 5 micron size and there are other filters on the tank suctions. Constant circulation and remixing of the blend and the returning fuel prevents separation.

The diesel is started and runs light on distillate fuel. As the load increases, heavy fuel is added.

Lubricating oil and treatment

Mineral oils for lubrication are, like fuel, derived from crude during refinery processes. Basic stocks are blended to make lubricants with the desired properties and correct viscosity for particular duties. Additives are used to enhance the general properties of the oil and these include oxidation and corrosion inhibitors, anti-corrosion and rust prevention additives, foam inhibitors and viscosity index (VI) improvers. The latter lowers the rate of change of viscosity with temperature.

Basic mineral oil is the term commonly used for oils with the additives mentioned above. These additives enhance the general properties of the oil.

HD or detergent type oils are derived in the same way as basic mineral oils by blending and the use of additives to enhance general properties but additional additives are used to confer special properties. Thus detergent-dispersant ability and the use of alkaline additives make these oils suitable for use in diesel engines.

Detergent type or HD lubricating oils

The main function of detergent-dispersant additives in a lubricating oil is to pick up and hold solids in suspension. This capability can be applied to other additives such as the acid neutralizing alkaline compounds as well as solid contaminants. Thus detergent oils hold contaminants in suspension and prevent both their agglomeration and deposition in the engine. This function reduces ring sticking, wear of piston rings and cylinder liners, and generally improves the cleanliness of the engine. Other functions include reduction of lacquer formation, corrosion and oil oxidation. These functions are achieved by the formation of an envelope of detergent oil round each particle of solid contaminant. This envelope prevents coagulation and deposition and keeps the solids in suspension in the oil.

In engines of the trunk piston type with a combined lubrication system for bearings and cylinders, in addition to the deposition of the products of incomplete combustion which occurs on pistons, piston rings and grooves, some of these products can be carried down into the crankcase, contaminating the crankcase oil with acid products and causing deposit build up on surfaces and in oil lines. Detergent oils are, therefore, widely used in this type of engine.

The detergent additives used today are, in most cases, completely soluble in the oil. There is a tendency for the detergent to be water soluble, so that an emulsion may be formed particularly if a water-washing system is used while purifying.

Manufacturers of centrifuges have carried out a considerable amount of research work in conjunction with the oil companies on the centrifuging of basic mineral and detergent lubricating oils using three different methods of centrifuging. These are purification, clarification and purification with water washing. The following is a summary of the findings and recommendations based on the results which were obtained.

When operating either as a purifier (with or without water washing) or as a clarifier, all particles of the order of 3–5 microns and upwards are completely extracted, and when such particles are of high specific gravity, for example iron oxide, very much smaller particles are removed. The average size of solid particles left in the oil after centrifuging are of the order of only 1–2 microns. (One micron is a thousandth part of a millimetre.) Particles left in the oil are not in general of sufficient size to penetrate any oil film in the lubricating oil system.

A centrifuge should be operated only with the bowl set up as a purifier, when the rate of contamination of the lubricating oil by water is likely to exceed the water-holding capacity of the centrifuge bowl between normal bowl cleanings.

When the rate of water contamination is negligible the centrifuge can be operated with the bowl set up as a clarifier. No sealing water is then required and this reduces the risk of emulsification of HD oils. Any water separated will be retained in the dirt-holding space of the bowl.

For basic mineral oils in good condition, purification with water washing can be employed to remove water soluble acids from the oil, in addition to solid and water contaminants. This method may be acceptable for some detergent lubricating oils but it should not be used without reference to the oil supplier.

Continuous bypass systems for diesel-engine and steam-turbine installations are illustrated in Figures 2.29 and 2.30.

Batch and continuous lubricating systems

For small or medium units without a circulatory lubricating system, the oil can be treated on the batch system. As large a quantity of oil as possible is pumped from the engine or system to a heating tank. The heated oil is passed through the purifier and back to the sump.

For removing soluble sludge, a system combining the batch and continuous systems is effective. The oil is pumped to a tank, where it is allowed to settle for 24 or 48 hours. The oil may be heated by steam coils and basic mineral oils in good condition may be water washed. After settling sludge is drawn off, and the oil is run through the purifier and back to the tank on a continuous system before being finally delivered back through the purifier to the sump.

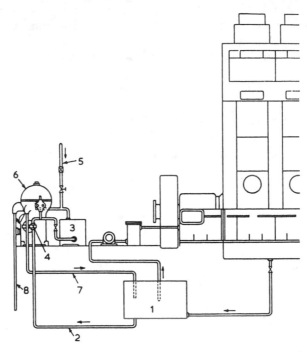

Figure 2.29   *Continuous by-pass purification for a diesel engine (Alfa-Laval Co. Ltd.)*

1. Sump tank for dirty oil from engine
2. Dirty oil to purifier
3. Heater
4. Pump
5. Hot water piping
6. Purifier
7. Purified oil
8. To waste

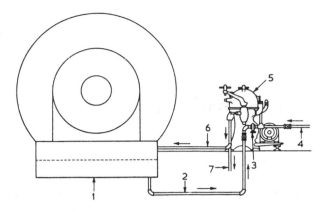

Figure 2.30 *Continuous pour steam-turbine*

1. Turbine oil tank
2. Dirty oil to purifier
3. Oil pump
4. Hot water piping
5. Purifier
6. Purified oil to turbine
7. To waste

Further reading

The Merchant Shipping Act, 1894 Report of Court (No. 8022) m.v. 'Capetown Castle' O.N. 166402.

3

Ship service systems

Some of the equipment in the machinery space is dedicated to servicing the ship in general and providing amenities for personnel or passengers. Thus the bilge system is available to clear oil/water leakage and residues from machinery and other spaces as well as to provide an emergency pumping capability. The domestic water and sewage systems provide amenities for personnel.

Bilge systems and oily/water separators

The essential purpose of a bilge system, is to clear water from the ship's 'dry' compartments, in emergency. The major uses of the system, are for clearing water and oil which accumulates in machinery space bilges as the result of leakage or draining, and when washing down dry cargo holds. The bilge main in the engine room, has connections from dry cargo holds, tunnel and machinery spaces. Tanks for liquid cargo and ballast are served by cargo discharge systems and ballast systems respectively. They are not connected to the bilge system unless they have a double function, as for example with deep tanks that are used for dry cargo or ballast. Spectacle blanks or change over chests are fitted to connect/isolate spaces of this kind, as necessary. Accommodation spaces are served by scuppers with non-return valves which are fitted at the ship's side.

Bilge system regulations

Regulations prescribe the requirements for bilge systems and the details of a proposed arrangement must be submitted for approval to the appropriate government department or classification society. The number of power operated bilge pumps (usually three or four) that are required in the machinery spaces is governed by the size and type of ship. For smaller vessels one of the pumps may be main engine driven but the other must be independently driven. A bilge ejector is acceptable as a substitute provided that, like the pumps, it is capable of giving an adequate flow rate. At least 120 m/min (400 ft/min) through the pipe is a figure that has been required. Pipe cross section is also governed by the rules, which means that this, combined with linear flow, dictates a discharge rate. Bilge ejectors are supplied with high pressure sea water from an associated pump.

The diameters of bilge main and branch pipes, are found as stated above from formulae based on ship size and the Classification Societies generally prescribe the bore of the main bilge line and branch bilge lines and relate the bilge pump capacity of each pump to that required to maintain a minimum water speed in the line. Fire pump capacity is related to the capacity of the bilge pump thus defined:

Bilge main dia. $d_1 = 1.68 \sqrt{L(B+D)} + 25$ mm
Branch dia. $d_2 = 2.16 \sqrt{C(B+D)} + 25$ mm
d_2 not to be less than 50 mm and need not exceed 100 mm.
d_1 must never be less than d_2

where

L = length of ship in m;
B = breadth of ship in m;
D = moulded depth at bulkhead deck in m;
C = length of compartment in m.

Each pump should have sufficient capacity to give a water speed of 122 m/min through the Rule size mains of this bore. Furthermore each bilge pump should have a capacity of not less than

$$\frac{0.565}{10^3} d_1^2 \, \text{m}^3/\text{h}$$

The fire pumps, excluding any emergency fire pump fitted, must be capable of delivering a total quantity of water at a defined head not less than two-thirds of the total bilge pumping capacity. The defined head ranges from 3.2 bar in the case of passenger ships of 4000 tons gross or more to 2.4 bar for cargo ships of less than 1000 tons gross.

Pumps installed for bilge pumping duties must be self-priming or able to be primed. The centrifugal type with an air pump is suitable and there are a number of rotary self-priming pumps available. Engine driven pumps are usually of the reciprocating type and there are still in use many pumps of this kind driven by electric motors through cranks.

The bilge pumps may be used for other duties such as general service, ballast and fire-fighting, which are intermittent. The statutory bilge pumps may not be used for continuous operation on other services such as cooling, although bilge injections can be fitted on such pumps and are a requirement on main or stand-by circulating pumps.

Common suction and discharge chests permit one pump to be used for bilge and ballast duties. The pipe systems for these services must, however, be separate and distinct. The ballast piping has screw lift valves so as to be able to both fill and empty purpose-constructed tanks with sea water. The bilge system is designed to remove water or oily water from 'dry' spaces throughout the vessel and is fitted with screw-down non-return valves to prevent any flooding back to the compartment served. The two could not be connected because they are incompatible. At the pump suction chest, the bilge valve must be of the screw down non-return type to prevent water from entering the bilge line from sea water or ballast suctions.

Materials which can be used are also given in the construction rules. When steel is used, it requires protection inside and out and both surfaces should be galvanized. The preparation of the surfaces for galvanizing is important as is the continuity of the coating. The external painting of steel pipes may be the only protection used to prevent rust arising from contact with water in the bilges. Flanged joints are made between sections of pipe and support must be adequate. Branch, direct and emergency bilge suctions are provided to conform with the regulations and as made necessary by the machinery space arrangement.

Bilge and ballast system layout

In the system shown, (Figure 3.1) the bilge main has suctions from the port and starboard sides of the engine room, from the tunnel well and from the different cargo holds. There are three pumps shown connected to the bilge main. These are the fire and bilge pump, the general service pump and the auxiliary bilge pump. These pumps also have direct bilge suctions to the engine room port side, starboard side and tunnel well respectively. The ballast pump (port side for'd) could be connected to the bilge main but is shown with an emergency bilge suction only. The main sea-water circulating pump at the starboard side of the machinery space also has an emergency suction. This emergency suction or the one on the ballast pump is required by the regulations. The ballast pump is self-priming and can serve as one of the required bilge pumps as well as being the stand-by sea-water circulating pump.

The auxiliary bilge pump is the workhorse of the system and need not be one of the statutorily required bilge pumps. For this installation, it is a low capacity, smooth flow pump which is suited for use in conjunction with the oily/water separator. All bilge suctions have screw down non-return valves with strainers or mud boxes at the bilge wells. Oily bilges and purifier sludge tanks have suitable connections for discharge to the oily water separator or ashore.

The system is tailored to suit the particular ship. Vessels with open floors in the machinery space may have bilge suctions near the centre line and in such cases, wing suctions would not be necessary provided the rise of floor was sharp enough.

The essential safety role of the bilge system means that bilge pumps must be capable of discharging directly overboard. This system is also used when washing down dry cargo spaces.

When clearing the water and oil which accumulates in machinery space bilges, the discharge overboard must be via the oily/water separator and usually with the use of the special bilge pump, i.e. the auxiliary bilge pump of the system shown.

The following paragraphs are extracted from the International Convention for the Safety of Life at Sea 1974 Chapter 11-1 Regulation 18 which relates to passenger ships:

The arrangement of the bilge and ballast pumping system shall be such as to prevent the possibility of water passing from the sea and from water ballast spaces into the cargo and machinery spaces, or from one compartment to another. Special provision shall be made to prevent any deep tank having bilge and ballast connections being inadvertently run up from the sea when containing cargo, or pumped out through a bilge pipe when containing water ballast.

Provision shall be made to prevent the compartment served by any bilge suction pipe being flooded in the event of the pipe being severed, or otherwise damaged by collision or grounding in any other compartment. For this purpose, where the pipe is at any part situated nearer the side of the ship than one-fifth the breadth of the ship (measured at right angles to the centre line at the level of the deepest subdivision load line), or in a duct keel, a non-return valve shall be fitted to the pipe in the compartment containing the open end.

All the distribution boxes, cocks and valves in connection with the bilge pumping arrangements shall be in positions which are accessible at all times under ordinary circumstances. They shall be so arranged that, in the event of flooding, one of the bilge pumps may be operative on any compartment; in addition, damage to a pump or its pipe connecting to the bilge main outboard of a line drawn at one-fifth of the breadth of the ship shall not put the bilge system out of action. If there is only one system of pipes common to all the pumps, the necessary cocks or valves for controlling the bilge suctions must be capable of being operated from above the bulkhead deck. Where in addition to the main bilge pumping system an emergency bilge pumping system is provided, it shall be independent of the main system and so arranged that a pump is capable of operating on any compartment under flooding condition; in that case only the cocks and valves necessary for the operation of the emergency system need be capable of being operated from above the bulkhead deck.

All cocks and valves mentioned in the above paragraph of this Regulation which can be operated from above the bulkhead deck shall have their controls at their place of operation clearly marked and provided with means to indicate whether they are open or closed.

Oil/water separators

Oil/water separators are necessary aboard vessels to prevent the discharge of oil overboard mainly when pumping out bilges. They also find service when deballasting or when cleaning oil tanks. The requirement to fit such devices is the result of international legislation. Legislation was needed because free oil and oily emulsions discharged in a waterway can interfere with natural processes such as photosynthesis and re-aeration, and induce the destruction of the algae and plankton so essential to fish life. Inshore discharge of oil can cause damage to bird life and mass pollution of beaches. Ships found discharging water containing more than 100 mg/litre of oil or discharging more than 60 litres of oil per nautical mile can be heavily fined, as also can the ship's Master.

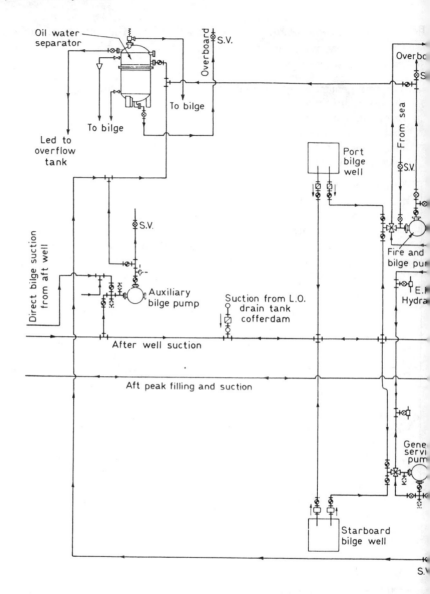

Oil water separator

Overboard S.V.

Overbc S

To bilge

To bilge

Led to overflow tank

From sea

Port bilge well

S.V.

Direct bilge suction from aft well

S.V.

Fire and bilge pu

Auxiliary bilge pump

Suction from L.O. drain tank cofferdam

E. Hydra

After well suction

Aft peak filling and suction

Gene servi pum

Starboard bilge well

S. V

Overboar

Figure 3.1   *Bilge, ballast and fuel main*

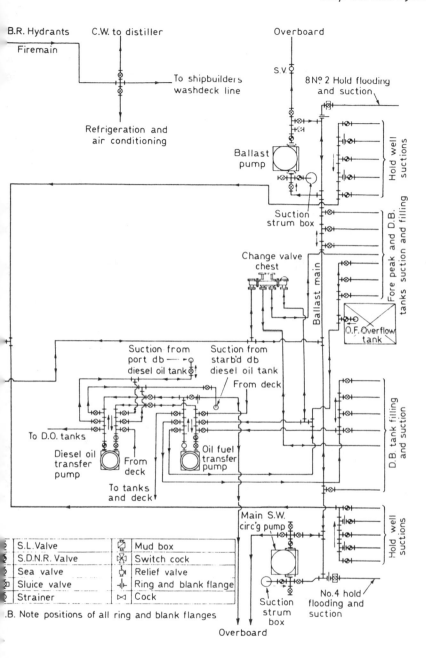

B.R. Hydrants

C.W. to distiller

Firemain

To shipbuilders
washdeck line

Refrigeration and
air conditioning

Overboard

S.V.

8 N⁰ 2 Hold flooding
and suction

Hold well
suctions

Ballast
pump

Suction
strum box

Change valve
chest

Ballast main

Fore peak and D.B.
tanks suction and filling

O.F. Overflow
tank

Suction from
port db
diesel oil tank

Suction from
starb'd db
diesel oil tank

From deck

To D.O. tanks

Diesel oil
transfer
pump

From
deck

Oil fuel
transfer
pump

D.B. tank filling
and suction

To tanks
and deck

Main S.W.
circ'g pump

Hold well
suctions

No.4 hold
flooding and
suction

Suction
strum
box

Overboard

S.L.Valve		Mud box	
S.D.N.R. Valve		Switch cock	
Sea valve		Relief valve	
Sluice valve		Ring and blank flange	
Strainer		Cock	

.B. Note positions of all ring and blank flanges

In consequence it is important that an oil/water separator is correctly installed, used and maintained. It is generally accepted that oil is less dense than water and this is the basis of the design of devices to separate the two liquids. Some of the modern heavy fuels however, have a density at 15°C which approaches, is the same as or is even higher than that of water and this has added to the problems of separation in oil/water separators and in centrifuges. The operation of oil/water separators relies heavily on gravity and a conventional difference in densities. Centrifuges by their speed of rotation, exert a force many times that of gravitational effect and the heater (see previous chapter) reduces density in comparison with that of water.

Oil/water separators and centrifuges are both employed for the purpose of separating oil and water but there are major differences. Oil/water separators are required to handle large quantities of water from which usually, small amounts of oil must be removed. Various features are necessary to aid removal of the oil from the large bulk of water particularly when the difference in densities is small.

Centrifuges are required to remove (again usually) small quantities of water from a much larger amount of oil. Additionally the centrifuge must separate solids and it must, with respect to fuel, handle large quantities at the rate at which the fuel is consumed.

Principle of operation

The main principle of separation by which commercially available oil/water separators function, is the gravity differential between oil and water.

In oily water mixtures, the oil exists as a collection of globules of various sizes. The force acting on such a globule, causing it to move in the water is proportional to the difference in weight between the oil particle and a particle of water of equal volume. This can be expressed as:

$$F_s = \frac{\pi}{6} D^3 (\rho_w - \rho_o) g \tag{1}$$

where:

F_s = separating force
ρ_w = density of water
ρ_o = density of oil
D = diameter of oil globule
g = acceleration due to gravity.

The resistance to the movement of the globule depends on its size and the viscosity of the fluids. For small particles moving under streamline flow conditions, the relationship between these properties can be expressed by Stoke's Law:

$$F_r = 3 \pi v \mu d \tag{2}$$

where:

F_r = resistance to movement
μ = viscosity of fluid
v = terminal velocity of particle
d = diameter of particle.

When separation of an oil globule in water is taking place F_s will equal F_r and the above equations can be worked to express the relationship of the terminal (or in this case rising) velocity of the globule with viscosity, relative density and particle size:

$$v = \left(\frac{g}{18\mu}\right)(\rho_w - \rho_o)\, d^2 \tag{3}$$

In general, a high rate of separation is encouraged by a large size of oil globule, elevated temperature of the system (which increases the specific gravity differential of the oil and water and reduces the viscosity of the oil) and the use of sea water. Turbulence or agitation should be avoided since it causes mixing and re-entrainment of the oil. Laminar or streamlined flow is beneficial.

In addition to the heating coils provided to optimize separation, there are various other means used to improve and speed up operation. The entrance area in oil/water separators is made large so that flow is slow and large slugs of oil can move to the surface quickly. (The low capacity pump encourages slow and laminar flow.) Alternation of flow path in a vertical direction continually brings oil near to the surface, where separation is enhanced by weirs which reduce liquid depth. Angled surfaces provide areas on which oil can accumulate and form globules, which then float upwards. Fine gauze screens are also used as coalescing or coagulating surfaces.

Pumping considerations

A faster rate of separation is obtained with large size oil globules or slugs and any break up of oil globules in the oily feed to the separator should be avoided. This factor can be seriously affected by the type and rating of the pump used. Tests were carried out by a British government research establishment some years ago on the suitability of various pumps for separator feed duties and the results are shown in Table 3.1.

It follows that equal care must be taken with pipe design and installation to avoid turbulence due to sharp bends or constrictions and to calculate correctly liquid flow and pipe size to guarantee laminar flow.

The Simplex-Turbulo oil/water separator

The Simplex-Turbulo oil/water separator (Figure 3.2) consists of a vertical cylindrical pressure vessel containing a number of inverted conical plates. The oily water enters the separator in the upper half of the unit and is directed

Table 3.1   *Pump suitability for oil/water separator duty*

Type	Remarks
Double vane Triple screw Single vane Rotary gear	Satisfactory at 50 per cent derating
Reciprocating Hypocycloidal	Not satisfactory: modification may improve efficiencies to 'satisfactory' level
Diaphragm Disc and shoe Centrifugal Flexible vane	Unsatisfactory

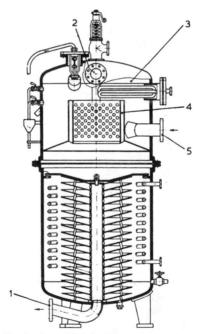

Figure 3.2   *Simplex-Turbulo oil/water separator*

1. Clean water run-off connection	3. Oil accumulation space
	4. Riser pipes
2. Outlet	5. Inlet connection

downwards to the conical plates. Large globules of oil separate out in the upper part of the separator. The smaller globules are carried by the water into the spaces between the plates. The rising velocity of the globules carries them upwards where they become trapped by the under-surfaces of the plates and coalesce until the enlarged globules have sufficient rising velocity to travel along the plate surface and break away at the periphery. The oil rises, is caught underneath an annular baffle and is then led up through the turbulent inlet area by risers to collect in the dome of the separator. The water leaves the conical plate pack via a central pipe which is connected to a flange at the base of the separator.

Two test cocks are provided to observe the depth of oil collected in the separator dome. When oil is seen at the lower test cock, the oil drain valve must be opened. An automatic air release valve is located in the separator dome. An electronically operated oil drainage valve is also frequently fitted. This works on an electric signal given be liquid level probes in the separator. Visual and audible oil overload indicators may also be fitted. To assist separation steam coils or electric heaters are fitted in the upper part of the separator. Where high viscosity oils are to be separated additional heating coils are installed in the lower part.

Before initial operation, the separator must be filled with clean water. To a large extent the conical plates are self-cleaning but periodically the top of the vessel should be removed and the plates examined for sludge build-up and corrosion. It is important that neither this separator nor any other type is run at over capacity. When a separator is overloaded the flow becomes turbulent, causing re-entrainment of the oil and consequent deterioration of the effluent quality.

To meet the requirement of legislation which came into force in October 1983 and which requires that the oil content of bilge discharges be reduced in general to 100 ppm and to 15 ppm in special areas and within 12 nautical miles of land, a second stage coalescer (Figure 3.3) was added in some designs. Filter elements in the second stage remove any small droplets of oil in the discharge and cause them to be held until they form larger droplets (coalesce). As the larger globules form, they rise to the oil collecting space.

Oil content monitoring

In the past, an inspection glass, fitted in the overboard discharge pipe of the oil/water separator permitted sighting of the flow. The discharge was illuminated by a light bulb fitted on the outside of the glass port opposite the viewer. The separator was shut down if there was any evidence of oil carry over, but problems with observation occurred due to poor light and accumulation of oily deposits on the inside of the glasses.

Present-day monitors are based on the same principle. However, whilst the eye can register anything from an emulsion to globules of oil a light-sensitive photo-cell detector cannot. Makers may therefore use a sampling and mixing pump to draw a representative sample with a general opaqueness more easily registered by the simple photo-cell monitor. Flow through the sampling chamber is made rapid to reduce deposit on glass lenses. They are easily removed for cleaning.

Bilge or ballast water passing through a sample chamber can be monitored

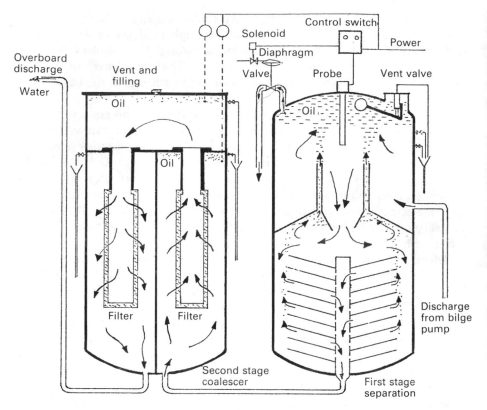

Figure 3.3   *Simplex-Turbulo oil/water separator with coalescer*

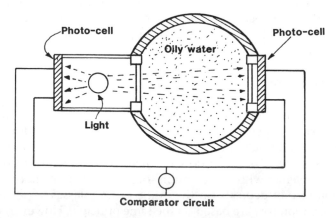

Figure 3.4   *Monitor for oily water using direct light*

by a strong light shining directly through it and on to a photo-cell (Figure 3.4). Light reaching the cell decreases with increasing oil content of the water. The effect of this light on the photo-cell compared with that of direct light on the reference cell to the left of the bulb, can be registered on a meter calibrated to show oil content.

Another approach is to register light scattered by oil particles dispersed in the water by the sampling pumps (Figure 3.5). Light reflected or scattered by any oil particles in the flow, illuminates the scattered light window. This light when compared with the source light increases to a maximum and then decreases with increasing oil content of the flow. Fibre optic tubes are used in the device shown to convey light from the source and from the scattered light window to the photo-cell. The motor-driven rotating disc with its slot, lets each light shine alternately on the photo-cell and also, by means of switches at the periphery, causes the signals to be passed independently to a comparator device.

These two methods briefly described, could be used together to improve accuracy, but they will not distinguish between oil and other particles in the flow. Methods of checking for oil by chemical test would give better results but take too long in a situation where excess amounts require immediate shut down of the oily water separator.

Tanker ballast

Sampling and monitoring equipment fitted in the pump room of a tanker can be made safe by using fibre optics to transmit light to and from the sampling chamber (Figure 3.6). The light source and photo-cell can be situated in the cargo control room together with the control, recording and alarm console. The sampling pump can be fitted in the pumproom to keep the sampling pipe short and so minimize time delay. For safety the drive motor is fitted in the machinery space, with the shaft passing through a gas-tight seal in the bulkhead.

Oil content reading of the discharge is fed into the control computer together with discharge rate and ship's speed to give a permanent record. Alarms, automatic shutdown, back-flushing and recalibration are incorporated.

Ballast arrangements

The ballasting of a vessel which is to proceed without cargo to the loading port is necessary for a safe voyage, sometimes in heavy weather conditions. On arrival at the port the large amount of ballast must be discharged rapidly in readiness for loading. Ballast pump capacity is governed by the volume of water that has to be discharged in a given time. The ballast pump is often also the stand-by sea-water circulating pump (Figure 3.1) but very large ballast discharge capacity is necessary for some ships. Vessels with tanks available for either ballast or oil fuel are fitted with a change-over chest or cock (see Chapter 4) designed to prevent mistakes. An oily water separator on the ballast pump discharge would prevent discharge of oil with the ballast from a tank that had been used for fuel or oil cargo.

Ballast carried in the empty cargo tanks of crude oil carriers has potential for pollution when discharged, particularly if cargo pumps are used for the purpose. Only very large oil/water separators have the capacity to reduce this

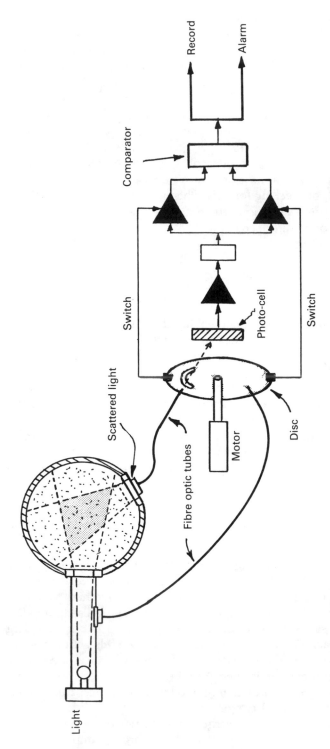

Figure 3.5 *Monitor based on scattered light (courtesy Sofrance)*

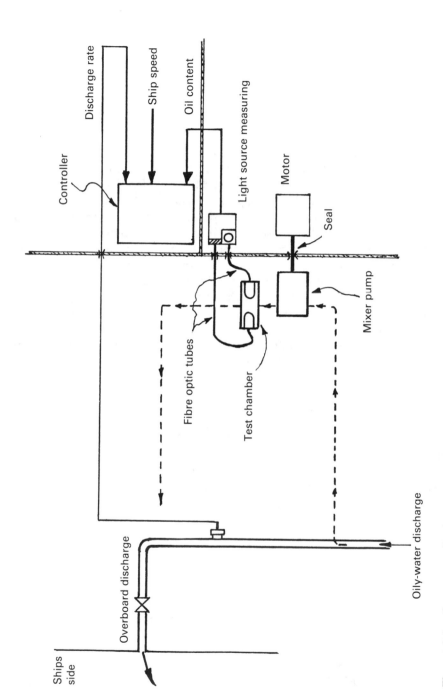

Figure 3.6 Seres monitoring system for tanker ballast

pollution. Segregated ballast tanks with dedicated ballast pumps prevent the problem. An example of a ballast pump for a segregated ballast tank, is given in Chapter 6.

Fore and aft peak tanks, double bottom and deep tanks used for ballast in dry cargo vessels as well as ballast spaces in bulk liquid carriers, can be dangerous due to lack of oxygen or the presence of harmful gases. Oxygen may be depleted by corrosion and harmful gases may be produced by organisms or pollutants in the water. The ballast water from some areas has been found to carry dangerous bacteria.

Ballast tank air and overflow pipes must be of the required size relative to the filling lines, that is, 25% greater in area and, in any case, not less than 50 mm bore. They are fitted at the highest part of the tank or at the opposite end to the filling connection. Tanks used for fuel storage also have to fulfil the requirements for fuel tanks. Nameplates are attached to the tops of all air pipes and sounding pipes must have means of identification. The latter are to be of steel with a striker plate at the bottom and must conform to the various rulings. The pipelines for ballasting must be of adequate strength and if of steel, protected by galvanizing or other means. The ballasting of some tanks, such as those in the double bottoms, is carried out by running up by opening appropriate valves, rather than by pumping. Remotely operated valves are installed with modern ballast systems. Pump and valve controls are then centrally located.

Centrifugal pumps with water ring primers, used for ballast pumping, are suitable for use as statutory bilge pumps.

Domestic water systems

Systems using gravity tanks to provide a head for domestic fresh and sanitary water, have long been superseded by schemes where supply pressure is maintained by a cushion of compressed air in the service tanks (Figure 3.7). The trade name Pneupress is commonly used to describe the tanks and system.

Fresh water

The fresh water is supplied to the system, by one of two pumps which are self-priming or situated at a lower level than the storage tanks. The pump starters are controlled by pressure switches which operate when pressure in the service tank varies within pre-determined limits as water is used. The pump discharges through filters to a rising main, branched to give cold and hot supplies, the latter through a calorifier. A circulating pump may be fitted in circuit with the steam or electrically heated calorifier. An ultra-violet light sterilizer is fitted adjacent to the Pneupress tank of some systems. Ultra-violet light acts in such an arrangement, as a point of use biocide. Although effective as a means of killing bacteria, it does not apparently provide protection in the long term. The Department of Transport requirement for protection of fresh

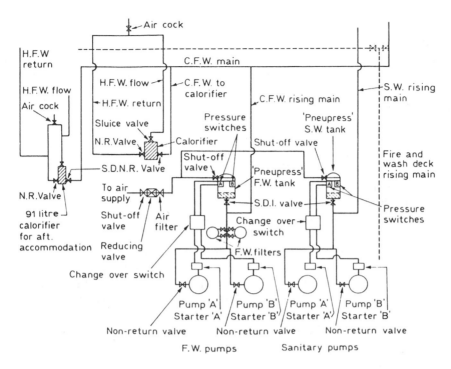

Figure 3.7 *Domestic fresh and sanitary water system*

water in storage tanks, is that chlorine dosing or the Electro-Katadyn method, be used. Guidance on the procedures to ensure that fresh water is safe for consumption is provided by M notices listed at the end of the chapter.

Sanitary water

The sanitary system operates on the same Pneupress principle as that described for fresh water. Pumps, if supplying sea water, are protected by filters on the suction side which require regular cleaning. A few sanitary systems use fresh or distilled water to reduce corrosion in pipes and flushing valves, particularly in vacuum systems where water consumption is minimal. Treated liquid effluent is recirculated in the chemical sewage treatment system described later in the chapter (p. 108); this also operates with a Pneupress system.

Water production

A considerable amount of fresh water is consumed in a ship. The crew uses on average about 70 litre/person/day and in a passenger ship, consumption can be as high as 225 litre/person/day. Water used in the machinery spaces as make up for cooling system losses may be fresh or distilled but distilled water is essential for steam plant where there is a water tube boiler. Steamship consumption for the propulsion plant and hotel services can be as high as 50 tonnes/day.

It is now common practice to take on only a minimal supply of potable water in port and to make up the rest by distillation of sea water. The saved storage capacity for water, is available for cargo and increases the earning power of the ship. A vessel which carries sufficient potable water for normal requirements is required, if ocean-going, to carry distillation plant for emergency use.

Modern low pressure evaporators and reverse osmosis systems give relatively trouble-free operation particularly in comparison with the types that were fitted in older ships. They are sufficiently reliable to provide, during continuous and unattended operation, the water needed for the engine room and domestic comsumption. An advantage of low pressure evaporators is that they enable otherwise wasted heat from diesel engine jacket cooling water to be put to good use.

Reverse osmosis systems were installed to give instant water production capacity without extensive modifications (as with vessels commandeered for hostilities in the Falklands War). They are used to advantage on some passenger cruise vessels and are fitted in ships which may remain stopped at sea for various reasons (tankers awaiting orders – outside 20 mile limit).

Warning is given in M Notice M620 that evaporators must not be operated within 20 miles of a coastline and that this distance should be greater in some circumstances. Pollution is present in inshore waters from sewage outfalls, disposal of chemical wastes from industry, drainage of fertilizers from the land and isolated cases of pollution from grounding or collision of ships and spillage of cargo.

Low pressure evaporators

The main object of distillation is to produce water essentially free of salts. Potable water should contain less than 500 mg/litre of suspended solids. Good quality boiler feed will contain less than 2.5 mg/litre. Sea water has a total dissolved solids content in the range 30 000–42 000 mg/litre, depending on its origin but the figure is usually given as 32 000 mg/litre.

Low pressure evaporators for the production of water can be adapted for steamships but operate to greatest advantage with engine cooling water on motorships. The relatively low temperature jacket water entering at about 65°C and leaving at about 60°C will produce evaporation because vacuum conditions reduce the boiling temperature of sea water from 100°C to less than 45°C.

The single effect, high vacuum, submerged tube evaporator shown in Figure 3.8 is supplied with diesel engine cooling water as the heating medium. Vapour evolved at a very rapid rate by boiling of the sea-water feed, tends to carry with it, small droplets of salt water which must be removed to avoid contamination of the product. The demister of knitted monel metal wire or polypropylene collects the salt-filled water droplets as they are carried through by the air. These coalesce forming drops large enough to fall back against the vapour flow.

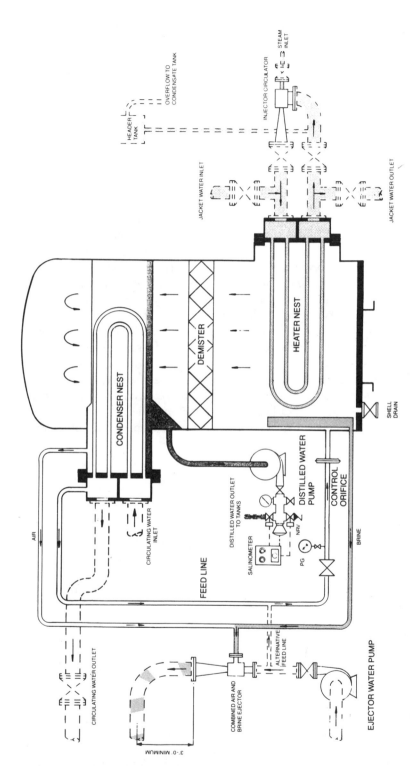

Figure 3.8 *High vacuum, submerged tube evaporatory (movac Mk2 – Caird & Rayner)*

Evaporation of part of the sea water leaves a brine the density of which must be controlled by continual removal through a brine ejector or pump. Air and other gases released by heating of the sea water, but which will not condense, are removed by the air ejector. The evaporator shown has a single combined ejector for extraction of both brine and air.

One of the gases liberated is CO_2 from calcium bi-carbonate in the sea water. Loss of carbon dioxide from calcium bi-carbonate, leaves plain calcium carbonate which has poor solubility and a tendency to form soft, white scale. Other potential scale-forming salts are calcium sulphate and magnesium compounds.

Scale is not a major problem where submerged heating coils reach a temperature of only 60°C. This heat is too low for formation of magnesium scales and provided brine density is controlled, calcium sulphate will not cause problems. Continuous removal of the brine by the brine pump or ejector, limits density. Approximately half of the sea-water feed is converted into distilled water, the quantity of brine extracted is equivalent to the remainder of the feed delivered. The level of water in the evaporator is maintained constant by means of a brine weir, over which excess passes to the ejector.

The small quantity of soft calcium carbonate scale can be removed by periodic cleaning with a commercially available agent or the evaporator can be continually dosed with synthetic polymer to bind the scale-forming salts into a 'flocc' which mostly discharges with the brine. Use of continuous treatment will defer acid cleaning to make it an annual exercise. Without continuous treatment, cleaning may be necessary after perhaps two months. Steam heated evaporators with their higher heating surface temperature, benefit more from chemical dosing, because magnesium scales form when surfaces are at 80°C or more.

Salinometer

The condensate or product, if of acceptable quality, is delivered to the appropriate tanks by the distilled water pump. Quality is continuously tested by the salinometer both at start up and during operation. If the device registers an excess of salinity it will dump the product and activate the alarm using its solenoid valves. The product is recirculated in some installations.

The electric salinometer

Pure distilled water may be considered a non-conductor of electricity. The addition of impurities such as salts in solution increases the conductivity of the water, and this can be measured. Since the conductivity of the water is, for low concentrations, related to the impurity content, a conductivity meter can be used to monitor the salinity of the water. The instrument can be calibrated in units of conductivity (micromhos) or directly in salinity units (older instruments in grains/gall., newer instruments in ppm or mg/litre) and it is on this basis that electric salinometers (Figure 3.9) operate. The probe type electrode cell (Figure 3.10) is fitted into the pipeline from the evaporator, co-axially through a retractable valve which permits it to be withdrawn for examination and cleaning. The cell cannot be removed while the valve is open and consists of two stainless steel concentric electrodes having a temperature compensator located within the hollow inner electrode. It operates within the limits of water pressure up to 10.5 bar and water temperatures between 15° and 110°C.

The incoming a.c. mains from control switch S2 through fuses FS, feed transformer T. A pilot lamp SL1 on the 24 V secondary winding indicates the circuit is live.

The indicating circuit comprises an applied voltage across the electrode cell and the indicator. The indicator shows the salinity by measuring the current which at a preset value actuates the alarm circuit warning relay. The

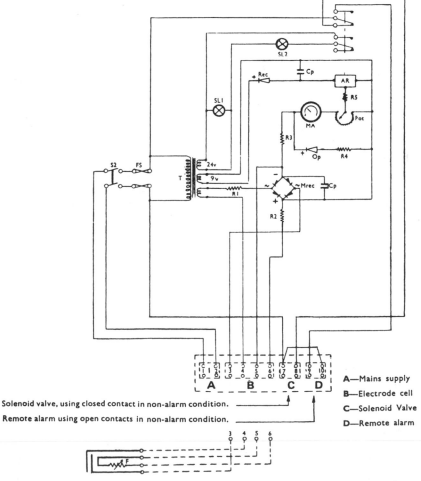

Solenoid valve, using closed contact in non-alarm condition.

Remote alarm using open contacts in non-alarm condition.

A—Mains supply

B—Electrode cell

C—Solenoid Valve

D—Remote alarm

Figure 3.9 *Schematic diagram of salinometer (W. Crockatt & Sons Ltd)*

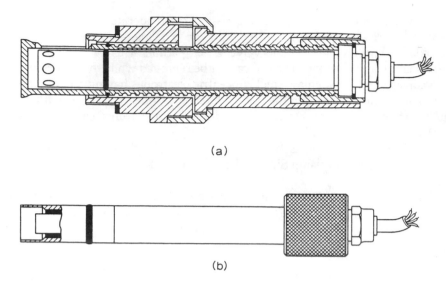

(a)

(b)

Figure 3.10 *Probe type electrode cell*

transformer cell tapped voltage is applied across a series circuit comprising the bridge rectifier Mrec, the current limiting resistor R1 and the electrode cell.

The current from rectifier Mrec divides into two paths, one through the temperature compensator F via resistor R2 and the other through the alarm relay potentiometer (Pot) indicator MA and resistor R3, the two paths joining in a common return to the low potential side of the rectifier.

The indicator is protected from overload by a semi-conductor in shunt across the indicator and potentiometer. When the water temperature is at the lower limit of the compensated range the total resistance of the compensator is in circuit and the two paths are as described above. As the temperature of the water rises, the resistance of the compensator device drops progressively, the electrical path through the compensator now has a lower resistance than the other and a large proportion of the cell current. The compensator therefore ensures that the alteration in the balance of the resistances of the two paths corresponds to the increased water conductivity due to the rise in temperature and a correct reading is thus obtained over the compensated range.

The alarm setting is adjustable and the contacts of the warning relay close to light a lamp or sound a horn when salinity exceeds the acceptable level.

The salinometer is also arranged to control a solenoid operated valve which dumps unacceptable feed water to the bilge or recirculates. The salinometer and valve reset automatically when the alarm condition clears.

Corrosion

The shell of the evaporator may be of cupro-nickel or other corrosion resistant material but more commonly, is of steel. The steel shell of evaporators is prone to corrosion. Protection is provided in the form of natural rubber, rolled and bonded to the previously shot-blasted steel. The adhesive is heat cured and the integrity of the rubber checked by spark test.

Reason for distillate treatment

The low operating temperature of the evaporator described, is not sufficient to sterilize the product. Despite precautions near the coast, harmful organisms may enter with the sea water and pass through to the domestic water tank and system. Additionally there is a likelihood that while in the domestic tank, water may become infested with bacteria, due to a build up of a colony of organisms from some initial contamination. Sterilization by the addition of chlorine, is recommended in Merchant Shipping Notice M1214. A later notice, M1401, states that the Electro-Katadyn process in use since the 1960s, has also been approved.

Another problem with distilled water is that having none of the dissolved solids common in fresh water it tastes flat. It also tends to be slightly acidic due to its ready absorption of carbon dioxide (CO_2). This condition makes it corrosive to pipe systems and less than beneficial to the human digestive tract.

Chlorine sterilization and conditioning

Initial treatment (Figure 3.11) involves passing the distillate through a neutralite unit containing magnesium and calcium carbonate. Some absorption of carbon dioxide from the water and the neutralizing effect of these compounds, removes acidity. The addition of hardness salts also gives the water a better taste. The sterilizing agent chlorine, being a gas, is carried into the water as a constituent of sodium hypochlorite (a liquid) or in granules of calcium chloride dissolved in water. The addition is set to bring chlorine content to 0.2 ppm. While the water resides in the domestic tank, chlorine should preserve sterility. In the long term, it will evaporate so that further additions of chlorine may be needed.

The passage of water from storage tanks to the domestic system, is by way of a carbon filter which removes the chlorine taste.

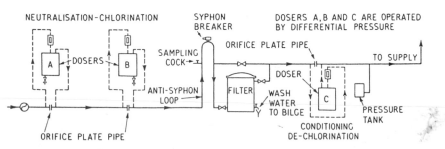

Figure 3.11 *Chlorine sterilization and conditioning*

Electro-katadyn method of sterilization

The Electro-katadyn process (Figure 3.12) accepted as an alternative to chlorination (see M1401) involves the use of a driven silver anode to inject silver ions (Ag^+) into the distilled water product of the low temperature evaporator. Silver is toxic to the various risk organisms. Unlike the gas chlorine, it will not evaporate but remains suspended in the water.

The sterilizer is placed close to the production equipment with the conditioning unit being installed after the sterilizer and before the storage tank.

The amount of metal released to water passing through the unit, is controlled by the current setting. If a large volume has to be treated, only part is bypassed through and a high current setting is used to inject a large amount of silver. The bypassed water is then added to the rest in the pipeline. With low water flow, all of the water is delivered through the device and the current setting is such as to give a concentration of 0.1 ppm of silver. The silver content of water in the domestic system, should be 0.08 ppm maximum.

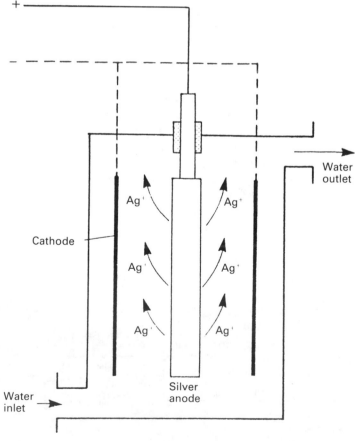

Figure 3.12   *Electro-katadyn sterilization*

Ultra-violet sterilizer

A means for sterilizing potable water at the point of use, is provided on many offshore installations and ships, by an ultra-violet radiation unit which is positioned after the hydrophore tank and as close as possible to the tap supply points. The stainless steel irradiation chamber contains low pressure mercury vapour tubes, housed in a quartz jacket. Tubes are wired in series with a transformer for safety. A wiper is fitted within the chamber to clean the jackets and lamp observation window. Units of a similar type are used for pretreatment disinfection in some reverse osmosis plant.

Flash evaporators

The evaporator described above, boils sea water at the saturation temperature corresponding to the uniform pressure through the evaporation and condensing chambers. With flash evaporators (Figure 3.13) the water is heated in one compartment before being released into a second chamber in which the pressure is substantially lower. The drop in pressure changes the saturation temperature below the actual temperature, so that some of the water instantly flashes off as vapour.

Steam in the chamber at sub-atmospheric pressure is condensed by contact with tubes circulated with the salt feed and is removed by a distillate pump. Suitably placed baffles and demisters, similar to those already described, prevent carry-over of saline droplets. The arrangements for continuous monitoring of distillate purity are similar to those described above.

If two or more vessels in series are maintained at progressively lower absolute pressures, the process can be repeated. Incoming salt feed absorbs the latent heat of the steam in each stage, with a resultant gain in economy of heat and fuel. This is known as cascade evaporation, a term which is self-explanatory. Figure 3.13 shows a two stage flash evaporator distiller. The flash chambers are maintained at a very low absolute pressure by ejectors, steam or water operated; the salt feed is heated initially by the condensing vapour in the flash chambers, subsequently in its passage through the ejector condenser (when steam-operated ejectors are used) and is raised to its final temperature in a heater supplied with low pressure exhaust steam. Brine density is maintained, as in the case of the evaporator-distillers described previously, by an excess of feed over evaporation and the removal of the excess by a pump. The re-circulation of brine may be provided for in plant.

It should be noted that when distillate is used for drinking it may require subsequent treatment to make it potable.

Reverse osmosis

Osmosis is the term used to describe the natural migration of water from one side of a semi-permeable membrane into a solution on the other side. The

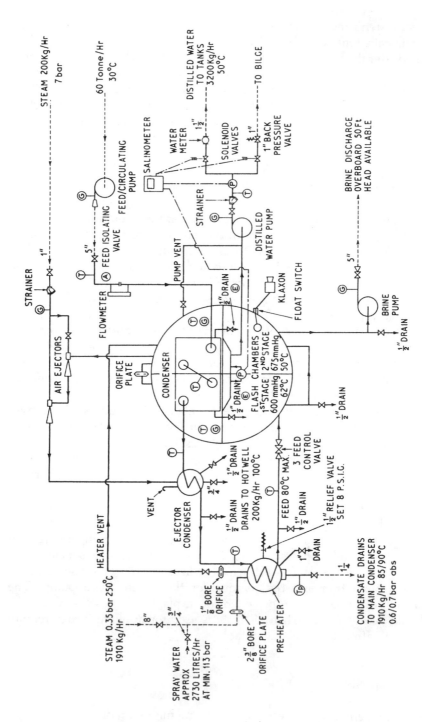

Figure 3.13 *Flow diagrams – cascade evaporator (Caird & Rayner Ltd)*

phenomenon occurs when moisture from the soil passes through the membrane covering of the roots of plants, with no loss of nutrient liquid from the plant. The membrane acts as a one way barrier, allowing the passage of water but not of the nutrients dissolved in the liquid within the root. Osmosis can be demonstrated in a laboratory with a parchment-covered, inverted thistle funnel partly filled with solution and immersed in a container of pure water. The liquid level in the funnel rises as pure water passes through the parchment and into the solution. The action will continue despite the rise of the head of the salt solution relative to that of the pure water. Osmotic pressure can be obtained by measuring the head of the solution when the action ceases.

The semi-permeable membrane and the parchment are like filters. They allow the water molecules through but not the larger molecules of dissolved substances. The phenomenon is important not only for the absorption of water through the roots of plants but in animal and plant systems generally.

Reverse osmosis is a water filtration process which makes use of semi-permeable membrane-like materials. Salt (sea) water on one side of the membrane (Figure 3.14) is pressurized by a pump and forced against the material. Pure water passes through but the membrane is able to prevent passage of the salts. For production of large amounts of pure water, the membrane area must be large and it must be arranged in a configuration which makes it strong enough to withstand the very high pump pressure needed.

The man-made membrane material used for sea-water purification is produced in the form of flimsy polyamide or polysulphonate sheets, which without backing would not be strong enough. The difficulty of combining the requirements of very large area with adequate reinforcement of the thin sheets is dealt with by making up spirally wound cartridges (Figure 3.15b). The core of the cartridge is a porous tube to which are attached the open edges of a large number of envelopes each made of two sheets of the membrane material. The envelopes, sealed together on three sides, contain a sheet of porous substance which acts as the path to the central porous tube for water which is squeezed

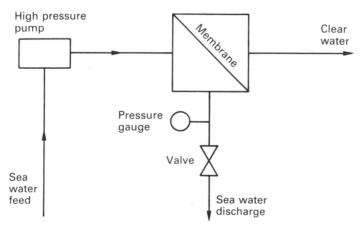

Figure 3.14 *Reverse osmosis principle*

through the membranes. The envelopes are separated by coarse gauze sheets. Assembled envelopes and separators initially have the appearance of a book opened so that the covers are in contact, the spine or binding forming a central tube. The finished cartridge is produced by rotating the actual central tube, so that envelopes and separators are wrapped around it in a spiral, to form a cylindrical shape. Cartridges with end spacers, are housed in tubes of stainless steel (Figure 3.15a) or other material. Output of the reverse osmosis plant is governed by the number of cartridge tubes in parallel. Quality is improved by installing sets of tubes in series.

One problem with any filtration system, is that deposit accumulates and gradually blocks the filter. Design of the cartridges is therefore such that the sea-water feed passes through the spiral windings and over the membrane sheets with a washing action that assists in keeping the surfaces clear of deposit. A dosing chemical, sodium hexametaphosphate, is also added to assist the action.

The pump delivery pressure for a reverse osmosis system of 60 bar (900 lb/in²) calls for a robust reciprocating or gear pump. The system must be protected by a relief arrangement.

Pre-treatment and post-treatment

Sea-water feed for reverse osmosis plant, is pretreated before being passed through. The chemical sodium hexametaphosphate is added to assist the wash through of salt deposit on the surface of the elements and the sea water is sterilized to remove bacteria which would otherwise become resident in the filter. Chlorine is reduced by the compressed carbon filter while solids are removed by the other filters.

Treatment is also necessary to make the water product of reverse osmosis potable. The method is much the same as for water produced in low temperature evaporators.

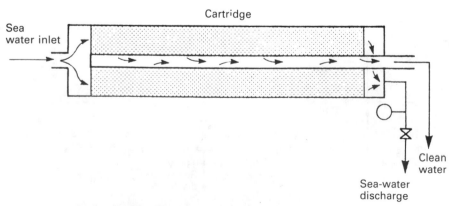

Figure 3.15a *Cartridge for reverse osmosis*

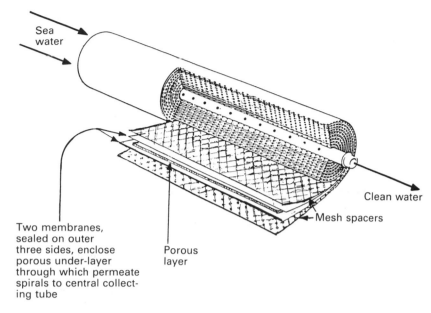

Sea
water

Clean water

Mesh spacers

Two membranes,
sealed on outer
three sides, enclose
porous under-layer
through which permeate
spirals to central collecting tube

Porous
layer

Figure 3.15b *Spirally wound cartridge for reverse osmosis*

Treatment of water from shore sources

There is a risk that water supplied from ashore may contain harmful organisms which can multiply and infect drinking or washing water storage tanks. All water from ashore, whether for drinking or washing purposes, is to be sterilized. When chlorine is used, the dose must be such as to give a concentration of 0.2 ppm. The Department of Transport recommends in Merchant Shipping Notice number M1214 that because of the risk from legionella bacteria entering the respiratory system by way of fine mist from a shower spray, all water including that for washing only, should be treated by sterilization.

The transfer hose for fresh water is to be marked and kept exclusively for that purpose. The ends must be capped after use and the hose must be stored clear of the deck to reduce the risk of contamination.

Domestic water tanks

Harmful organisms in drinking water storage tanks have caused major health problems on passenger vessels and in general to ship's crews and personnel working on oil platforms. To eliminate this problem, water storage tanks should be pumped out at six-month intervals and, if necessary, the surfaces should be hosed to down clean them. At the 12-month inspection, recoating may be needed in addition to the cleaning. Washing with a 50 ppm solution of chlorine is suggested. Super-chlorinating when the vessel is drydocked, consists of leaving a 50 ppm chlorine solution in the tank over a four hour period, followed by flushing with clean water.

The steel tank surfaces may be prepared for coating by wire brushing and priming. Subsequently a cement wash is applied or an epoxy or other coating suitable for use in fresh water tanks.

Sewage systems

The exact amount of sewage and waste water flow generated on board ship is difficult to quantify. European designers tend to work on the basis of 70 litres/person/day of toilet waste (including flushing water) and about 130–150 litres/person/day of washing water (including baths, laundries, etc.). US authorities suggest that the flow from toilet discharges is as high as 114 litres/person/day with twice this amount of washing water.

The breakdown of raw sewage in water is effected by aerobic bacteria if there is a relatively ample presence of oxygen, but by anaerobic bacteria if the oxygen has been depleted. When the amount of sewage relative to water is small, dissolved oxygen in the water will assist a bio-chemical (aerobic) action which breaks down the sewage into simple, clean components and carbon dioxide. This type of action is produced in biological sewage treatment plant in which air (containing 21% oxygen) is bubbled through to sustain the aerobic bacteria. The final discharge from an aerobic treatment plant has a clean and clear appearance.

The discharge of large quantities of raw sewage into restricted waters such as those of inland waterways and enclosed docks, will cause rapid depletion of any oxygen in the water so that aerobic bacteria are unable to survive. When the self-purification ability of the limited quantity of water is overwhelmed in this way, breakdown by putrefaction occurs. Anaerobic bacteria, not reliant on oxygen for survival are associated with this action which results in the production of black, turgid water and gases which are toxic and flammable. The process is used deliberately in some shore sewage treatment works to produce gas which is then used as fuel for internal combustion engines on the site.

The very obvious effects of sewage discharge in waterways and enclosed docks prompted the Port of London Authority and others to establish regulations concerning sewage discharge and to provide facilities ashore for ships' crews. The lavatories were vandalized and the scheme was found to be impractical. Legislation imposed nationally by the USA (through the Coast Guard) and the Canadian Government was more effective and together with the anticipation of the ratification of Annexe IV of the 1973 IMCO Conference on Marine Pollution was probably more responsible for the development of holding tanks and on board sewage treatment plant.

Some plants are designed so that the effluent is retained in the vessel for discharge well away from land, or to a receiving facility ashore; others are designed to produce an effluent which is acceptable to port authorities for discharge inshore. In the former type, the plant consists of holding tanks which receive all lavatory and urinal emptyings, including flushing water, while wash-basins, showers and baths are permitted to discharge overboard. Some are designed to minimize the amount of liquid retained by flushing with recycled effluent. It is claimed that such a system only requires about 1% of the retaining capacity of a conventional retention system.

Effluent quality standards

To discharge sewage in territorial waters the effluent quality may have to be within certain standards laid down by the local or national authorities. These will usually be based on one or more of three factors, namely the bio-chemical oxygen demand (BOD), suspended solids content and e-coliform count of the discharge.

Bio-chemical oxygen demand

The bio-chemical oxygen demand (BOD) is determined by incubating at 20°C, a sample of sewage effluent which has been well-oxygenated. The amount of oxygen absorbed over a five-day period is then measured. The test is used in this context to evaluate the effectiveness of treatment as it measures the total amount of oxygen taken up as final and complete breakdown of organic matter by aerobic bacteria in the effluent occurs. The quantity of oxygen used equates to the amount of further breakdown required.

Suspended solids

Suspended solids are unsightly and over a period of time can give rise to silting problems. They are usually a sign of a malfunctioning sewage plant and when very high will be accompanied by a high BOD. Suspended solids are measured by filtering a sample through a pre-weighed pad which is then dried and re-weighed.

Coliform count

The e-coliform is a family of bacteria which live in the human intestine. They can be quantified easily in a laboratory test the result of which is indicative of the amount of human waste present in a particular sewage sample. The result of this test is called the e-coli. count and is expressed per 100 ml.

Holding tanks

Simple holding tanks may be acceptable for ships which are in port for only a very brief period. The capacity would need to be excessively large for long stays because of the amount of flushing water. They require a vent, with the outlet suitably and safely positioned because of gas emissions. A flame trap reduces risk. Inhibiting internal corrosion implies some form of coating and, for washing through of the tank and pump after discharge of the contents at sea, a fresh water connection is required.

Elsan holding and recirculation (zero discharge) system

A retention or holding tank is required where no discharge of treated or untreated sewage is allowed in a port area. The sewage is pumped out to shore reception facilities or overboard when the vessel is proceeding on passage at sea, usually beyond the 12 nautical mile limit.

Straight holding tanks for retention of sewage during the period of a ship's stay in port were of a size large enough to contain not only the actual sewage but also the flushing water. Each flush delivered perhaps 5 litres of sea water. Passenger vessels or ferries with automatic flushing for urinals required very large holding tanks.

Problems resulting from the retention of untreated wastes relate to its breakdown by anaerobic bacteria. Clean breakdown by aerobic organisms occurs where there is ample oxygen, as described previously. In the conditions of a plain retention tank where there is no oxygen, anaerobic bacteria and other organisms thrive. These cause putrefaction, probably with corrosion in the tank and production of toxic and flammable gases.

The Elsan type plant (Figure 3.16) has an initial reception chamber in which separation of liquid and solid sewage takes place. Wastes drop on to a moving perforated rubber belt (driven by an electric motor) which the liquid passes through but solids travel with the belt to fall into a caustic treatment tank. Solids are then transferred by a grinder pump to the sullage or holding tank. The liquid passes via the perforated belt to treatment tanks which contain chlorine and caustic based compounds. These chemicals make the liquid effluent acceptable for use as a flushing fluid. The Pneupress arrangement which supplies liquid for flushing the toilets can deliver recirculated fluid or, when the vessel is on passage, sea water.

Capacity of the holding tank is 2 litres per/person/day. The tank is pumped out at sea, or to shore if the ship is in port for a long period. Tank size is small because liquid effluent passes mainly to the flushing system (excess overflows to the sullage tanks).

Biological sewage treatment

A number of biological sewage treatment plant types are in use at sea but nearly all work on what is called the extended aeration process. Basically this consists of oxygenating by bubbling air through or by agitating the surface. By so doing a family of bacteria is propagated which thrives on the oxygen content and digests the sewage to produce an innocuous sludge. In order to exist, the bacteria need a continuing supply of oxygen from the air and sewage

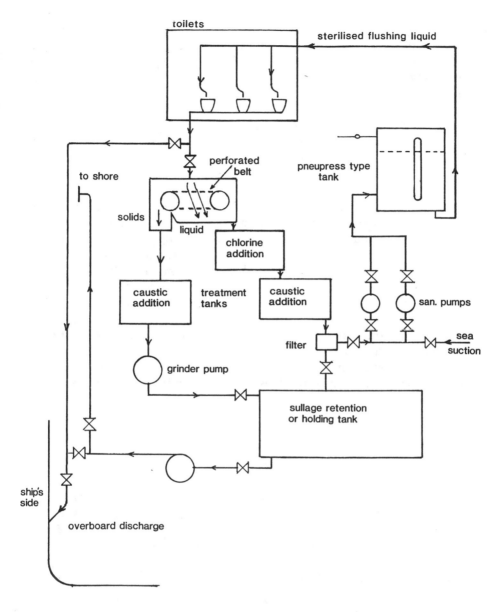

Figure 3.16 *Elsan type sewage plant*

wastes. If plant is shut down or bypassed or if the air supply fails, the bacteria die and the plant cannot function correctly until a new bacteria colony is generated. Change of flushing liquid – as when a ship moves from a sea-water environment to fresh water – drastic change of temperature or excess use of lavatory cleaning agents can also affect the bacteria colony. The process of regeneration can take several days depending on the level of harm caused.

Bacteria which thrive in the presence of oxygen are said to be aerobic. When oxygen is not present, the aerobic bacteria cannot live but a different family of bacteria is generated. These bacteria are said to be anaerobic. Whilst they are equally capable of breaking down sludge, in so doing they generate gases such as hydrogen sulphide and methane. Continuing use of a biological sewage system after a failure of the air supply, could result in propagation of anaerobic bacteria and processes. The gases produced by anaerobic activity are dangerous, being flammable and toxic.

Extended aeration plants used at sea are package plants consisting basically of three inter-connected tanks (Figure 3.17). The effluent may be comminuted (i.e. passed through a device which consists of a rotating knife-edge drum which acts both as a filter and a cutter) or simply passed through a bar screen from where it passes into the first chamber. Air is supplied to this chamber via a diffuser which breaks the air up into fine bubbles. The air is forced through the diffuser by a compressor. After a while a biological sludge is formed and this is dispersed throughout the tank by the agitation caused by the rising air bubbles.

The liquid from the aeration tank passes to a settling tank where under quiescent conditions, the activated sludge, as it is known, settles and leaves a clear effluent. The activated sludge cannot be allowed to remain in the settling tank since there is no oxygen supplied to this area and in a very short time the collected sludge would become anaerobic and give off offensive odours. The sludge is therefore continuously recycled to the aeration tank where it mixes with the incoming waste to assist in the treatment process.

Over a period of time the quantity of sludge in an aeration tank increases due to the collection of inert residues resulting from the digestion process, this

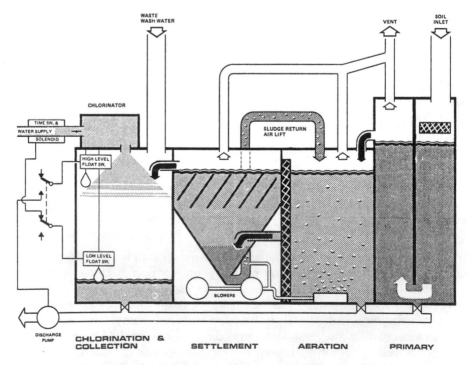

Figure 3.17   *Biological sewage treatment plant (Hamworthy)*

build up in sludge is measured in ppm or mg/litre, the rate of increase being a function of the tank size. Most marine biological waste treatment plants are designed to be desludged at intervals of about three months. The desludging operation entails pumping out about three quarters of the aeration tank contents and refilling with clean water.

The clear effluent discharged from a settling tank must be disinfected to reduce the number of coliforms to an acceptable level. Disinfection is achieved by treating the clean effluent with a solution of calcium or sodium hypochlorite, this is usually carried out in a tank or compartment on the end of the sewage treatment unit. The chlorinator shown in Figure 3.17 uses tablets of calcium hypochlorite retained in perforated plastic tubes around which the clean effluent flows dissolving some of the tablet material as it does so. The treated effluent is then held in the collection tank for 60 minutes to enable the process of disinfection to be completed. In some plants the disinfection is carried out by ultra-violet radiation.

Further reading

Allanson, J. T. and Charnley, R. (1987) Drinking water from the sea: reverse osmosis, the modern alternative, *Trans I Mar E, **88***.

Gilchrist, A. (1976) Sea Water Distillers, *Trans I Mar E, **88***.

Hill, E. C. (1987) *Legionella and Ships' Water Systems*, MER

Merchant Shipping Notice No. M1214 Recommendations to Prevent Contamination of Ships' Fresh Water Storage and Distribution Systems.

Merchant Shipping Notice No. M1401 Disinfection of Ships' Domestic Fresh Water.

The Merchant Shipping (Crew Accommodation) Regulations 1978, HMSO.

4

Valves and pipelines

The various pipe systems for commercial ships must comply with any applicable rules of the responsible government department and those of the designated classification society. Guidance is provided in government and classification society publications and it is required that plans for principal systems are submitted for approval. The safety and reliability of critical individual fittings is ensured by a requirement that they are made to specification by an approved manufacturer. Materials are tested, welds are inspected, major fittings are tested and marked, systems are pressure tested by or in the presence of a representative of the appropriate authority. Every effort is made to ensure safety and reliability. Replacement components for pipe systems must be of the same standard and obtained if necessary, from an approved maker. Some accidents have been the result of replacement valves and other components being of inferior quality.

Materials – corrosion – erosion

Galvanic corrosion is a major challenge for any pipes which carry sea water. Rust is a particular corrosion problem for steel pipes exposed to contact with sea water or moisture generally and air. Pipe runs along tank tops or on deck, are examples of the latter. Steel pipes in these areas require external as well as internal protection.

Sea water is an electrolyte and therefore a conductor of electricity, because the molecules of its dissolved salts split into positive and negative ions which are available as current carriers. Electrolytic action can result if there are different metals or even differences in the same metal in a pipeline. Galvanic corrosion can occur if the different metals are connected electrically and mutually in contact with the sea water. A corrosion cell formed between steel and brass in contact with sea water results in wastage of the less noble steel. A list is given in the galvanic series, in which the more noble metals are placed in order after the less noble thus: zinc, aluminium, carbon steels, cast iron, lead–tin alloys, lead, brass, copper, bronze, gunmetal, copper-nickel iron, monel metal. A metal in contact with one occurring later in the series, as with steel and brass, may corrode rapidly in sea water. Because the action is galvanic, less noble sacrificial anodes can give protection.

Steel

Steel being subject not only to galvanic corrosion but also to rusting, appears to be a poor material to select for sea water pipes or for installation in tank top or deck areas. Mild steel pipes for sea water are protected by being galvanized or rubber lined. Welding and pipe bending should be completed before galvanizing or application of a lining, so that weld spatter and deposits from manufacture can be removed. The mild steel, electric resistance welded (ERW) or hot rolled pipes are galvanized by hot dipping. Inadequate protection of steel, results if there are pinholes or discontinuities in protective linings. Linings should always be carried over the flange faces. Mild steel welded fabrications, similarly lined, are also used for large ship side fittings.

Seamless mild steel is used for steam, high pressure air, feed discharges and all oil fuel pressure piping. Its strength reduces however, at about 460°C and above this figure, steels require small additions of alloying materials such as molybdenum and chromium. Flanges are secured to steel pipes by fusion welding or by screwing and expanding.

Cast iron

Cast iron has poor corrosion resistance in sea water, being especially vulnerable to graphitization. This form of attack gradually removes the iron from the surface in contact with sea water to leave soft, black graphite. The weakness of ordinary grey cast iron in tension and under shock loading limits its use to low pressure applications, and the brittle nature of ordinary grey cast iron excludes its use for side shell fittings where failure could result in flooding of the machinery space. Ease of casting makes the material ideal for the production of fittings and fortunately techniques for improving strength have been developed. Spheroidal graphite cast iron (SG iron) and meehanite are examples of high strength versions of the material. These are suitable for use in ship side valves if made to specification by an approved manufacturer. SG iron may be used for high pressure services and for steam below 461°C.

Cast iron with its high carbon content and consequent low melting temperature is ideal for the production of fittings by casting.

Copper

Copper pipes are suitable for moderate pressures and temperatures. Flanges are secured to copper and its alloys by brazing or sweating.

Non-ferrous alloys

Basically, brass is an alloy of copper and zinc; bronze an alloy of copper and tin. In both cases there may be additions of other metals and there is some confusion of nomenclature; some high-tensile brasses are called 'bronze' and the practice has prevailed for so long as to be accepted.

Aluminium brass and other non-ferrous pipelines, are considered very resistant to corrosion in sea water, but concentrated galvanic corrosion can occur if some part of the pipe system has a different make up. A localized corrosion cell can be set up when a fitting, such as a thermometer pocket, is of a brass, bronze or other material which is different to the parent material. Pipe systems are ideally of the same material throughout but non-ferrous alloys are protected against corrosion by the deposition of iron ions so that use of iron or steel fittings is beneficial. Iron ion protection can alternatively be supplied from sacrificial or driven iron anodes or by dosing with ferrous sulphate.

Dezincification of brasses is a particular type of corrosion that occurs in the presence of sea water. The attack removes zinc from the alloy, leaving porous copper which is soft. The problem is marked by a patch of copper colour in the brass. Dezincification is inhibited in brasses which are intended for sea-water contact, by additions of a very small amount of arsenic (0.04%) or other elements.

Some brasses are prone to corrosion-stress cracking but this is a phenomenon associated chiefly with brass tube which has been stressed by expanding or by being worked in the unsoftened condition and which is also in contact with corrosive fluids, such as sea water. Splitting can occur suddenly, or even violently as a result of stress corrosion cracking.

Stainless steel

A different problem is presented by corrosive liquids and those that contain hard particles and are therefore likely to cause erosion. These can cause differing rates of wastage in conventional metal pipes or cargo tanks. With some corrosive liquids wastage is slow enough, lasting over a period of years, to permit the use of common metals. Expensive stainless steel is widely used for the cargo pipes of chemical tankers intended for carriage of very corrosive cargoes.

Erosion

Erosion of metal may be the result of abrasives or of high water speeds, entrained air, turbulence and cavitation. The latter are often caused by protuberances, tight bends or an abrupt change of pipe cross sectional area. Erosion from turbulent flow and cavitation also aids corrosion (corrosion/erosion) by removing the oxide film that assists in the protection of metal surfaces. The exposed metal surfaces can form galvanic corrosion cells with adjacent areas where oxide film is still present. Erosion is reduced by limiting speed of flow, avoiding sharp bends, changes of section and impediments to flow such as incorrectly cut jointing or weld deposits. Speed of liquid flow should be no greater than 1 m/s for copper; 3 m/s for galvanised steel and aluminium brass; 3.5 m/s for 90/10 cupro-nickel: 4 m/s for 70/30 cupro-nickel.

Strength of materials

The strength of materials used for pipes and fittings must be adequate for the system pressures and possible over-pressures. Pipelines and valves, for example, used to carry and control the flow of high temperature, high pressure steam must obviously be made to very exacting specifications by approved manufacturers.

Various and often varying pressures and temperatures pose problems. Temperatures of about 450°C can cause recrystallization and creep in iron and steels. Very low temperatures as with liquefied natural gas, can result in brittle failure. Varying temperatures give problems with stress due to expansion and contraction.

The term fittings covers valves, cocks, branch and bulkhead pieces, reducers, strainers and filters, separators and expansion pieces, in short, everything in a system which is not a pipe. Couplings and unions are used only in small bore pipes.

Cast iron and gunmetal fittings are used freely in small sizes at moderate pressures. Large fittings, those for high pressure and temperature and for oil fuel under pressure, are cast or fusion welded (fabricated) mild steel or SG iron. For temperatures above 460°C they are usually of 0.5% molybdenum steel. The addition of 0.5% molybdenum, inhibits recrystallization and therefore the resulting creep.

Pipe installation

Vibration is the frequent cause of eventual pipe failure but supports and clips to prevent this problem must permit free expansion and contraction. A pipe which has to be twisted or bowed when being connected, has inbuilt stress which can lead to ultimate failure. Pipes should be accurately made (particularly replacement sections) and installed with simple supports before being permanently clipped. If pumps are designed so that the driving motor or turbine is mounted upon an extension to the pump casing proper, the tendency for mal-alignment, due to pipeline stresses, is practically eliminated. Nevertheless, it is essential that the pipe systems and heavy valve chests, are separately supported and stayed during installation, the flanged connection to the pumps being the last to be coupled after the faces are correctly aligned. This can contribute materially to the life of the unit.

Horizontal pumps should be laid down on suitable chocks, accurately fitted to ensure that the couplings, with their bolts removed, are in correct alignment and with their faces parallel. This alignment should be checked after tightening the holding down bolts and again after the pipes are coupled and preferably full of liquid.

Colour coding

It is usual to identify pipes by a colour code for the individual system or by bands of paint at intervals on pipes of a common colour. There are standard codes but individuals or companies may prefer variations. Frequently pipes are incorrectly coloured. Before working on or using a pipe system, it should be traced and verified.

Cleaning the system

It is often found, in new ships, that the bilges and bilge systems have not been thoroughly cleaned with the result that wood, nuts, bolts, rags and other debris are found inside valves and pipes after initial bilge pumping. These choke the valve-chests and prevent the valves from being properly closed. They also block strainers. It is vital to clean before the bilge system is tested to ensure that all suction pipes, joints, valves and glands are free from air leaks. Pipes too must be cleaned and checked as being clear before and after assembly. Blockage has sometimes been found due to failure to cut apertures in metal or joints. Obviously with hydraulic or pneumatic pipe systems, foreign objects or residues from manufacture can cause serious malfunction.

Drains

Disastrous explosions have been caused by accumulations of oil or oil vapour in diesel engine air lines which were not regularly drained. Severe damage has been caused by 'water hammer' when steam has been admitted to pipes containing water, especially when a slight inclination of the pipe from the horizontal allowed the water to have a large free surface area.

On steam being admitted, condensation occurs on the cool water surface or in a cold section of the pipe, a partial vacuum develops and the water moves along the pipe at great speed. The impact of this water at a bend or valve, can cause fracture of the pipe. Water hammer is indicated by severe and often repeated banging in the pipe. Steam pipes are fitted with drains which should be left open so that water will not accumulate otherwise drains must be opened before admitting steam. Steam master valves are first opened very slightly or 'cracked open' when a line is being brought into use until the pipe is thoroughly warmed. Only then should the valve be opened fully.

Expansion arrangements

Provision must be made in pipe systems to accommodate changes in length due to change of temperature, and so prevent undue stress or distortion as pipes expand or contract. One type of expansion joint (Figure 4.1) has an anchored sleeve with a stuffing box and gland in which an extension of the joining pipe can slide freely within imposed limits. Simpler schemes (Figure 4.2a and 4.2b) allow for change of length with a right angle bend arrangement or a loop. For high pressures and temperatures with associated greater pipe diameter and thickness other methods may be more appropriate.

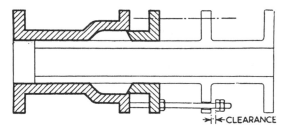

Figure 4.1 *Tie rod expansion joint*

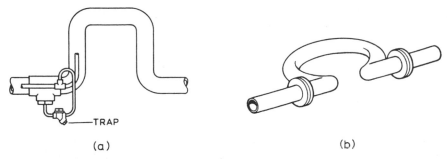

(a)   (b)

Figure 4.2 *Steam-line expansion arrangements (a) Expansion loop upwards. Large bore drain pocket fitted before loop (b) Expansion loop horizontal, no drainage required*

Stainless steel bellows expansion joints (Figure 4.3) are commonly used since they will absorb some movement or vibration in several planes, eliminate maintenance, reduce friction and heat losses.

Maximum and minimum working temperatures must be considered when choosing a bellows piece, which must be so installed that it is neither over-compressed nor over-extended. Its length must be correct for the temperature change. Stainless steel is the usual material for temperatures up to 500°C. Beyond that and for severe corrosive conditions, other materials are required.

Normally the bellows has an internal sleeve, to give smooth flow, to act as a heat shield and to prevent erosion. If exposed to the possibility of external damage, it should have a cover. In usual marine applications, bellows joints are designed and fitted to accommodate straight-line axial movement only and the associated piping requires adequate anchors and guides to prevent misalignment. It will be apparent that, in certain cases, the end connections will act adequately as anchors and that well designed hangers will be effective guides.

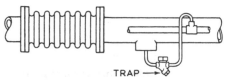

Figure 4.3 *Bellows type expansion fitting*

An axial bellows expansion joint can accommodate compression and extension, usually stated as plus or minus X mm, i.e. it will compress or extend X mm from the free length, at which it is supplied. It is most important that the unit be installed at its correct length as extension or compression outside its specified limits will cause premature fatigue.

Watertight bulkheads

Pipes are carried through watertight bulkheads with the use of special fittings (Figure 4.4) to avoid impairment of their integrity. The large flange of the fitting, covers the necessary clearance in the bulkhead.

Joints

Joints between flanges should be impervious to damage from the fluids carried and a variety of materials are available to suit the different requirements. Rubber for example, with or without cotton insertion, is suitable for water but not for oil. High pressure can force a joint out of a flange so that the thinnest joints are used for the highest pressures. Some jointing fabrics are sheathed with copper or stainless steel, which may be grooved finely and lightly in the

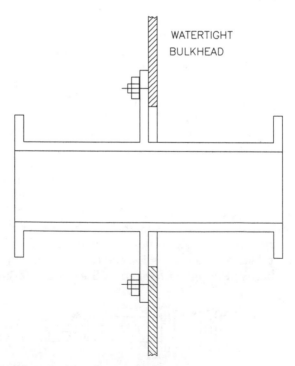

WATERTIGHT
BULKHEAD

Figure 4.4   *Bulkhead piece for use when a pipe passes through a watertight bulkhead*

area adjacent to the pipe bore. Most materials deteriorate with time and temperature so that periodic replacement may be necessary. Graphite compounds assist flexibility.

Mating flanges should be parallel and accurately machined. Bolts should fit reasonably well and have good threads.

Cocks and valves

Cocks and valves are designed to control or interrupt flow. This is done in cocks by rotating the plug, and in valves by lowering, raising or rotating a disc in relation to a seating surface or by controlling the movement of a ball. These fittings have bodies furnished with flanged or screwed ends (or ends prepared by welding) for connection to the joining pipes.

Cocks

A cock may be straight-through, right-angled or open-bottomed as required by its situation in a pipe system. Its plug may be tapered or parallel with tightness achieved by lapping in or by resilient packing material (Figure 4.5) often in the form of a ready made sleeve.

In machinery spaces, the short sounding pipes for fuel or lubricating oil tanks, must be fitted with cocks having parallel as opposed to tapered plugs. This, together with the requirement for weighted handles which will automatically close the cock when released, is for safety. Tapered plugs, when tightened to hold the cock open for sounding and then forgotten, have contributed to fires when tanks have overflowed.

Boiler blowdown cocks on the ship's shell, are constructed so that the handle can be removed only when the cock is closed.

Globe valves

The globe valve (Figure 4.6) has a bulbous body, housing a valve seat and screw down plug or disc arranged at right angles to the axis of the pipe. For the valve shown, both seat and disc faces are stellited and almost indestructible. Alternatively, the seat may be renewable and screwed into the valve chest or given a light interference fit and secured by grub screw. The seatings may be flat or more commonly mitred. The spindle or stem may have a vee or square thread, below or above the stuffing box. If the latter it will work in a removable or an integral bridge (bonnet).

The spindle may be held in the valve disc (or 'lid') by a nut as shown or the button may locate in a simple horseshoe. Leakage along the valve spindle is prevented by a stuffing box, packed with a suitable material and a gland. If there is a change of direction, as in a bilge suction, the valve is referred to as an angle valve. Flow is from below the valve seat, so that the gland is not subject

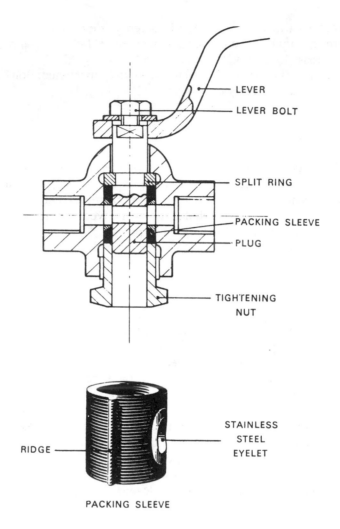

LEVER

LEVER BOLT

SPLIT RING

PACKING SLEEVE

PLUG

TIGHTENING NUT

RIDGE

STAINLESS STEEL EYELET

PACKING SLEEVE

Figure 4.5   *Example of a sleeve-packed cock (Richard Klinger Ltd)*

to higher static pressure when the valve is closed. The disc must be guided by wings or a stem on the underside for location, or by a piston as shown.

The type of valve with the disc attached to the spindle is of the screw lift type. When the disc is not attached to the spindle (inset Figure 4.6) it is a screw-down non-return (SDNR) valve, as used for bilge systems, to prevent back flooding. They are also used as feed check and boiler stop valves. The disc requires guide vanes or a stem to keep it concentric with the seat when open. The greatest lift required is one-quarter of the bore; guides must be of a greater length than the lift.

A free-lifting non-return valve (Figure 4.7) is fitted in the compartment served by a bilge suction line, when the pipe is nearer to the ship side than one fifth of the ship's breadth. Such valves are intended to prevent flooding of the compartment in the event of collision damage.

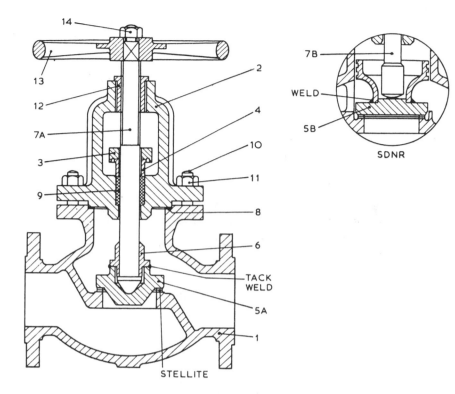

Figure 4.6 *Example of a globe valve with (inset) detail of the valve arranged as a screw-down non-return (SDNR) unit (Hattersley Newman Hender Ltd.)*

1. Body
2. Bonnet
3. Gland flange
4. Gland
5A Disc – stop type
5B Disc – piston SDNR
6. Disc stem nut
7A Stem – stop type
7B Stem – piston SDNR
8. Bonnet gasket
9. Gland packing
10. Bonnet studs
11. Bonnet stud nuts
12. Yoke bush
13. Handwheel
14. Handwheel nut

Gate valves

Unlike the globe valve, gate (or sluice) valves (Figure 4.8) give full bore flow without change of direction. The valve disc known appropriately as a gate, is moved at right angles into the flow by a screwed spindle working in a nut. It rests when closed, between circular openings furnished with seats. Valves and seats may be tapered or parallel on their facing sides.

Such a valve is not suitable to partially open operation since wire-drawing of the seat will occur. The bonnets of these valves are frequently of cast iron and care should be taken when overhauling. To ensure tightness, some parallel gates are fitted with twin discs, dimensioned similarly to the chest seats but pressed against the seats by a spring when closed.

Where change of direction is required, a full bore angle valve (Figure 4.9) may be used.

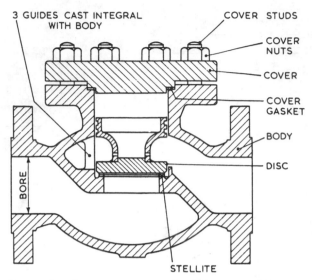

Figure 4.7 *Example of a non-return valve (Hattersley Newman Hender Ltd.)*

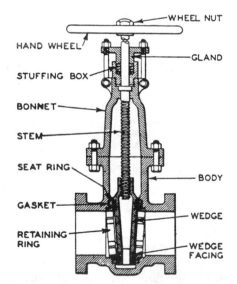

Figure 4.8 *Gate (or sluice) valve*

Butterfly valves

A butterfly valve (Figure 4.10) consists basically of a disc pivoted across the bore of a ring body having the same radial dimensions as the pipe in which it is fitted. The full bore straight through flow arrangement of this type of valve, especially if combined with a carefully streamlined disc profile, gives excellent

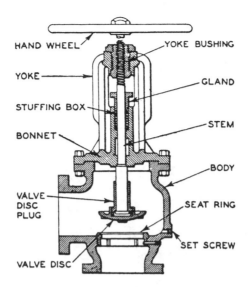

Figure 4.9 *Full bore angle valve*

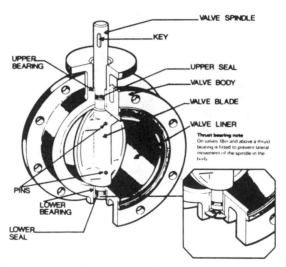

Figure 4.10 *Type B A butterfly valve (Hindle Cockburns Ltd)*

flow characteristics and low pressure drop. The valve is quick-acting if required, as only a quarter of a turn of the spindle is required to move the valve from the fully open to the fully closed position. Sizes range from 6 mm to over 1000 mm bore.

For fine control of cooling water temperature a special type of ganged butterfly valve may be used to bypass coolers. Known as a diverter valve it consists of a Y or T casting with butterfly valves in two of the legs (Figure 4.11). A pneumatic actuator working from a signal provided by a temperature sensor opens one valve while closing the other. This gives precise control of

Figure 4.11   *A diverter valve (Cockburn-Rockwell Ltd.)*

the flow rate in the main and branch lines. In the event of a temperature controller failure, a built-in return spring opens or closes the main and branch lines (as appropriate to the system of operation) to provide maximum cooling flow. Manual control is available for emergencies.

Flap valves

Scupper pipes from accommodation spaces are fitted with non-return valves. Those scuppers from spaces below the bulkhead deck, are required to be fitted with non-return valves which can be positively closed from above the bulkhead deck or, if this is not practical, with two non-return valves. A common type of non-return valve (Figure 4.12) has a hinged flap which is pushed open by outward flow and closed by its own weight. The flap prevents inward passage of sea water.

Change-over valve chests

Dual purpose tanks such as those for either oil or water ballast require exclusive connections to separate systems. Special valve chests (Figure 4.13) with interchangeable blanks and connecting passages are installed for this duty. The example shown has two suction valves flanked by a blank on one side and a dome on the other. In the position shown, the two suction valves have access

Figure 4.12 *Flap check valve (Blakeborough and Sons Ltd.)*

1. Cover
2. Body
3. Hinge shaft
4. shaft bearing

5. Seat ring
6. Face ring
7. Door

via the dome, to the water ballast main but not to the oil fuel main. By simply changing over the blank and the dome the situation is reversed.

Valve actuators

A variety of valve actuators to control the opening and closing of globe, gate and butterfly valves are available. In some types an electric motor, fitted with limit switches is used to turn a threaded stem through a yoke, purely substituting the action of a handwheel. Most remotely operated valves have pneumatic or hydraulic actuators. These give linear motion to a piston which for a globe or gate valve moves the valve stem axially up or down. The globe valve disc may be given a slight turn on landing to clean the seat. The piston actuator for a butterfly valve rotates the valve disc through 90° directly or through a scroll arrangement (Figure 4.14).

Relief valves

Excess pressure is eased by a relief valve (Figure 4.15). This consists of a disc held closed by a spring loaded stem. The compression on the spring can be

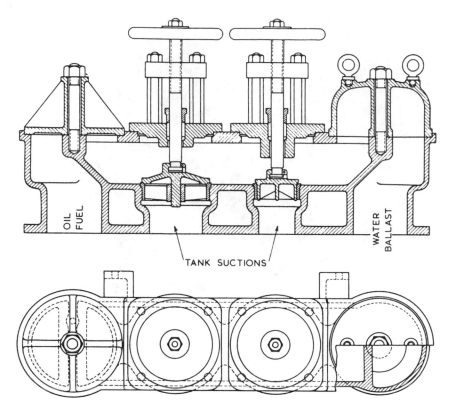

Figure 4.13   *2-valve change-over chest for oil and ballast suctions as arranged when filling or discharging ballast*

adjusted so that the valve opens at the desired pressure. The special case of boiler safety valves is dealt with in *Marine Steam Boilers* by J. H. Milton and R. M. Leach. Selection of a valve of the correct size and loading is important since they have a narrow pressure range.

Under normal conditions a relief valve should operate consistently within reasonable limits of its set pressure. Incorrect function may be due to the setting, valve seat deposit or damage. Relief valve seats should be checked whenever the pump is overhauled.

Pressure reducing valves

If it is necessary to provide steam or air at a pressure less than that of the boiler or compressed air reservoir, a reducing valve is fitted. This will maintain the downstream pressure within defined limits over a range of flow, despite any changes in supply pressure.

In the reducing valve shown (Figure 4.16) the higher inlet pressure (P_1) acts in an upward direction on the main valve and in a downward direction on the

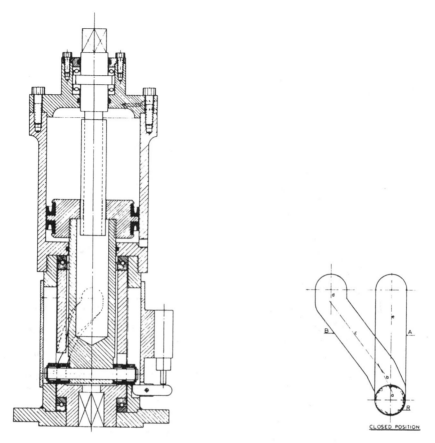

Figure 4.14 *Pneumatic butterfly valve actuator showing scroll cam arrangement*

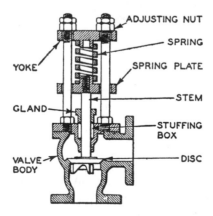

Figure 4.15 *Relief valve*

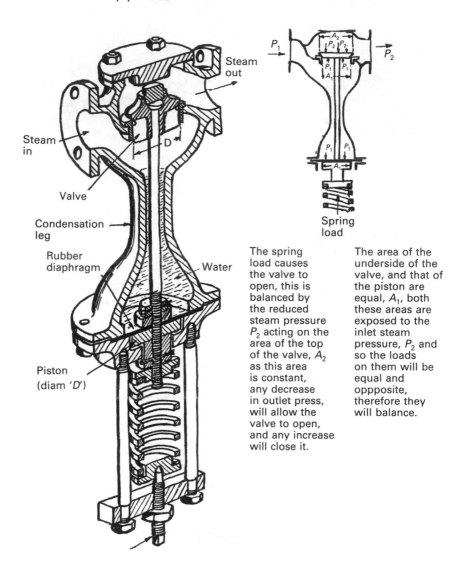

The spring load causes the valve to open, this is balanced by the reduced steam pressure P_2 acting on the area of the top of the valve, A_2 as this area is constant, any decrease in outlet press, will allow the valve to open, and any increase will close it.

The area of the underside of the valve, and that of the piston are equal, A_1, both these areas are exposed to the inlet steam pressure, P_2 and so the loads on them will be equal and oppposite, therefore they will balance.

Figure 4.16   *Pressure reducing valve*

controlling flexible diaphragm and the piston beneath it. These two parts are in a state of balance. The large spring pushes against the spindle, tending to open the valve against the reduced steam pressure (P_2) acting on the area A_2 at the top of the valve. Any decrease in pressure on the outlet side, will allow the valve to be pushed open by the spring. Any increase will close it. It is important that this type of valve is installed in the vertical position.

The self-regulating valve can be replaced by an automatic process control valve for fluid pressure control as used for control functions in unmanned machinery spaces. This permits remote control of the set-point of the valve and by careful selection of valve trim (the control industry's term for the internal parts of the valve which come in contact with the controlled fluid and form the actual control portion) a variety of flow characteristics can be achieved. The subject is dealt with further in Chapter 16.

Quick closing valves

Fuel oil service and some other tanks must be fitted with valves that can be closed rapidly and remotely in the event of an emergency such as fire. Wire operated valves (Figure 4.17) are commonly fitted, with wire pull levers located externally to the machinery space. The type shown is a Howden Instanter valve. As an alternative a hydraulically operated quick-closing valve (Figure 4.18) can be fitted.

Quick-closing valves are examined and tested when installed and then periodically when the tank is not in use, to ensure that the mechanism functions correctly. Wires are sometimes found to be slack or hydraulic systems empty.

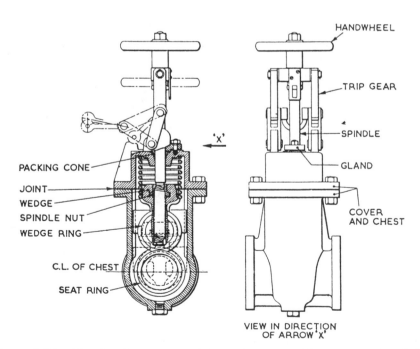

Figure 4.17 *Howden Instanter quick closing valve (James Howden & Co. Ltd.)*

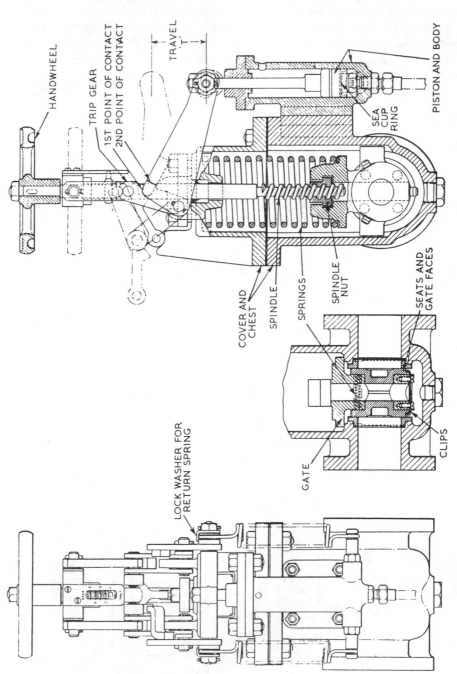

HANDWHEEL

TRIP GEAR

1ST POINT OF CONTACT
2ND POINT OF CONTACT

TRAVEL 'T'

SEA
CUP
RING

PISTON AND BODY

COVER AND
CHEST

SPINDLE

SPRINGS

SPINDLE
NUT

SEATS AND
GATE FACES

GATE

CLIPS

LOCK WASHER FOR
RETURN SPRING

Figure 4.18 Arrangement of Instanter quick closing parallel slide valve with oil operating cylinder

Steam traps

A steam trap is a special type of valve which prevents the passage of steam but allows condensate through. It works automatically and is used in steam heating lines to drain condensate without passing any steam. The benefit gained with a steam trap, is that steam is contained in the heating line until it condenses, thus giving up all of its latent heat. There are three main types which are the mechanical, the thermostatic and the thermodynamic. There is also the vacuum trap or automatic pump.

Mechanical traps

Mechanical traps have been installed with ball floats (Figure 4.19) or open floats (Figure 4.20) for control of a needle valve to release condensate.

Thermostatic traps

Thermostatic traps (Figures 4.21a, 4.21b and 4.21c) use the expansion of an oil-filled element, a bimetallic strip or flexible bellows to actuate a valve.

As the condensate temperature rises in the oil filled element type (Figure 4.21a) element A expands to close the valve D. An adjustment screw E permits the valve to be set up for condensate release at a specific temperature. Clearly in an application where the pressure varies, there could be a broad band of operation in which the trap would be either waterlogged or passing steam.

With the bimetallic strip type (Figure 4.21b) deflection of the bimetallic strip when temperature increases, closes the valve. The device will work over a

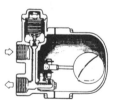

Figure 4.19 *Ball float type mechanical trap*

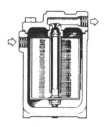

Figure 4.20 *Open float mechanical trap*

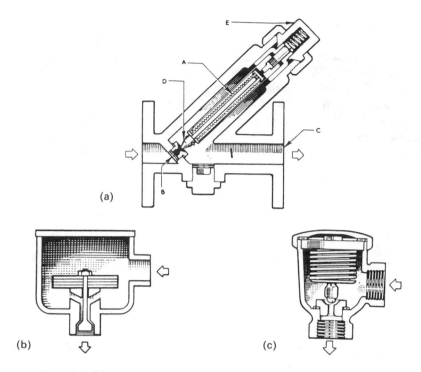

Figure 4.21 *(a) Oil filled thermostatic steam trap; (b) Bi-metallic steam trap; (c) Bellows type steam trap (Spirax-Sarco Ltd.)*

A. Element   D. Valve head
B. Orifice   E. Adjuster
C. Outlet

range of pressures without the need for re-adjustment and will operate satisfactorily under superheat conditions. It is not particularly prone to damage by water hammer or vibration.

In the flexible bellows type (Figure 4.21c) the bellows is filled with a mixture which boils at a lower temperature than does the steam. The trap self-compensates for operating pressure. It will be damaged if water hammer occurs and will burst if subjected to superheated steam.

Thermodynamic steam trap

This type of trap uses the pressure energy of the steam to close the valve which consists of a simple metal disc. The sequence of operation is shown in Figure 4.22. In (i), disc A is raised from seat rings C by incoming pressure allowing discharge of air and condensate through outlet B. As the condensate approaches steam temperature it flashes to steam at the trap orifice. This means that the rate of fluid flow radially outwards under the disc is greatly increased. There is thus an increase in dynamic pressure and a reduction in static pressure. The disc is therefore drawn towards the seat. Due to this alone the disc will never seat.

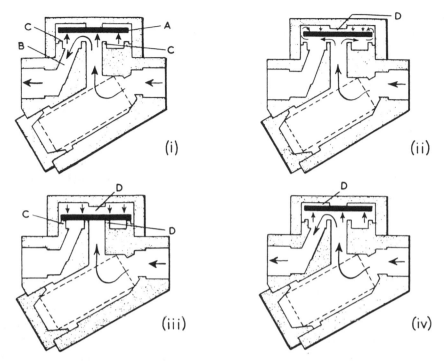

Figure 4.22 *Operation of thermodynamic steam trap (Spirax-Sarco Ltd.)*

However, steam can flow round the edge of the disc resulting in a pressure build up in the control chamber D as shown in (ii). When the steam pressure in chamber D acting over the full area of the disc (iii) exceeds the incoming condensate/steam pressure acting on the much smaller inlet area, the disc snaps shut over the orifice. This snap action is important. It removes any possibility of wire-drawing the seat, while the seating itself is tight, ensuring no leakage. As shown in (iv) the incoming pressure will eventually exceed the control chamber pressure and the disc will be raised, starting the cycle all over again.

The rate of operating will depend on the steam pressure and on the ambient air temperature. In practice, the trap will usually open after 15–25 s, the length of time open depending on the amount of condensate to be discharged. If no condensate has been formed, then the trap snaps shut immediately. From the foregoing it will be seen that the trap is never closed for more than 15–25 s, so condensate is removed virtually as soon as it is formed.

Vacuum or pumping traps

The layout of drain systems can often be improved by the use of automatic pumps, sometimes referred to as vacuum traps. Condensate can be drained by gravity more readily to local receivers and then pumped back to the engine room as required. Similarly engine room drains can be relieved of much back pressure by taking them to a low level hot well before pumping to a high level boiler feed tank. A high level feed tank will enable the pumps to handle warm feed water and can improve the overall efficiency of the plant. A typical pumping trap is shown in Figure 4.23.

When the trap is empty, the exhaust valve C is open and the steam valve D is closed. Water flowing into the trap through the non-return valve A raises the float E until it compresses the spring H, exerting a force which plucks the spindle J away from the magnet G so that the exhaust valve is projected into the closed position and the steam valve D is opened through the movement of the lever. Steam is therefore admitted to the trap and drives out the water through the non-return valve B. The float falls with the level of the water inside the trap and engages the collar F, pulling the spindle J down so that the exhaust valve opens and the steam valve closes. The cycle of operation is then repeated. Being float controlled, the trap operates only when water is flowing into it.

Maintenance of traps

A defective trap wastes steam and fails to ensure adequate heating. Poor heater performance can sometimes be traced to a defective steam trap. One remedy is to close in the steam outlet valve to perhaps half or quarter of a turn. Most traps can be regularly opened for inspection. The only exception is the thermodynamic type. When operating correctly this gives a characteristic 'click', usually at intervals of between 20 and 30 s so that its performance can be checked simply by listening.

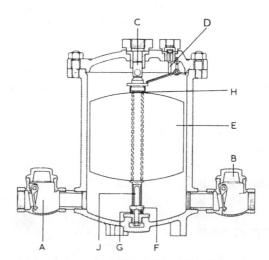

Figure 4.23   *Automatic pumping or vacuum trap (Royles Ltd.)*

A, B. Non-return valves   F. Collar
C. Exhaust valve   G. Magnet
D. Steam valve   H. Spring
E. Float   J. Spindle

Before inspecting any trap, it is advisable to check the strainer which should always be provided at the inlet of the trap. The contents of the strainer screen will give an indication of the cleanliness of the system. Fine dirt passing through the strainer is one of the chief causes of defective steam traps. Pieces of pipe scale or dirt jammed across the seat can prevent proper closure and if left for a period of time can give rise to wire drawing.

In most cases cleaning and reassembling should be sufficient to ensure satisfactory operation. If the thermodynamic type fails to operate after cleaning, then it could be that the disc and seat require lapping. Mechanical traps should be checked for defective floats and buckets and wear on any linkages, while the valves and seats should be renewed if necessary. In the case of thermostatic traps, the elements should be checked to ensure that they are sound. Particular care should be taken with the elements of bimetallic steam traps. Incorrect adjustment can give rise to excessive waterlogging or cause the traps to blow steam, while the elements themselves can sometimes assume a permanent 'set'.

Strainers and filters

The term strainer is sometimes used specifically for a simple device made up with a single layer of coarse gauze, a very coarse wire mesh or a drilled or perforated plate. The strainer in this sense is installed to hold large foreign objects which could cause damage or blockage. The term filters also describes a device designed to prevent the passage of unwanted solids into or further along a system, but the implication is that the particles are very small and the filter cartridge or element has a more complex or elaborate make up.

The simplest strainer consists of a box with a removable lid in which a flat perforated plate is inserted such that the fluid must pass through the perforations. Such strainers are found in bilge systems immediately before bilge valves. Perforation sizes vary according to duty and manufacture but are usually in the range 3–12 mm.

Strainer plates corrode and erode and when cleaning due attention should be paid to the condition of the plate. If a gap has formed at the top or bottom of the plate (so that solid objects or rags could pass through) or if it is in danger of breaking up then it should be renewed. These strainers protect the pump. If strainer covers are hinged the hinges are made loose to avoid the problem of poor seating. The state of the gasket should be checked, mating surfaces must be clean and care should be taken when closing the cover. Since these covers are frequently of cast iron and are secured by lugs and thumb screws care must be taken not to use undue force when closing them.

For high pressure water, steam or oil services simple basket strainers are used. These consist of a cylindrical container in which a perforated metal or wire basket is suspended (shown left Figure 4.24). Flow through these units is from the top, into the basket and out from the outside of the basket. They may be installed as duplex units with three way cocks at inlet and outlet so that one or both baskets can be in use, but one can be shut down for cleaning. Close to

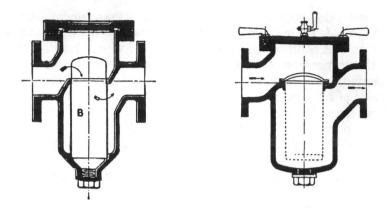

Figure 4.24 *Single strainers for (left) high pressure water, steam or oil service and (right) low pressure water or oil service*

sea-water suction valves, similar basket strainers (shown right in Figure 4.24) having air release cocks, are installed.

Lubricating oil systems are fitted with a wide variety of strainers some of which can be cleaned *in situ*. The knife edge strainer, (Figure 4.25) has a series of discs ganged to a shaft. Interspaced between the discs are a number of thin fingers. The solid particles are trapped on the edges of, and between, the discs. By rotating the disc shaft the particles are cleared by the fingers and fall to a sump, which is drained periodically. It is essential to operate these strainers regularly to prevent clogging when rotation may be difficult. Some have been fitted with electric motors for continuous operation.

Cartridge filters (Figure 4.26) are usually of the duplex type and have various types of element. Some cartridges can be removed for cleaning; others

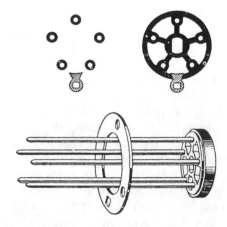

Figure 4.25 *The Auto-Klean strainer showing circular straining plates, separation washers and spider (Auto-Klean Strainers Ltd.)*

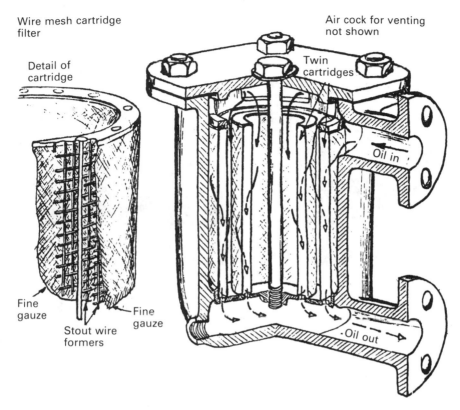

Wire mesh cartridge filter

Detail of cartridge

Air cock for venting not shown

Twin cartridges

Oil in

Fine gauze

Fine gauze

Stout wire formers

Oil out

Figure 4.26 *Cartridge filters*

have renewable elements. Cartridges can be of gauze layers with different mesh sizes others are made up with layers of mesh and felt, some use membrane type materials.

Magnetic filters (Figure 4.27) provide extra protection for engines and gearboxes where iron or steel wear particles are likely to be present.

Further reading

Milton, J. H. and Leach, R. M. (1980) *Marine Steam Boilers*, 4th edn. Butterworths.

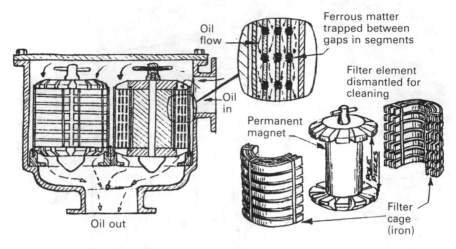

Figure 4.27 *Magnetic filter*

5

Pumps and pumping

The centrifugal pump is now used for most applications and systems on ships. In the machinery space it provides a much more reliable service than the steam reciprocating pumps that were still being installed in the 'fifties as auxiliary boiler feed and fuel pumps for example. These reciprocating pumps required regular maintenance and, if neglected, they needed constant attention to keep them functioning. The general use of the centrifugal pump helped to make the unmanned machinery space viable.

General pumping system characteristics

A pump divides its pipe system into two distinct parts, each with different characteristics. These are the suction and discharge sides. On the suction side the drop in pressure that can be produced by a pump is limited to that of an almost perfect vacuum. On the discharge side there is theoretically, no limit to the height through which a liquid can be raised.

Suction conditions

If a liquid to be pumped is in a tank which is open to atmosphere and it is also at a height above the pump (Figure 5.1) then the liquid will flow into the pump because of its head and due to the effect of atmospheric pressure on its surface. The pump in this case only adds to the energy of the system.

When an open tank containing the liquid to be transferred is at a lower level than the pump, the energy required to bring the liquid to the pump, is provided by atmospheric pressure on its surface but the pump must create the drop in pressure which makes the atmospheric pressure effective.

A discussion of the relevant features of the suction side of a pumping system must include not only the height of the liquid surface above or below the pump and the effect of atmospheric pressure, but also the effect of a vacuum or zero pressure on the liquid surface, vapour pressure and the characteristics of the suction pipe. Liquid flow through a pipe is impeded by friction over its length, by valves or other restrictions and by changes of direction.

Figure 5.1 shows pressure head H_{es} acting on the liquid surface at the suction inlet. The vertical distance of the pump centre H_s from the surface of the liquid will affect the head available at the pump and must be added algebraically to

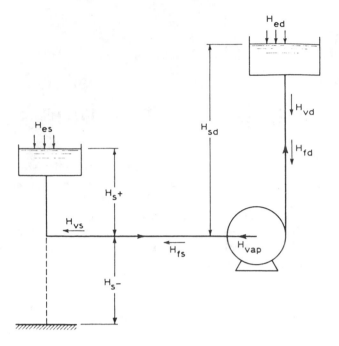

Figure 5.1   *A simple pumping system*

H_{es}. If the pump is below the liquid level then H_s will be positive; if it is above the liquid level H_s will be negative. The pipe will have some frictional resistance resulting in a loss of pressure head H_{fs}. A further head loss H_v, due to the velocity of the liquid will also occur but, except for very high velocities, it is negligible.

Providing that the sum of these head losses $H_v + H_{fs} \pm H_s$ is less than H_{es}, the suction condition at the pump might be thought to be adequate. There are two further factors to take into consideration however. These are the vapour pressure of the liquid being pumped and the amount of remaining positive suction head required at the pump suction to effect the designed delivery rate. This factor is known as the required NPSH (net positive suction head).

Every liquid has a pressure at which it will vaporize and this pressure varies with temperature. If the combination of pressure and temperature within the suction pipe is such that vaporization occurs, the efficiency of the pump deteriorates and a condition can be reached where the pump will cease to function. The vapour pressure H_{vap} is thus usually shown as a suction head loss.

The summation $H_{es} \pm H_s - H_{fs} - H_{vs} - H_{vap}$ is known as the available NPSH (net positive suction head). In application to systems and neglecting the velocity head the expression becomes:

$$\text{available NPSH} = \frac{10.2}{\rho}[P_{bar} + P_{es} - P_{vap}] - H_{fs} \pm H_s,$$

where:

ρ = density of liquid at max operating temp, kg/litre,

P_{bar} = barometric pressure at the pump, bar,

P_{es} = minimum pressure on the free liquid level at the suction inlet (negative when under vacuum), bar gauge,

P_{vap} = vapour pressure of the liquid at the maximum operating temperature, bar abs,

H_s = height of liquid free surface above the centre line of the pump (negative when level is below pump), m,

H_{fs} = friction head losses in suction piping system, m.

In application, the available NPSH must always be greater than the required NPSH. The former may be calculated knowing the details of the suction piping while the latter may be obtained from the pump manufacturer.

The significance of vapour pressure is most easily seen when considering a pump drawing from a negative suction head (usually referred to as a suction lift).

The theoretical suction lift of a pump at sea level with water at 15°C is $1.013 \times 10.2 = 10.3$ m where the barometric pressure is 1.013 bar (1 atm) and 10.2 m is the head of water equivalent to 1 bar (1 bar = 10^5 N/m² = 14.51 lb/in²). In practice the suction lift will exceed 7 m only under very favourable conditions. This is because of friction losses in the suction pipe and because of the limitations of the pump design. Any increase in water temperature above 15°C will have a detrimental effect on the vapour pressure. As an example, at 50°C water will boil at an absolute pressure of 0.14 bar, so that the lift reduces to $10.2(1.013 - 0.14) = 9$ m, drastically reducing the available NPSH. It follows that suction lift should be as small as conditions allow and that for water temperatures above about 75°C the suction head must be positive or if this is impossible the suction pipe must be short, straight, free from interference and the speed of flow must be low, say less than 1 m/s.

Discharge conditions

Some of the energy fed into the pump will be dissipated as heat due to mechanical inefficiencies. The remainder will be converted into pressure rise and fluid velocity. Some of the pressure head generated will be lost in overcoming the friction of the discharge pipe H_{fd}, some in the static head of the pipe system H_{sd}, and some in the pressure head acting on the free surface at the terminal point H_{ed}. There will also be a velocity head loss but, as in the case of the suction line, for most practical purposes this can be neglected.

Pump power

The total work done by the pump, neglecting losses within the pump itself, will be proportional to the equivalent head difference between the points of suction and discharge. This is known as total head H_{tot}:

$$H_{tot} = H_{fs} + H_{fd} + H_{vap} + H_{sd} \pm H_s$$

The power absorbed by the pump, P_a, then becomes:

$$P_a = \frac{Q \times H_{tot} \times w}{K}$$

where:

P_a = Power absorbed (kW),
Q = Quantity delivered in litres/s
H_{tot} = Total head in metres,
w = Density of liquid in gm/ml (1 for fresh water),
K = 101.9368 (102).

The input power P_i to the pump required from the prime mover is

$$P_a \times \frac{1}{\text{pump erfficiency}}$$

For an electrically driven pump, the power consumed is

$$P \times \frac{1}{\text{pump efficiency}} \times \frac{1}{\text{motor efficiency}} \text{(kW)}.$$

Where the head available is small, the suction line, passages and valves are specially designed and of a large enough area to reduce the suction losses to a minimum. This increases the cost of the pump and installation and reduces efficiency, but is unavoidable for duties such as the extraction of condensate from a condenser where the head available is frequently a matter of millimetres.

Generally speaking, suction heads require to be greater for high speed or large capacities than for low speed or small capacities. Condensate pumps, heater drain pumps and feed pumps operating with direct-contact feed heaters, must be arranged below the water level as the static head of water is the only force available to cause the water to flow into the pump, because the water and the steam on the surface are at the same temperature.

Before liquid can flow into a pump, the air or vapour in the suction line must be evacuated sufficiently to cause the liquid to flow into the suction chamber.

Some pumps (known as self-priming pumps) do this automatically when they are started. In others special priming devices must be used to withdraw the air and lower the pressure in the pump sufficiently to cause flow.

Friction losses

The sum of these losses depends upon the sectional area and the internal condition of the pipes and fittings, the velocity and viscosity of the liquid being pumped and the friction caused by bends, valves and other fittings. Frictional resistance to the flow of water varies approximately as the square of the velocity. Thus, if the frictional resistance of a condenser and system of piping is equivalent to a head of 5 m when 800 litres/s are passing, the frictional resistance would rise to 11.25 m with 1200 litres/s and to 20 m with

1600 litres/s. Tables are provided by pump manufacturers to find the head in metres required to overcome the friction of flow in pipes of different materials and size.

The general law of frictional resistance due to the flow of water in a straight circular pipe running full of water may be expressed accurately enough for practical purposes as

$$Hm = \frac{KLV^2}{2GR} \text{ if } R = \frac{\text{area of pipe bore}}{\text{wetted perimeter}} = \frac{D}{4}$$

or
$$Hm = \frac{KLV^2}{2GD}$$

where:

Hm = head loss (m),
L = length of pipe (m),
V = speed of flow (m/s),
D = bore of pipes (m),
G = gravitational constant = 9.81 m/s² = 9.81 N/kg.

To this must be added the loss due to bends each equivalent to from 3 to 6 m of straight pipe, depending upon the radius of the bend. Tables (for example Table 5.1) are provided by manufacturers to find losses due to bends and fittings.

Drawings or prints are supplied with pumps by the manufacturers, giving sizes and particulars of flanges, positions of foundation bolts and other information necessary for the arrangements of pipe connections. These must be exactly adhered to; it is little use installing a highly efficient pump if the power is dissipated and increased by the use of unsuitable pipes and fittings or by poor layout.

Pipe connections

Connecting pipes of the correct bore should be fitted and this is of special importance for the suction pipe where restriction causes cavitation. Pipe connections should be as direct as possible, sharp bends and loops such as that shown in Figureure 5.2 must be avoided. The loop could increase turbulence and be the location of an air pocket.

A common fault, causing a great deal of trouble, is shown in Figure 5.3. Here the suction pipe connected to the pump slopes downwards to the pump, with the result that an air pocket is formed.

A similar fault is found when reducing pipes are used to connect an oversize pipe to the pump inlet. The type of reducing pipe shown at A (Figure 5.4) in conjunction with a 90° bend, creates an air pocket. Such a pipe is quite satisfactory on the discharge side of a pump when used to increase the diameter of the piping. At the suction side the type of reducer shown at B, could be used.

Table 5.1 Loss for fittings in equivalent lengths of straight pipe (m)

Bore of pipe mm	Vel. head for ordinary pipe	Vel. head for bell-mouthed entry	Bend	Foot valve	Non-return valve	Delivery valve full open	Strainer	Tess and elbows
25	1.37	0.82	0.76	0.24	0.305	0.24	0.091	0.83
38	2.2	1.31	1.13	0.36	0.49	0.36	0.152	1.31
50	3.0	1.8	1.52	0.52	0.7	0.52	0.214	1.8
65	3.8	2.31	1.95	0.64	0.85	0.64	0.275	2.31
75	4.75	2.86	2.44	0.79	1.06	0.79	0.305	2.87
100	6.4	3.96	3.3	1.10	1.43	1.10	0.427	3.96
125	8.5	5.2	4.27	1.43	1.86	1.43	0.58	5.2
150	10.7	6.4	5.27	1.76	2.31	1.76	0.70	6.4

Figure 5.2 *Loop to be avoided*

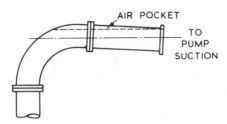

Figure 5.3 *Suction pipe sloping upwards from pump, forming air pocket*

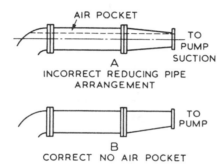

Figure 5.4 *Reducing pipes*

Types of pump

Marine pumps fall into the two broad classes of being either displacement or rotodynamic:

1. Displacement. The liquid or gas is displaced from the suction to the discharge by the mechanical variation of the volume of a chamber or chambers. Displacement pumps can be subdivided into two classes: reciprocating pumps, in which a plunger or piston is mechanically reciprocated in a liquid cylinder, and rotary pumps, where the liquid is forced through the pump cylinder or casing by means of screws or gears.

2. Rotodynamic. The centrifugal pump operates on a rotodynamic principle in that flow through the pump is induced by the centrifugal force imparted to the liquid by rotation of an impeller. The axial flow pump relies on rotation and the screw propeller form of the rotor.

Centrifugal pumps

Rotation of a centrifugal pump impeller (Figure 5.5) causes the liquid it contains to move outwards from the centre to beyond the circumference of the impeller. The revolving liquid is impelled by centrifugal effect. It can only be projected into the casing around the periphery of the impeller if other liquid in the casing can be displaced. Displaced liquid in moving from the casing to the delivery pipe, causes flow in the discharge side of the system.

The liquid in the impeller and casing of a centrifugal pump is also essential to its operation. In moving out under the influence of the centrifugal effect, it drops the pressure at the centre, to which the suction or supply pipe delivers the liquid to be pumped. The moving liquid acts in the same way as a reciprocating pump piston on its suction stroke. Provided that a centrifugal pump is filled initially with liquid and that flow is maintained, the suction stroke action will continue. If such a pump contains no liquid initially, it is as though an essential part is missing.

The volute pump (Figures 5.5 and 5.6) is so called because of the shape of the casing. The object of the volute is gradually to reduce the velocity of the water after it leaves the impeller, and so convert part of its kinetic energy to pressure energy. For general purposes the volute pump is commonly used. The diffuser pump (Figure 5.6) is so called because of the ring of guide passages around the impeller. The design is used for high pressure as in multi-stage boiler feed pumps. The diffuser passages are able to convert a larger amount of the kinetic energy of the liquid as it leaves the impeller into pressure energy. A single stage diffuser pump is able to deliver to a much greater head than an ordinary

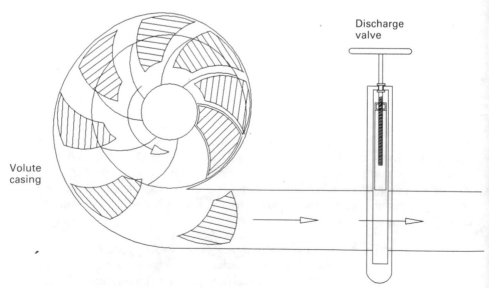

Discharge
valve

Volute
casing

Figure 5.5   *Centrifugal impeller action*

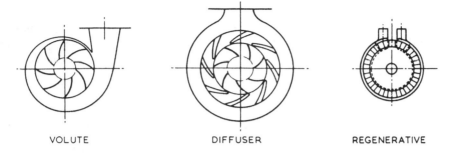

VOLUTE   DIFFUSER   REGENERATIVE

Figure 5.6   *Types of centrifugal pumps (Hamworthy Engineering Ltd)*

volute pump. The term turbine pump is sometimes used to describe the diffuser pump. Multi-stage deepwell pumps for bulk liquid cargoes may be referred to as turbine pumps.

The regenerative pump (Figure 5.6) is used where a relatively high pressure and small capacity are required.

Centrifugal pump discharge characteristics

From a mathematical consideration of the action of a centrifugal pump it can be shown that the theoretical relationship between head, H, and throughput, Q, is a straight line (Figure 5.7), with minimum throughput occurring when the head is maximum. Because of shock and eddy losses caused by impeller blade

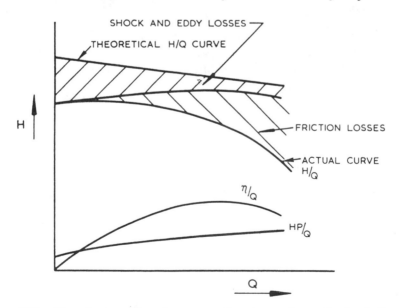

Figure 5.7   *Showing centifugal pump action by mathematical relationship (Hamworthy Engineering Ltd.)*

thickness and other mechanical considerations there will be some head loss, increasing slightly with throughput. These losses, together with friction losses due to fluid contact with the pump casing and inlet and impact losses, result in the actual H/Q curve shown in the figure. The shape of this curve, shows the discharge characteristic which varies according to the design and features of the particular pump. The discharge characteristic is obtained for a pump type, by measuring throughput (Q) through increase of head (H) during a test at constant speed. The actual discharge characteristic provides important information for the designer of a pumping system; it also explains why the throughput of a centrifugal pump alters with discharge head or back pressure. A slow rate of discharge by a centrifugal cargo pump can be explained by increasing head due to a restricted or very long discharge pipe, high viscosity of the liquid, discharge to a storage tank sited at a high level or even a partly open valve on the discharge line.

Depending on application, centrifugal pumps can be designed with relatively flat H/Q curves or if required the curve can be steep to give a relatively large shut-off head.

From Figure 5.7 and from HP/Q, the power curve, that minimum power is consumed by the pump when there is no flow and when the discharge head is at its highest. This equates to the discharge valve being closed. Because maximum pressure with the discharge closed is only moderately above working pressure, a relief valve is not necessary for a centrifugal pump.

It will be noticed that the efficiency curve for the pump is convex which means that maximum efficiency occurs at a point somewhere between maximum and minimum discharge head and throughput conditions.

In the case of a variable speed pump:

1   Head varies as the square of the speed.
2   Capacity varies directly as the speed.
3   Power varies as the cube of the speed since it is a function of head and capacity.

In the case of a constant speed pump:

1   Head varies as the square of the diameter.
2   Capacity varies as the diameter.
3   Power varies as the cube of the diameter.

Where the head in a given installation is known, the following formula could be used to calculate the necessary speed of the pump:

$$N = \frac{95\ H.C}{D}$$

where:

N = rev/min,
D = diameter of impeller over blade tips in m,
H = total head in m,
C = constant.

The value of C varies considerably with pump shape but is generally between 1.05 and 1.2; the higher value being taken for pumps working considerably beyond their normal duty, or for pumps with impellers having small tip angles.

Construction and installation

Marine pumps are usually installed with the shaft vertical and the motor above the pump. This positions the pump as low as possible for the best NPSH, takes up the least horizontal space and leaves the electric motor safer from gland or other leakage.

Shaft sealing

It is preferable on vertical pumps to have shaft sealing at the pump upper end only. This permits observation and adjustment of the shaft seal and ensures that the pump does not drain through a leaking gland during idle periods (i.e. the pump will remain free because deposits will not harden).

Mechanical seals (Figures 5.8a and 5.8b) are spring loaded to hold the sealing faces together. It is important that cooling and lubricating liquid is led

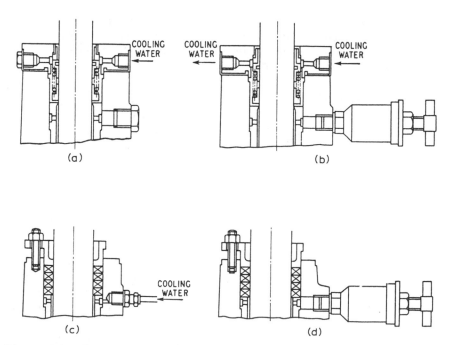

Figure 5.8 *Typical arrangements of pump seals (Hamworthy Engineering Ltd.)*

(a) Water lubricated bearing   (c) Water lubricated bearing, soft packed
(b) Grease lubricated bearing   (d) Grease lubricated bearing, soft packed

to mechanical seals from the lowest point on the pressure side of the pump, to ensure that some liquid reaches them, even when priming. Special header tanks have been installed for the seals in some applications. They must not run dry and care must be taken to prevent ingress of foreign matter. Many mechanical seals incorporate a carbon face and there is a possibility of electrolytic action in the presence of sea water. Soft packing may be preferred in sea-water pumps.

Stuffing box type glands (Figures 5.8c and 5.8d) may be packed with soft or metal foil type packing.

Pump internal bearings may be lubricated and cooled by the pumped liquid in situations where liquid is always available when the pump is running. Lubricators for the application of grease are fitted in some circumstances. Pumps used for slurries or those for suction dredgers, have external bearings and a nozzle located to exclude solids from the gland area.

Materials

Pumps like pipelines, are used for high or low temperature liquids, those which are corrosive and some that carry abrasive particles. The materials chosen for pump construction must be suitable.

(a)   Pumps for engine cooling water, fresh or potable water: high grade cast iron casings with bronze internals − shaft of bronze or stainless steel (EN57. 18 Cr/2Ni) the latter material gives better wear life.

(b)   Sea-water pumps: (these must also handle harbour, river and canal water) may have gunmetal casing with aluminium bronze impeller (BS 1400 AB2 9: 5AI − 5 Fe − 5.5 Ni) − shaft either stainless steel (EN57) for soft packed stuffing boxes or EN58J (18Cr − 10Ni − 3 Mo) under mechanical seals or bearings

(c)   Boiler feed pumps: because of the high pressures and temperatures casings are of cast steel − shafts and impellers of stainless steel.

(d)   Cargo pumps: stainless steel casings, impeller and shaft suitable for most chemicals − nickel steels for low temperature liquefied gas.

Priming

Pumps may be mounted above the level of the liquid to be pumped even though placed low in the ship. Bilge pumps for this reason, must be self-priming or equipped with a means of priming, to create a vacuum in the suction pipeline. Ballast and other pumps (which may also be statutory bilge pumps) may need to be self-priming or equipped with a means of priming. Cargo pumps for oil tankers (see Chapter 6) are likely to be arranged for stripping the maximum amount of liquid from tanks.

A centrifugal pump placed above the liquid to be pumped, is not self-priming because it cannot exhaust the air contained in its casing and suction pipe. A displacement pump can and is self-priming. A centrifugal pump must be placed below the level of the liquid to be pumped (when it will flood if valves are opened) or it must be provided with an external device for the removal of air.

Air handling methods

The removal of air from pump suction pipes is usually achieved with a liquid ring primer. This is necessary in order to produce vacuum conditions, so that atmospheric pressure on the surface of the liquid to be pumped will promote flow into and priming of the pump.

The liquid ring air pump (Figure 5.9) consists of a bladed circular rotor, shrouded on the underside, which rotates in an oval casing. Sealing water is drawn into the oval casing through a make-up supply pipe (in older types is was added through a plug). The water, thrown out to the casing periphery by the turning rotor, whirls around, to form a moving layer against the oval casing. The water seals the rotor blades and also recedes from and re-approaches the rotor boss twice in each revolution. The effect is to produce a series of reciprocating water pistons between the blades.

As the water surface moves out from the rotor boss, it provides a suction stroke and, as it moves in, a discharge stroke. The shaped suction and discharge ports, provided above the elliptical core formed by the rotating water, permit air to be drawn in from the main pump suction pipe float chamber and expelled through the discharge ports, to atmosphere.

A continuous supply of sealing water is circulated from the primer reservoir to the whirlpool casing, and discharged with the air back to the reservoir. The air passes to atmosphere through the overflow pipe. This circulation ensures that a full water-ring is maintained and the cooling coil incorporated in the reservoir, limits the temperature rise of the sealing water during long periods of operation. The supply for the cooling coil can be taken from any convenient sea-water connection. About 0.152 litres/s is required at a pressure not exceeding 2 bar.

The air handling capacity of water ring primers is good, with air gulps being quickly cleared and small air leakages being handled without any fall in pump performance. This type of primer, replaced the now obsolete reciprocating pump primers. Eccentric vane primers have good air handling capability but vanes wear and they sometimes jam in the rotor slots. Ejectors are effective if sized correctly but their efficiency is low.

Float chamber

The water ring primer draws air from the pump suction pipe, through a float chamber (Figure 5.9). The float rises as liquid replaces air and as the level rises well above the pump the impeller and casing are flooded and the float spindle closes off the suction. This ensures that the primer itself is not flooded.

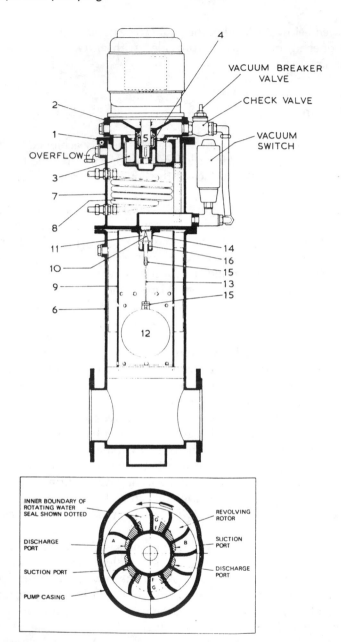

Figure 5.9 *Section arrangements of liquid ring primer (Weir Pumps Ltd.)*

1. Air pump casing
2. Air pump top cover
3. Rotor
4. Mechanical pump seal spring
   Stationary sealing ring
   Shroud 'O' ring
5. Pump shaft and coupling Flange (long coupled

pumps only) Motor half coupling (long coupled pumps only) Combined pump and motor shaft (close coupled pumps only)
6. Separating chamber
7. Sealing water tank
8. Cooling coil

9. Float gear cage
10. Needle valve
11. Needle valve seat
12. Ball float
13. Pendant
14. Bridge piece
15. Roll pins
16. Spring

Central priming system

This system may be used when more than four pumps require priming facilities. It gives a large air exhausting reservoir as well as capacity greater than individual pumps can carry. Pump casings can be filled with liquid before starting. The air exhausting units are usually of the liquid ring type. A typical schematic arrangement is shown in Figure 5.10.

A float chamber arrangement is also used with central primers to prevent flooding of the priming unit.

General purpose pumps

Single entry general purpose pumps (Figure 5.11) are used for salt and fresh water circulating and also for bilge and ballast duties. The impeller is suspended from the shaft with no bottom support. A neck bush provides lateral location. The eye of the impeller faces downwards, to the suction inlet below the impeller. There are renewable wear rings, usually of aluminium bronze, located at the top and bottom around the collars or boss of the impeller. The clearance between the wear rings and collars is minimal to restrict the flow of liquid from the discharge side. A short circuit flow would reduce efficiency and pressurize the shaft seal. In this design access for maintenance is via the top cover. A distance piece arranged in the shaft, is removed to permit the impeller, shaft and cover, to be lifted out without disturbing the motor or the pipework.

Figure 5.12 shows a different design of a pump intended for similar duties. The impeller is arranged with its eye uppermost, with the suction branch being elevated. This arrangement is claimed to give better venting to eliminate any possibility of vapour locking. Another significant design difference is that the casing is split vertically so that the impeller and shaft can be removed sideways. The wear rings and neck bush fitted in this design are stepped to abut with the removable part of the casing, to prevent them from turning.

Where a single entry pump is to be employed to supply a large pressure head, an impeller of a greater diameter (Figure 5.13) can be used. The model shown also has a vertically split casing and an impeller eye which faces upwards. The added lower guide bush is deemed necessary for the larger diameter impeller.

A novel design of single entry pump (Figure 5.14) was produced for ease of maintenance and adaptability. In this pump the impeller eye faces downwards but the impeller is open-sided, with the bottom of the pump casing effectively shrouding the vanes. This design allows the motor and cover of the pump to be tilted on a hinge so that the operation of a simple screw jack exposes the internal parts. A mechanical seal prevents water leakage or air ingress. This type of pump was designed for a wide range of capacities through simply fitting an impeller of suitable diameter and tip width. As previously stated, the performance of a centrifugal pump is dictated by speed of rotation, the impeller diameter and the area of the flow passage through the impeller (or width). The

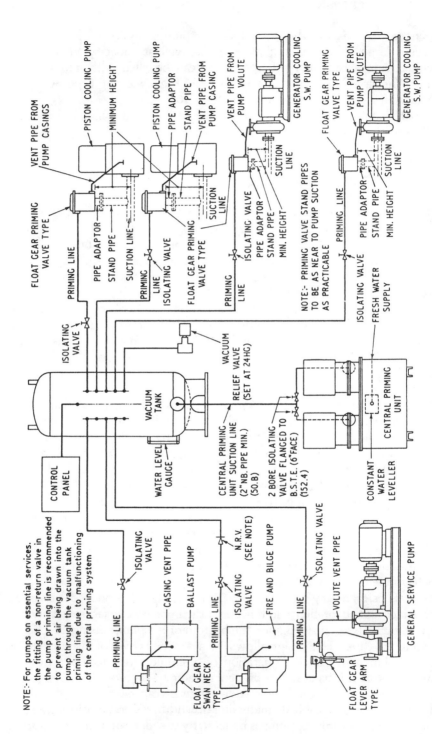

NOTE:- For pumps on essential services. the fitting of a non-return valve in the pump priming line is recommended to prevent air being drawn into the pump through the vacuum tank priming line due to malfunctioning of the central priming system

Figure 5.10 *Layout showing central priming system (Hamworthy Engineering Ltd.)*

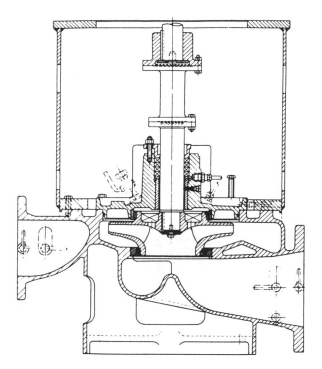

Figure 5.11   *Typical single stage centrifugal pump (Hamworthy Engineering Ltd.)*

first two variables basically control the pressure generated and the last the quantity of liquid delivered.

For a constant speed pump, a set of performance curves (Figure 5.15) can be obtained, which reflect operation with different impeller diameters. A similar effect would be produced by using different speeds.

Variation of capacity at constant speed and diameter, facilitated by fitting impellers of different widths (Figure 5.16) produces a different type of performance variation. These curves also have points where performance is at its highest. Obviously, by altering impeller diameters and widths, a pump can be tailored to requirements.

A two-stage pump (Figure 5.17) may be installed as a fire and general service pump. Because both low and high head are available from the one pump, it can be readily used for a double duty. Lower head is obtained by pumping through the first stage impeller only and a higher head when pumping through both impellers.

Erosion by abrasives

A pump handling liquids which contain abrasives, will suffer erosion on all internal surfaces, including bearings and shaft seals. The sea-water circulating pumps of ships operating in waters that contain large quantities of silt and sand

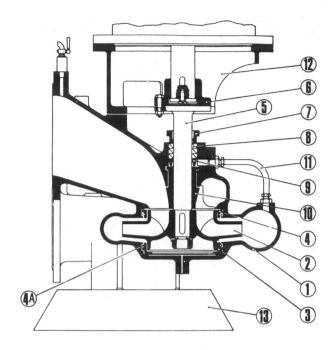

Figure 5.12 *Single-entry pump giving throughput of 425 m³/hr against heads of up to 54m (Weir Pumps Ltd.)*

1.	Pump casing and cover	7.	Gland
2.	Impeller	8.	Packing
3.	Casing ring (bottom)	9.	Lantern ring (split)
4.	Casing ring (top)	10.	Neck bush
4A.	Locking pins	11.	Water service pipe to stuffing box
5.	Pump spindle	12.	Motor stool
6.	Coupling (motor half)	13.	Pump foot

require frequent renewal of shaft seals or packing, also of shaft sleeves in way of the gland and bearings. Impellers are sometimes extensively damaged with resulting perforations and massive enlargement of wear ring clearance. Pump casings suffer erosion on all internal surfaces. Impellers and wear rings may have to be changed frequently and casings may need to be renewed at longer intervals. Special provision can be made in such pumps or those employed in suction dredgers (Figure 5.18) to safeguard bearings and shaft seals. The protection is provided by a water service to the shaft which washes solids away from the shaft seal area. Bearings can be protected by being mounted external to the casing. The pumps designed for use in suction dredgers require frequent casing repairs: a solution has been provided by one manufacturer with a casing which is built from renewable parts.

Erosion due to cavitation as opposed to the presence of abrasives is selective. The problem occurs, as with cavitation damage on propeller blades, in certain areas where cavitation pockets or bubbles collapse.

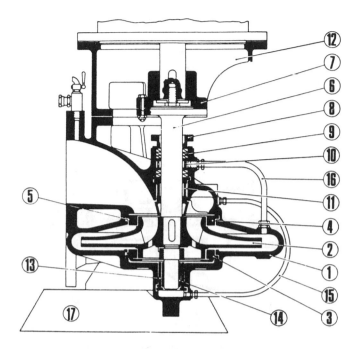

Figure 5.13 *Large diameter impeller for large pressure head (Weir Pumps Ltd.)*

1. Pump casing and cover
2. Impeller
3. Casing ring (bottom)
4. Casing ring (top)
5. Locking pins
6. Pump spindle
7. Coupling (motor half)
8. Gland
9. Packing
10. Lantern ring (split)
11. Neck bush
12. Motor stool
13. Bottom bush housing
14. Bottom bush liner
15. Water service pipe to bottom bush
16. Water service pipe to stuffing box
17. Pump foot

Centrifugal pump cavitation

During operation, if the drop in pressure created at the suction side of a centrifugal pump (by liquid moving radially outwards from the eye of the impeller) is greater than the vapour pressure for the temperature of the liquid being pumped, then vapour will be drawn from the liquid in this area. The phenomenon is likely to occur if there is a restriction in the suction pipe, if the liquid is volatile or has a higher temperature than anticipated, or if the impeller speed is excessive. A vapour cavity of this type is likely to cause loss of suction or erratic operation. A lesser cavitation problem occurs when NPSH required by the pump is only just matched by the NPSH available from the system, because centrifugal pumps have features which promote localized cavitation. The impeller entry on the side away from the shaft has a profile which resembles that of a hydrofoil, and local liquid flow creates a drop in pressure

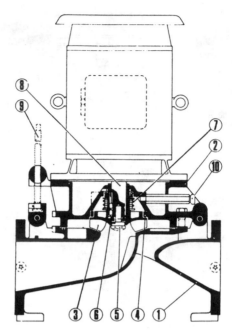

Figure 5.14 *Single-entry pump with open-sided impeller (Weir Pumps Ltd.)*

1. Pump casing
2. Casing cover
3. Impeller
4. Casing ring
5. Impeller locking screw

6. Shims
7. Mechanical seal
8. Combined pump and motor shaft
9. Screw jack
10. Air release plug

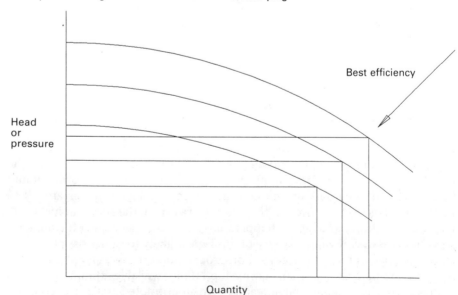

Figure 5.15 *Performance curves for centrifugal pump impellers of different diameters but running at the same speed*

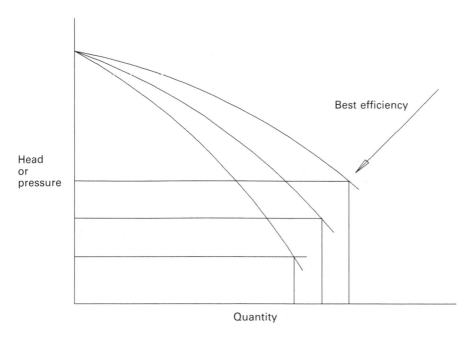

Figure 5.16 *Performance curves for different width impellers of the same diameter and constant speed*

which starts at the effective leading edge and extends along the surface. A vapour pocket created by a drop in pressure at this surface would collapse in an area of higher pressure and cause cavitation damage. A change of flow direction from axial to radial causes the fluid to experience different velocities and at the extremity, a drop in pressure could produce cavitation, again with subsequent bubble collapse and damage. The types of cavitation described could cause surface roughening and generate some tell-tale noise.

Corrosion wastage could cause impeller wastage but the cast iron casing of a sea-water pump is a prime target.

Centrifugal pumps for lubricating oil duties

Because of their self-priming ability, positive displacement pumps are widely used for lubricating oil duties. This practice is completely satisfactory in installations where the pump speed is variable but when the pump is driven by a constant speed a.c. motor it is necessary to arrange a bypass which can be closed in to boost flow. By using a centrifugal pump with an extended spindle, such that its impeller can be located at the bottom of the oil tank, the H/Q characteristics of the centrifugal pump can be utilized without the priming disadvantages. Known as the tank type pump this pump has a small open impeller. It can be driven directly by a high speed alternating current motor without capacity restrictions, whereas the permissible operating speed of the

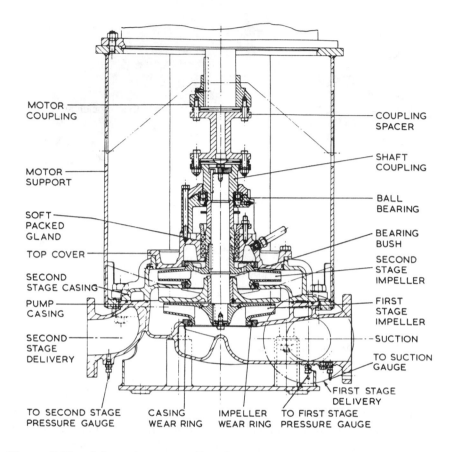

Figure 5.17   *A two-stage centrifugal pump, typical of a fire pump (Hamworthy Engineering Ltd.)*

positive displacement type, pumping an incompressible fluid, decreases as the capacity increases.

The pump will deliver an increased amount of oil as the system resistance is reduced, when the oil temperature rises and viscosity falls. In practice this is what the engine requires. A conventionally mounted self-priming pump is satisfactory for this duty if the suction pipe is short, direct and generously dimensioned. In a positive displacement pump the output varies little, if at all. The latter point is readily appreciated on reference to Figure 5.19 in which, for simplicity, the effect of a change in the viscosity of the oil on the pump characteristics has been omitted, although this would favour the centrifugal pump and enhance the comparison still further.

Centrifugal pumps – general

A centrifugal feed pump must not be operated unless it is fully primed. The pump casing should be filled before starting, the suction pipe and pipe branch

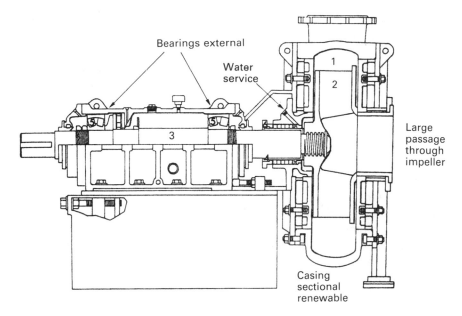

Figure 5.18 *Dredge pump (GEC-Elliot Mechanical Handling)*

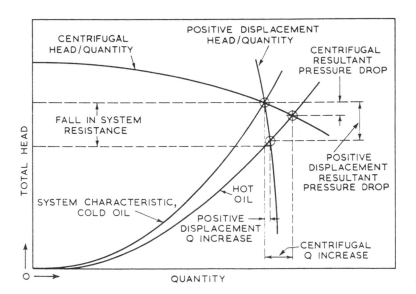

Figure 5.19 *Comparison of centrifugal and positive displacement pump characteristics with respect to lubricating oil duties*

to the discharge stop valve must also be full. If the liquid enters the pump suction by gravity, priming is usually unnecessary and the pump will remain full of liquid when shut down. To prime a pump initially, or one that has become filled with air, open the small air cock on the top of the pump casing to release the air and close it when liquid commences to flow.

If the pump is operating with a suction lift, it may be primed either from an independent supply, for example by opening the sea suction of a ballast pump, or by means of an exhauster or priming system, which will evacuate air from the pump and suction piping. The discharge valve is kept closed while priming the pump.

Axial flow pumps

An axial flow pump (Figure 1.12 Chapter 1) is one in which a screw propeller is used to create an increase in pressure by causing an axial acceleration of liquid within its blades. The incidental rotation imparted to the liquid is converted into straight axial movement by suitably shaped outlet guide vanes.

Axial flow pumps are sometimes classed with centrifugal pumps although centrifugal force plays no useful part in the pumping action. A comparison of discharge characteristics (Figures 5.20 and 5.21) shows that H/Q and working efficiency characteristics for the two pumps are quite different.

The discharge characteristics (Figure 5.20) drawn in each case for constant speed show those for the axial flow pump with a solid line and those for the centrifugal with a broken line. Starting from the point of normal duty, throttling of the discharge of an axial flow pump reduces the flow but also causes a rise in pressure and power. With the valve closed and zero discharge, the head can be about three times greater and absorbed power about doubled. The action of throttling to reduce throughput would overload the electric

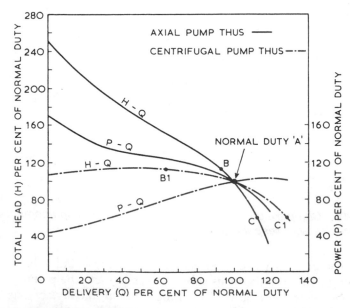

Figure 5.20   *Head/quantity curves, at constant speed, of axial-flow and centrifugal pumps*

motor and cause operation of protective devices. The throughput needs to be controlled in another way or the pump motor must be over-rated. In fact, the closing in of the discharge tends to cause water hammer. If the discharge of the centrifugal pump is closed in, as previously described, the pressure rises by a moderate amount and power demand actually drops.

Beyond the point of normal duty, the axial pump power requirement decreases with lower head and delivery increases by a modest amount. For the centrifugal pump operating at lower head, the greater throughput is matched by a rise in absorbed power (although there is a marginal fall off beyond normal duty). The centrifugal pump electric motor can be overloaded at low head and high throughput – the opposite condition for overload of an axial pump.

The axial pump (Figure 5.21) retains reasonable efficiency over a wider head range, than the centrifugal pump. There are three other features of the axial flow pump not indicated by the graph but of particular importance in their application. These are:

1   Under the low head (2.5 to 6.2 m) high throughput (2800–9500 m³/hr) conditions commonly required by main condensers, an axial flow pump with a higher speed than an equally matched centrifugal pump can be used. The electric motor for the pump, can be of smaller size.
2   The pump will idle and offer little resistance when a flow is induced through it by extremal means.
3   The pump is reversible.

This combination of characteristics makes the axial flow pump ideal for condenser circulating duties, especially in conjunction with a scoop injection (Figure 1.12) where the motion of the ship under normal steaming conditions is

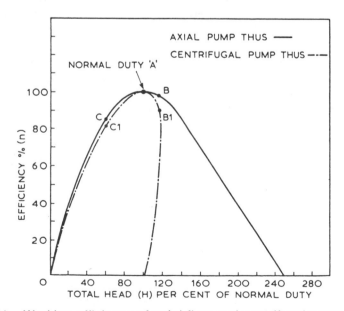

Figure 5.21 *Working efficiency of axial-flow and centrifugal pumps where the head varies within large limits*

sufficient to induce a flow through the idling pump and the condenser.

Reversibility and high throughput make it ideal for heeling and trimming duties. An axial flow pump may be fitted on the straight transfer pipe between tanks installed for this purpose.

When used for sea-water circulation the pump will normally have a gunmetal casing; for heeling and trimming applications it is more usual to find pumps with cast iron casings. Impellers are of aluminium-bronze, guide vanes of gunmetal and the shaft of stainless steel, with a renewable stainless steel sleeve in way of the bush.

Axial pump cavitation and supercavitating pumps

The hydrofoil section of conventional axial pump blades (upper right Figure 5.22), like those of propellers, promotes a severe drop in pressure at the leading edge and on the surface of the suction side, while producing positive pressure on the discharge surface. The blade shape enhances operation but cavitation can be a problem mainly on the suction surface of pumps rotating at very high speed. Cavitation occurs when a drop in pressure at the leading edge and across the blade surface is sufficient to produce vapour (and air if present) from the liquid. The pressure of the liquid increases as it passes through the pump and vapour pockets or bubbles, subjected to positive pressure, then collapse, as vapour is re-absorbed back into the liquid. Erosion is caused because vapour pocket collapse is very rapid and liquid rushes in with considerable force. The impact damages the metal surface.

An axial flow pump with hydrofoil shaped blades, which operates at excessive speed, also creates a demand for rapid inflow of the liquid being pumped. Flow may not match the demand with resulting loss of efficiency which may be compounded by cavitation and erosion on blade surfaces.

Supercavitating pumps have blades of a wedge cross section (lower right Figure 5.22) to produce a different cavitation pattern. A drop in pressure is not pronounced at the leading edge; it is minimal over the full area of the suction side of the blades and exaggerated only at the trailing edge. Vapour pockets persist until clear of the blade and collapse where they cause no damage. The supercavitating type of axial flow pump operates at high speed but generates low discharge head. As a small, light-weight, high-speed pump, it is used for fuel transfer in aircraft and rockets. It is also used as a suction booster or inducer for conventional centrifugal cargo pumps in liquefied gas carriers because it can handle liquids at, or near their vapour pressure, with higher speeds than can conventional pumps.

Rotary displacement pumps

Positive displacement rotary pumps rely on fine clearances between moving parts for their efficient operation. Excessive wear or erosion of parts, due to friction contact or the presence of abrasives, is avoided by employing this type

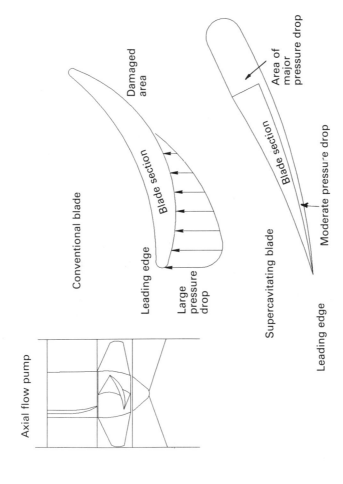

Figure 5.22 *Hydrofoil and supercavitating blades*

of pump for specialized rather than general duties. Contact between elements in some screw pumps (Figure 5.23) is made unnecessary by gear drives. When used for lubricating oil and hydraulic systems, rotary displacement pumps benefit from the sealing effect and provision of lubrication between parts.

Pumped volume

For a rotary positive displacement pump operating at zero pressure the volumetric efficiency should be 100% but as the differential pressure increases (Figure 5.24) the amount of leakage (slip) through clearances will increase. This

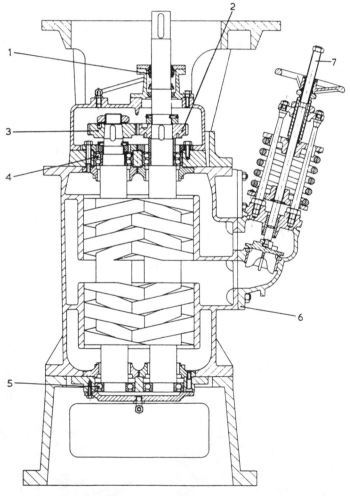

Figure 5.23 *A two-screw displacement pump (Weir Pumps Ltd.)*

1. Mechanical seal
2. Timing gear (driving)
3. Timing gear (driven)
4. Upper bearing
5. Lower bearing
6. Valve body
7. Relief valve spindle

slip (the terminology normally used) will naturally be less the more viscous the pumped liquid is, for any given pressure. It should also be noted that the slip, being a function of the clearance, viscosity and differential pressure, is constant irrespective of running speed. In practice, changes in flow conditions affect this slightly. Slip or leakage can cause erosion and increase of clearances, particularly when the liquid being pumped contains abrasives.

Pressure is limited by the torque available from the drive motor, the strength of parts, the amount of slip or leakage and considerations of overall efficiency.

Power

The power requirement of the pump may be split into two components, namely hydraulic power and frictional power. The hydraulic power is that

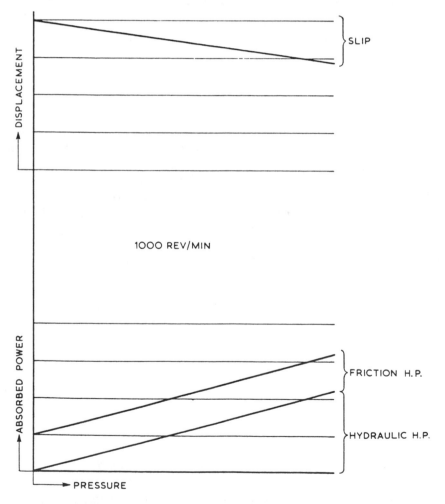

Figure 5.24 *Screw pump performance characteristics*

required for the pumped medium. Since slip is only 'slip' because it has been pumped and then leaked back via the clearances, it is only necessary to consider the pump displacement at its running speed, and the differential pressure through which the liquid is being raised. Frictional loss is minimal where rotors are not in contact.

Acceleration forces

These are by far the greatest losses in a rotary pump and the liquid properties are beyond the designers' control. This leaves only the distance of movement of the dynamic unit and speed within the designers' control (acceleration is a function of both).

Depth and form of tooth influence this in a gear pump. With the screw pump, the pitch of the screw is the major factor. Thus high helix angle screws can be used up to a relatively high speed on small pumps, and lower helix angles on larger pumps, to keep the pitch within limits set by the field requirements for suction performance.

Screw pumps

Both double-screw pumps, in which the screws are driven in phase by timing gears (Figure 5.23), and triple screw pumps (Figure 5.25), in which the centre screw is driven and the outer screws idle are used at sea especially for pumping high viscosity liquids such as oil and some liquid cargoes. Being self-priming and able to pump liquid and vapour without loss of suction they are particularly useful when draining tanks of high vapour pressure liquids. They are suitable for operation at high rotational speed (units are in operation with speeds of 3500 rev/min, delivering over 1000 litres/min) and can thus be easily matched with standard electric motors. Performance characteristics of screw pumps are illustrated in Figure 5.24.

In the IMO triple screw pump the centre screw is driven mechanically, through a flexible coupling. The two outer screws are driven by the fluid pressure and act purely as seals. The screws work in a renewable cast iron sleeve mounted in a cast iron pump casing. When the screws rotate, their close relation to each other creates pockets in the helices; these pockets move axially and have the same effect as a piston moving constantly in one direction. These pumps work well at high pressure and with high viscosity fluids (up to 4000 centistokes). The axial thrust on the power rotor is balanced hydraulically, that of the idlers by thrust washers.

Double-screw pump with timing gears

This type of pump can be mounted either horizontally or vertically. Pumping is effected by two intermeshing screws rotating within a pump casing. Each

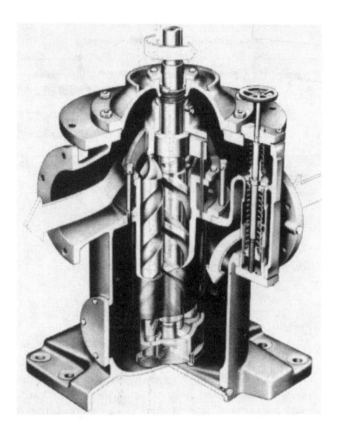

Figure 5.25 *A triple-screw displacement pump (IMO Industri)*

screwshaft has a right and a left hand screw which ensures axial hydraulic balance, there being no load imposed on the location bearing.

Metal contact is avoided by driving the screwshaft through hardened and ground timing gears. Once the casing has been filled with liquid, the pump is self-priming and ready for operation. Displacement on pumping occurs when the screws are rotated and liquid is drawn into the screws at the outer ends and pumped inwards to the discharge located at rotor mid-length. The discharge is without pulsation.

When pumps are installed for handling non-corrosive liquids of reasonable lubricity, it is normal to find units with internal bearings. For handling corrosive chemicals, liquids with a lack of lubricity and/or high very high viscosity, designs incorporating outside bearings (Figure 5.26) which can be independently lubricated would be used.

Pumps with inside bearings are shorter and lighter than their outside bearing counterparts and have only one shaft seal as against four.

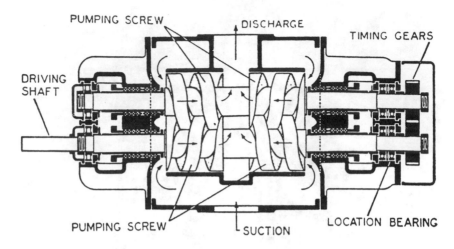

Figure 5.26 *Counterscrew pump for oil or water service*

Shaft sealing

The double-screw pump (Figure 5.23) shaft sealing is effected by either mechanical seals or packed glands. The shaft seal is at the suction end of the pump, so that seals are subjected only to low pressure or vacuum.

Shaft seals must ensure leak-free operation of cargo pumps for toxic and corrosive liquids. Provision is usually made for mechanical seals to be cooled and lubricated by the pumped liquid. For installations in pumprooms an external header may be fitted to service the seal.

Materials

For the general range of liquids handled, cast iron is suitable for the casing and bearing housings. Screwshafts are normally of high grade carbon steel or stainless steel. Materials having high corrosion resistant qualities such as stainless steel EN 58 J and Hastelloy are used. If screws are not in direct contact they will not be scuffed.

Protection

Since screw pumps are essentially displacement pumps and will produce increasing pressure until rupture or drive failure occurs it is necessary to safeguard the pump, prime mover and its associated pipework against the event of a discharge line valve closure. The pumps are equipped with full flow relief valves capable of bypassing the entire throughput of the pump. This is for safety purposes only however, and should only operate for a short time otherwise excessive liquid/pump temperatures will result. The valve may be fitted with either manual or automatic control to facilitate starting up under a no load condition. This is necessary where the discharge system is under pressure, to avoid excessive starting torque (electrical load) and long run-up times.

Relief valves are also often fitted with automatic volume control valves, which control the output of the pump in order to maintain either a constant pressure or a vacuum at a specific point in the system − as in diesel engine lubricating oil supply − to ensure constant pressure at engine inlet irrespective of oil viscosity.

When hot or viscous liquids are handled it may be necessary to preheat the casing of a pump already filled with liquid. Means for heating are available, for example electric immersion heaters or a coil through which low pressure steam or hot oil is circulated.

Single screw pumps

Oil and water separators require delivery from low capacity, smooth flow bilge pumps. The mono pump, with its corrosion resisting rotor and a stator of natural or synthetic rubber, is frequently used.

Gear pumps

Diesel engine and gearbox lubrication systems are normally supplied by gear pumps which are independently driven for large slow speed engines and stand-by duties but usually shaft driven for medium and high speed engines. Gear pumps are also used for fuel and oil transfer, boiler combustion systems and other duties.

The simple gear pump (Figure 5.27) has rotors and shafts which are integrally forged of nitralloy steel. They are hardened all over and ground finished after hardening. Close grained grey cast iron is used for the casing and bearing housings. The shafts of the pump shown are supported in white metal bearings but needle roller bearings are an alternative. Lubrication is provided by the oil passing through. The liquid being pumped, is forced out after being carried around between gear teeth and housing, as the teeth mesh together. It is certain that the centrifugal effect contributes to the pumping action. There is no side thrust with straight gear teeth.

Side thrust produced by single helical gears, causes severe wear and in one pump opened for examination, bronze bearing bushes exhibited wear to a depth of 3 mm so that gears had become displaced axially. Despite the excessive clearance due to wear, it was noted that the pump continued to be effective when repaired as far as possible (no spares being available).

To gain the benefit of using helical gears, a double helical is necessary. This larger pump is of the type used for large, slow speed engine lubrication

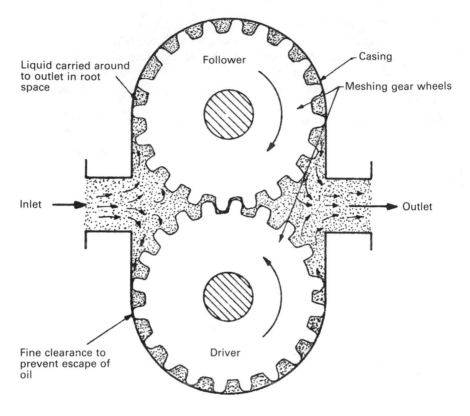

Figure 5.27 *Simple gear pump*

systems. Spring loaded relief valves are usually fitted and may be of the adjustable calibrated type to facilitate the setting of delivery pressure.

Lobe pumps

Lobe pumps as manufactured by Stothert and Pitt have inner and outer elements which rotate in a renewable liner fitted in the pump body. The inner rotor is eccentric to the outer and is fitted to a shaft located by bearings in the pump covers. The pump types are defined by the number of lobes and recesses, for example Three–Four and Seven–Eight (Figure 5.28). As the rotors turn, pockets of increasing and then decreasing capacity are created to give a pumping action.

The normal maximum pressure is 21 bar with capacities of up to 400 tonne/hour. Special designs for up to 83 bar have been made for hydraulic control applications.

The Three–Four types are particularly suited for handling high viscosity fluids and are set to run at comparatively slow speeds which may require a gear drive from the motor. The Seven–Eight types operate higher speeds, being

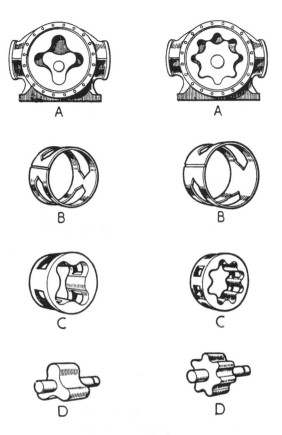

Figure 5.28 *Lobe displacement pumps. Pumping elements of (left)
Three-Four type for high-viscosity fluids and (right) Seven-Eight type for
low viscosity fluids (Stothert & Pitt Ltd.)*

most efficient at around 720 rev/min when handling fluids in the lower
viscosity range.

Metering pumps

A metering pump is designed primarily for measuring and dispensing a liquid,
rather than merely transferring it from one place to another. It has to be a
precisely designed piece of equipment, manufactured to close engineering
limits. It is frequently used to dispense concentrated and highly corrosive
liquids and for this reason considerable care has to be taken to ensure that the
materials used in construction, particularly those of parts in direct contact with
the liquid, will resist attack.

Most metering pumps are of the positive displacement type. They consist of
a prime mover, a drive mechanism and a pumphead. Pumpheads are usually of
the piston or plunger type (Figure 5.29a) where the pump is to be used against
high pressures. For lower pressure duties the diaphragm version (Figure 5.29b)

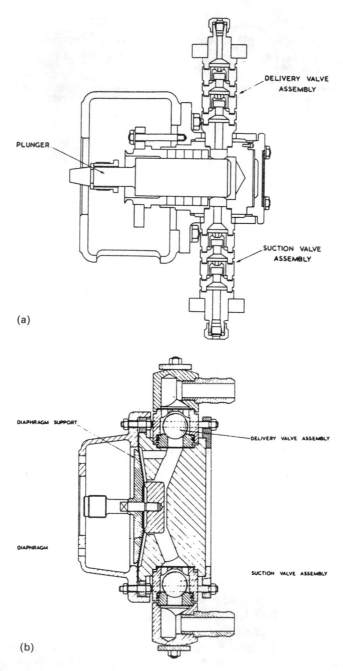

Figure 5.29   *Metering pump (a) Typical plunger head for MPL Type Q pump; (b) Typical diaphragm head for MPL Type Q pump*

is generally used. The plunger model is more exact in its performance than the diaphragm type, but the diaphragm type which requires no glands, is completely leak proof.

Pump output is controlled by varying the length of pump stroke. It is possible to adjust output progressively from zero to maximum whilst the pump is in operation as well as when it is at rest.

A further feature of many metering pumps is that arrangements are made in design so that more than one pumphead and mechanism can be accommodated, each with its own stroke control. Similarly each pumphead mechanism can be equipped with a different gear ratio, thus enabling one pump to meter various liquids in different ranges of flow rates. The metering pump is thus highly versatile and capable of adaptation to a wide range of duties.

Further reading

Pearsall, I. S. (1973) The supercavitating pump, *Proc I Mech E*, **187**, 54, 649–65.

6

Tanker and gas carrier cargo pumps and systems

The nominal time for complete discharge of cargo from a tanker is 24 hours and this is the figure that is normally written into the cargo pumping warranty. The form of the warranty in the tanker charter may be 'owners warrant the vessel is capable of discharging the entire cargo within twenty four hours or pumps are capable of maintaining 7 bar (sometimes $100\,lb/in^2$) at the ships rail'. The last part is important if cargo cannot be accepted ashore at the required rate for a 24 hour discharge or if there is resistance to flow because of the size or length of the shore discharge pipe, height of the storage facility or, possibly, excessively low ambient temperature. The time lost if a vessel is at fault, cannot be counted as laytime or as time on demurrage. Time lost due to restrictions imposed by the terminal can count as laytime or as time on demurrage but the ship's master must have immediately issued a letter of protest (ideally acknowledged) and informed the charterers.

It is important to realize that there is usually no allowance for fall off in performance of cargo pumps in a time charter. The 24 hour discharge or 7 bar ($100\,lb/in^2$) capability must be maintained. (There may be an allowance for fall in ship's speed.) Cargo pumps obviously have to be designed for a better performance than required by the warranty, to allow for deterioration. Deterioration and maintenance are the responsibility of the engineering staff on the ship.

Oil tankers (crude oil)

Crude oil is delivered to refineries in large crude oil carriers which may carry more than one grade. Refinery products are transported in smaller tankers which may be used for a variety of cargoes.

Cargo pumping

Before the advent of very large tankers, reciprocating pumps were commonly installed for cargo discharge and tank draining. These pumps require routine maintenance of suction and discharge valves, pistons and buckets, pump rod glands, slide valve chests and steam valves. Being positive displacement pumps, they have good tank draining capability.

Centrifugal cargo pumps with a double entry impeller (Figure 6.1) have replaced reciprocating pumps in modern oil tankers. These pumps are simple, have no suction and delivery valves, pistons, piston rings, steam control valve or linkages and therefore require much less maintenance. The compact centrifugal pump can be mounted horizontally or vertically in the pumproom with a turbine, or in some ships, electric motor drive from the machinery space. The drive shaft passes through the engine room bulkhead via a gas seal.

Rate of pumping is high (typically 2600 m³/hr) until a low level is reached,

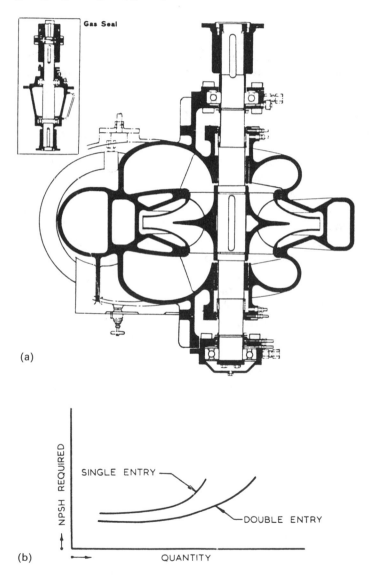

Figure 6.1 *(a) Typical double entry pump arranged for cargo pumping (Weir Pumps Ltd.) (b) Showing the improved NPSH characteristic available from a double entry pump (Hamworthy Engineering Ltd.)*

when a loss of suction head and impeded flow through frames and limber holes makes a slowdown in the rate of pumping necessary, if the use of a small stripping pump is to be avoided. Systems such as the Worthington-Simpson 'Vac-Strip' enable a faster general rate of discharge to be maintained, while reducing temporarily the rate of discharge when a low level is reached in the cargo tank to allow for draining.

Oil tanker discharge arrangements

Control equipment is available for managing the loading and discharge of oil tankers and instrumentation is used to monitor the state of the cargo.

There are three main systems of pipelines in use each requiring a different method of handling:

(a)   The ring system, in which a ring pipeline has separate suction tail pipes tapped off to each tank and controlled by a gate valve. By opening the appropriate gate valve, each tank in turn can be emptied by the main pumps. When loading, the main pumps are bypassed and the ring pipeline becomes a gravity flow line so that tanks can be loaded by opening the appropriate gate valve.

(b)   The direct system, in which the tanker is divided into three or more sections, each of which in turn is subdivided into individual tanks. One pump is assigned to each section by a main pipeline, which in turn is connected to tanks by tailpipes and gate valves. The main pipelines are interconnected through cross-over sluice valves.

(c)   The free-flow system avoids the use of extensive pipelines, but has the disadvantage that the tanks can only be emptied simultaneously or in a fixed sequence. Oil reaches the pump through a suction tail pipe in the after tank and from the remaining tanks through valves in the transverse bulkheads between the tanks. Sometimes the free flow and the direct system are combined so that the direct system is used for port and starboard wing tanks and free-flow for the centre tanks.

It will be seen that a large number of valves must be controlled in a logical sequence, and without centralized remote control this requires a great deal of effort. In many cases a mixed cargo of different grades of oil is involved and correct handling is important. To avoid dangerous stressing of the hull structure due to unfavourable bending moments as the buoyancy of individual tanks changes it is necessary to operate to a known safe sequence. It is also necessary to take trim and list into account.

Centralized cargo control and the employment of computers for cargo operations is common. Electrical operation of valves would involve danger of explosion; hydraulic systems are therefore favoured for this purpose. To give protection against incorrect operation some of the valves can be arranged with sequence interlocking. For control and instrumentation purposes, low voltage/low current intrinsically safe electrical and electronic circuits are permissible subject to classification society approval. Intrinsically safe (Ex i) circuits, are so designed that in no circumstances, including conditions involving one (Ex i(b)) or two (Ex i(a)) faults, can a spark be created which could cause explosion.

When the main pumps have emptied a tank, the sediment and sludge which settle on the bottom of the tank are removed by crude oil washing and the stripping action of the cargo pumps described below. Separate stripping pumps delivering through an alternative to the main pipeline may be used. Any separate stripping system is brought into use as soon as the cargo pump has finished emptying the tank as far as possible. When the level of oil is below that of the steam heating coils the remains in a tank can cool quickly and may become unpumpable.

Vac-strip system

Within its capacity, a pump will discharge at the rate at which fluid will flow to its suction orifice; this will be determined by viscosity, specific gravity, line friction, entrained air (as the tank level falls) and, in the case of gassed oils, the release of volatile light fractions if the pressure in the suction pipe falls below the vapour pressure of the oil. This is the highest practicable rate. It will fall under adverse pumping conditions but a useful rate will be obtained from a displacement pump while enough oil passes to keep it sealed. The centrifugal pump however, will not regain a lost suction until the pump fills with liquid after the vapour is released. This problem disappears if the non-liquid elements are separated from the fluid before it reaches the suction orifice and if the discharge rate is limited to that of flow to the suction. Figure 6.2 shows a typical system of this kind, for one pump.

A separator 10 fitted with a vapour outlet pipe, a sight glass and a level controller 1, is placed near the pump in the suction line. The vapour outlet is led to an interceptor tank 11 having a moisture eliminator, a drain to 10 and an outlet to a vapour extraction pump 7 through an exhaust control valve 2. This pump (of the water ring type) controlled by the pressure switches A and B, draws from the interceptor tank and discharges to atmosphere at a suitable height through tank 8, which has an internal cooling coil. The small quantities of liquid carried with the vapour, drop out in the tank and return to the pump by the pipe shown. This arrangement, by means described later, ensures that the pump is always fully primed. The discharge rate is regulated by a butterfly valve 6 in the cargo discharge line, controlled by a pneumatic positioner. A thermostat guards against overheating when the pump is working against a restricted or closed discharge.

As non-liquid elements separate out, the liquid level in the separator tank will fall, so long as the exhaust control valve is closed. The level controller is supplied with clean, dry, air (this is a prerequisite of satisfactory operation) at a suitable pressure and is so arranged that its outlet pressure increases progressively from say, 200 mbar to 1000 mbar, as the level falls. At 240 mbar the pressure switch A will close but the extraction pump 7 will not start. At 270 mbar the exhaust control valve 2 will open and if there is a head on the

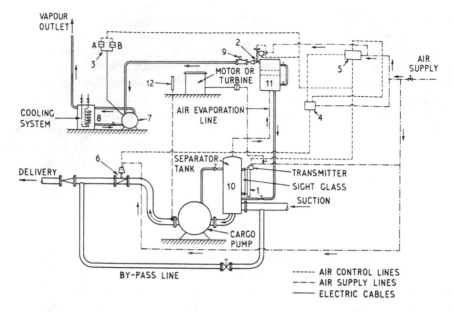

Figure 6.2   *Diagram of 'Vac-Strip' cargo system (Worthington-Simpson Ltd.)*

1. Leveltrol (level controller)
2. Exhaust control valve
3. Pressure switches
4. Auto/manual selector
   (panel loader)
5. High selector relays
6. Butterfly valve
7. Vacuum pump
8. Reservoir tank
9. Vacuum breaker
10. Separator
11. Interceptor tank
12. Thermostat

suction line, 10 will vent to atmosphere through 1, 2, 7 and 8. Following this, the level in 10 will rise, the level controller outlet pressure will fall, 2 will close and A will break.

As pumping continues and the entrained vapour and air increases, the level in 10 will fall again, the outlet pressure of 1 will rise, 2 will open, switches A and B will both close, 7 will start and 10 will be positively vented. The level in 10 will rise, 7 will stop, 2 will close, A and B will break. This sequence will be repeated until frequent snoring at the suction brings in so much air that 7 runs continuously. As pumping goes on, the situation becomes one in which the cargo pump is discharging at a faster rate than oil can flow to the suction and the level in the separator tank will fall continuously. As a result, the level controller outlet pressure will be high and will act to increase the discharge head, so reducing the pump capacity by closing the butterfly valve 6. The level controller outlet is led to the high selector relay 5 and thence to the butterfly valve positioner, through the auto-manual selector 4. This may be set by hand to limit the pump discharge pressure to that desired in the receiving system, or to operate in automatic response to the level controller outlet pressure fed to it from 5. The manual setting will override the automatic if the discharge pressure desired is below that consequent upon restricting the pump discharge rate to a figure not above the rate of flow to the suction. The butterfly valve will therefore be either fully open or partly closed, either to limit discharge pressure or to limit pump capacity. The pump speed may be reduced in a similar manner before the butterfly valve begins to close. It is important to bear in mind the difficulties of working centrifugal pumps in parallel at different speeds – the slower running pump may cease to pump at all.

The above description refers to a basic automatic system for one turbine-driven pump. Separators, interceptors and other fittings will increase in number with the pumps but one or more vapour extraction pumps may serve all the interceptors. It is possible to overload the motor of an electrically driven cargo pump and this may be prevented by the means shown in Figure 6.3). The motor armature current taken through transformer, rectifier and electro-pneumatic converter, produces pressure variations in the feed to a controller between no load and overload, of the same order as those arising from variations in separator tank level. The controller has an adjustable set point. These pressure variations are fed to the auto-manual selector through two high-selector relays, to vary appropriately the opening of the butterfly valve. Two high-selector relays are necessary because these instruments can only select the higher of two impulses transmitted to them. Here the highest of three pressures has to be selected from the level controller, the auto-manual selector manual setting and the electro-pneumatic converter.

The system described pre-supposes the absence of stripping pumps. One or more are sometimes fitted, either as a safeguard against unforeseen difficulty with the main pumps or to deal with tank washings.

The level controller comprises a vessel, common to the separator tank, in which is suspended a body connected by an arm to a torque tube in such a way that variations in liquid level (i.e. variations in the immersion of the suspended body) give rise to proportionate variations in the torque applied to the rotary shaft (Figure 6.4).

Figure 6.5 shows the arrangement which translates the degree of rotation of the rotary shaft into the controller outlet pressure operating the exhaust control valve, the pressure switches and the pump discharge butterfly valve.

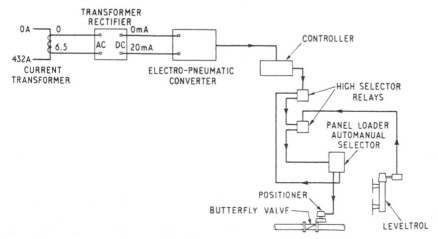

Figure 6.3 *'Vac-Strip' motor overload system (Worthington-Simpson Ltd.)*

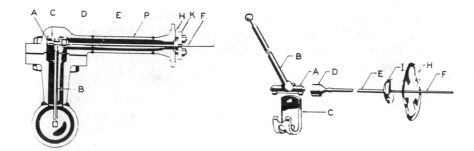

Figure 6.4 *Level controller (left) exploded view (right) plan view (G.E.C.-Elliott Control Valves Ltd.)*

A. Torque tube driver
B. Arm
C. Torque tube driver bearing
D,E. Torque tube assembly

F. Rotary shaft
H. Positioning plate
I. Torque tube flange
K. Retaining flange

Air at 1.5 bar is fed to the inlet valve O, to the upper diaphragm chamber L and to nozzle D through orifice J. D is large enough to bleed off all the air passing J if it is not restricted by the beam G (i.e. L is free of pressure and O is closed unless D is wholly or partly closed by G as a result of a fall in the liquid level). The double diaphragm assembly M and P is fully floating and pressure balanced. When the pressure in L rises, the assembly moves downwards, opens the inlet valve O and so admits pressure to the lower diaphragm chamber N; the assembly then rising, returns to its original position and O closes. When the pressure in L falls because of the rise in liquid level, the assembly moves upwards, opens exhaust valve K and so reduces the pressure in N until the assembly returns to its original position again and K closes. It will be seen that the pressure at the controller outlet is the same as that in N, varying with the degree of restriction of D by G and that there is a lead to the bellows A through the proportional valve T so that, as the outlet pressure rises, A extending, moves G away from D and limits the pressure rise. M and P are so proportioned that the pressure in N and therefore at the controller is always three times that in L.

High velocity vents

Tank vapours can be released and sent clear of the decks during loading through large, high velocity vents. The type shown (Figure 6.6) has a moving orifice, held down by a counterweight to seal around the bottom of a fixed cone. Pressure build up in the tank as filling proceeds causes the moving orifice to lift. The small gap between orifice lip and the fixed cone gives high velocity to the emitted vapour. It is directed upwards with an estimated velocity of 30 m/s. Air drawn in by the ejector effect dilutes the plume.

The conical flame screen fixed to the moving orifice to give protection against flame travel will, like the moving parts, require periodic cleaning to

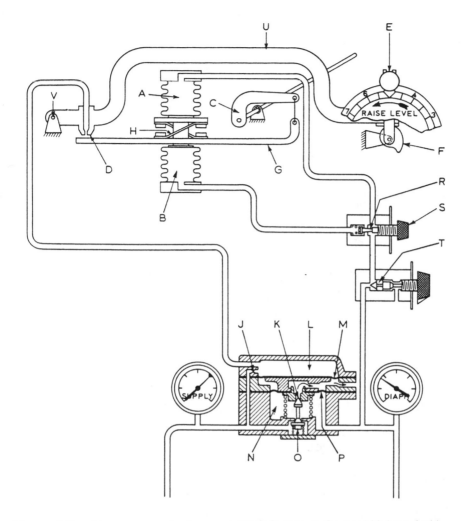

Figure 6.5   *Air pressure instrument (G.E.C.-Elliott Control Valves Ltd.)*

A, B. Bellows	G. Beam	M,P. Diaphragm assembly
C. Torque arm	J. Air supply orifice	L,N. Upper and lower
D. Nozzle	K. Exhaust valve	diaphragm chambers
E. Level adjustment	R. Reset valve	O. Inlet valve
F. Level adjustment arm	T. Proportional valve	

remove gummy deposit. The cover is closed (as shown) when the vessel is on passage.

A simpler design of vent (Figure 6.7), has two weighted flaps which are pushed open by pressure build up to achieve a similar nozzle effect. The gauze flame traps and vents tend to collect a sticky residue which should be cleaned off regularly to ensure unimpeded venting.

Tanks should be vented during loading through high velocity or masthead vents. The practice of venting through open tank hatches is dangerous particularly during thundery conditions.

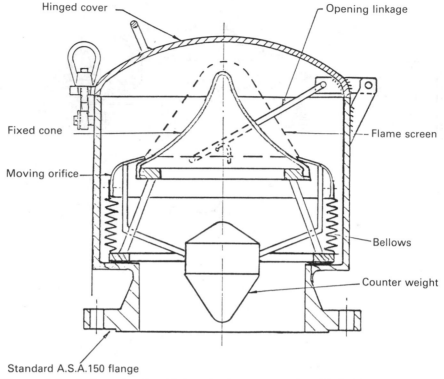

Figure 6.6 *Iotta valve for high velocity gas venting (F. R. Hughes)*

Pressure/vacuum valves

Moderate pressures of 0.24 bar (3.5 lb/in²) acting on the large surfaces in liquid cargo tanks, are sufficient to cause damage and rupture. The pressure on each unit of area multiplied by the total area gives a very large loading on the underside of the top of a tank or other surface. Distortion can result or the metal plate may be ruptured. Similarly, a vacuum within a tank, can result in damage in the form of inward collapse, due to greater atmospheric pressure on the outside. The hatches of at least one OBO (Ore-Bulk-Oil carrier) were severely damaged when condensing steam produced a vacuum. Pressure/vacuum valves (Figure 6.8) attached to tank vents or in the ventilation system, will prevent either over or under pressure. They are set usually so that tank pressure of about 0.14 bar (2 lb/in²) will lift the larger main valve (the smaller valve will lift with it) and release excess pressure. The vapour passes to atmosphere through a gauze flame trap. A drop in tank pressure compared to that of the outside atmosphere will make the small valve open downwards to admit air and equalize internal pressure with that outside.

Pressure vacuum valves can relieve moderate changes in tank pressure due to variations of temperature and vapour quantity. A drop towards vacuum conditions as the result of the condensation of steam will also be handled by the valve. A rapid pressure rise due to an explosion would not be relieved nor is the pressure/vacuum valve suitable as a vent when loading.

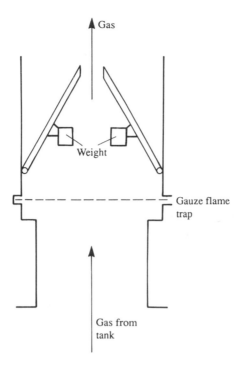

Figure 6.7 *Simple high velocity vent*

Inert gas systems

Hydrocarbon gases or vapours from crude oil form very flammable mixtures (Figure 6.9) with air when present in the proportion of between 1% and 10% hydrocarbon with 99% down to 90% normal air. Below the lower explosive limit (LEL) the mixture is too lean to burn rapidly, although a lean mixture will burn slowly in the presence of a naked flame or a spark, as is proved by the operation of explosimeters in this range. Over rich mixtures exist when the level of the hydrocarbon exceeds 10%. The risk of explosion or fire within the cargo tanks, pump rooms and other enclosed areas of an oil tanker is high as a consequence of the very small quantity of hydrocarbon gas or vapour which makes up an explosive mixture. Within tanks, vapour is readily evolved from cargo and immediately above the surface of the liquid there tends to be a layer of undiluted hydrocarbon vapour. Above this dilution by air, if present, increases towards the top of the tank and gives a mixture with a decreasing amount of hydrocarbon vapour. At the top of the tank, the larger quantity of air would be likely to produce a lean mixture. The free space within the cargo tank above any liquid petroleum cargo, or in an empty tank from which the cargo has been discharged, is likely to contain a middle flammable layer, sandwiched between lower over-rich and upper lean layers. Leakage of crude oil or products from pump glands, would produce the same effect in pump rooms and in any space where oil vapour could collect.

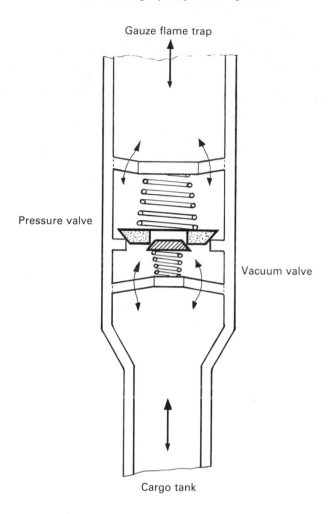

Figure 6.8   *Pressure/vacuum valve*

The effect of inert gas

Inert gas is delivered to the cargo tanks of crude oil tankers before loading commences, to purge the air content and to maintain a slightly higher than atmospheric pressure and so exclude air during operations. With no air present, an explosive mixture cannot form and tanker safety is enhanced. The inert gas can also be used to displace any air from a tank after an inspection or other period when the inert gas system has not been in use.

Inert gas cannot be used for pumprooms or other spaces where there is normal access for personnel. Crude oil vapours can collect in such areas and form a flammable mixture with air. In these places and on deck precautions to avoid the possibility of ignition are necessary.

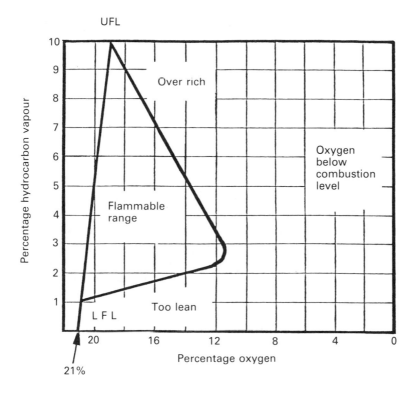

Figure 6.9 *Flammable air and hydrocarbon mixtures*

Inert gas system

Mention has already been made of the necessity for oxygen to be present for combustion to take place and of how carbon dioxide can be used to displace air from a compartment until combustion can no longer be supported. A gas such as CO_2, which is incapable of supporting combustion, is said to be inert. Another example of an inert gas is nitrogen and, under certain conditions, boiler flue gas with a low oxygen content, can also be considered inert. The latter is used aboard tankers to prevent explosions occurring in cargo tanks.

Safety

Combustible gases and vapours, such as petroleum vapour, when mixed in the correct proportion with air in an enclosed vessel will burn so rapidly that an explosion occurs. The burning can be initiated quite easily, ignition often being caused by a relatively small spark. For each gas or vapour there is, however, an upper and lower concentration of vapour in air between which an explosion can occur. These limits are referred to as the Lower Flammable and Upper Flammable limits (LFL and UFL) or, alternatively, the lower and upper explosive limits (LEL and UEL). Petroleum vapour, for instance, has an LEL of 1.4% and a UEL of 6.4% while hydrogen has an LEL of 4% and a UEL of 74%. Crude oil contains a variety of hydrocarbon mixtures and the vapours contained in crude oil tanks will exhibit a wide variety of upper and lower explosive limits.

By ensuring that the vapour/air mixture in the tank is maintained outside the explosive limits, an increased measure of safety can be achieved. Boiler flue gas can be used to maintain the environment below the LEL. In designing the inerting system the following must be borne in mind:

1   Boiler flue gas contains highly corrosive elements which must be removed.
2   The gas must be cooled and any sparks extinguished.
3   Measures must be taken to prevent the back flow of hydrocarbon gases to unsafe areas.
4   Interlocks must be provided to ensure that the inert gas system is not fed from a boiler undergoing soot-blowing.
5   On products tankers, which carry a variety of cargoes, it is necessary to ensure that contamination of one cargo by another cannot occur.

Operation of the inert gas system

The boiler flue gas is delivered to the scrubbing tower (Figure 6.10) via butterfly valves designed for remote operation with pneumatic actuators. Valve bodies are of cast iron and suitable for temperatures up to 450°C. To enter the scrubbing tower, the gas has to pass through a water seal which provides initial cooling and extinguishes any sparks. Sea water continuously pumped through the seal and scrubber sprays, cools the gas from the boiler uptake temperature (upwards of 135°C) to only a few degrees above the sea-water temperature. The sea water also removes most of the (corrosive) impurities in the gas.

In the F. A. Hughes system, the scrubbing tower (Figure 6.11) is a rectangular mild steel structure containing a number of polypropylene trays. Each tray is pierced by a number of shrouded slots. The shrouds are called tunnel caps and this type of scrubber is referred to as a tunnel cap scrubber.

The sea water enters the scrubber at the top and flows across each tray, a series of weirs being arranged to ensure that each tray is flooded to a depth of about 20 mm. Downcomers are arranged to conduct the water from one tray layer to the next.

The gas enters the scrubber at the bottom through a water seal and rises through the pack, bubbling up onto the trays through the tunnel caps which distribute the gas across the free water surface on the trays. At the top of the tray-pack a polypropylene mattress demister is arranged which removes the bulk of any entrained water. Above the demister a number of water sprays are located. These are used to wash down the scrubber after use.

The amount of sea water used varies to some extent with the design of the scrubber but units with sea-water flows of from 100 tonnes/hr to 350 tonnes/hr are in common use.

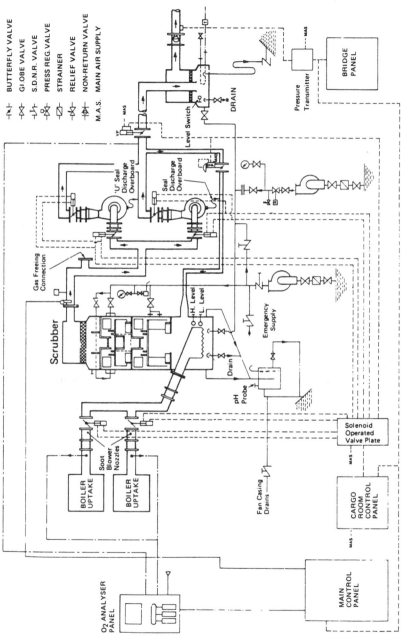

BUTTERFLY VALVE
GLOBE VALVE
S.D.N.R. VALVE
PRESS REG. VALVE
STRAINER
RELIEF VALVE
NON-RETURN VALVE
M.A.S. MAIN AIR SUPPLY

BRIDGE PANEL

Pressure Transmitter

DRAIN

Level Switch

'U' Seal Discharge Overboard

Seal Discharge Overboard

Gas Freeing Connection

Scrubber

H. Level
L. Level

Emergency Supply

Drain

pH Probe

Solenoid Operated Valve Plate

Soot Blower Nozzles

BOILER UPTAKE

BOILER UPTAKE

Fan Casing Drains

CARGO ROOM CONTROL PANEL

O₂ ANALYSER PANEL

MAIN CONTROL PANEL

Figure 6.10 *Schematic arrangement of a Peabody inert gas plant*

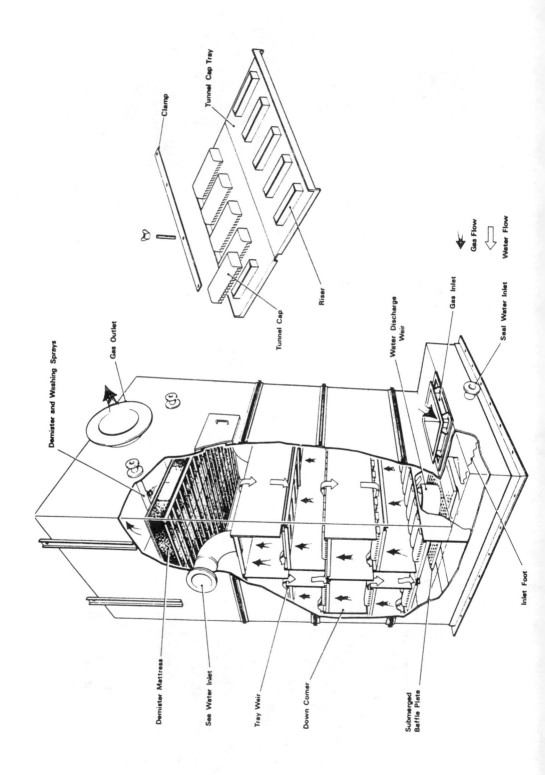

Clamp

Tunnel Cap Tray

Riser

Tunnel Cap

Gas Outlet

Demister and Washing Sprays

Water Discharge Weir

Gas Inlet

Seal Water Inlet

Gas Flow

Water Flow

Inlet Foot

Demister Mattress

Sea Water Inlet

Tray Weir

Down Corner

Submerged Baffle Plate

Because of the sulphur dioxide absorbed by the sea water its pH value is changed from around about 7 to about 2.5 as it passes through the scrubber – hence the selection of polypropylene for the tray tunnel caps and demister mattress. The tower itself is lined with ebonite rubber.

Other designs of scrubber used at sea include the impingement and agglomerating type such as the Peabody circular tower (Figure 6.12). In this type the incoming flue gas is first wetted by sea-water sprayers and then passes upwards through a venturi slot stage which agglomerates the solid particles. The gas then rises through slots in a series of trays. Above the slots are a number of baffle plates. The trays are covered in water introduced at the top of the tower in much the same way as in the tunnel cap tower previously described. A mesh type demister is arranged at the top of the tower.

Fans

The fans used in the inert gas system must be capable of providing a throughput equivalent to about 1.33 times the maximum cargo pumping rate since the tanks must be kept supplied with inert gas during cargo discharge. At full output, the fans must be capable of delivering at an over-pressure of

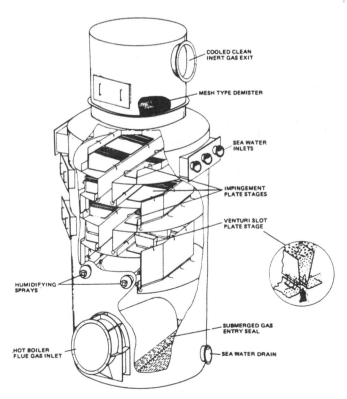

Figure 6.12   *Impingement and agglomerating type tower (Peabody)*

670–1000 mm w.g. which with pipeline losses equates to a static pressure at the fan of up to 1600 mm w.g.

Both electric motor-driven and steam turbine driven fans are used and it is usual to provide one running and one stand-by unit. These are normally both 100% duty units although some installations with a 100% duty and a 50% duty fan have been used. Because of the corrosive nature of the gas the fan materials must be carefully selected. Fan impellers of stainless steel or nickel-aluminium bronze are frequently used and the mild steel casings are internally coated with, for example, coal tar epoxy. Some problems have been encountered with bearing failures on inert gas fans and these have frequently been caused by blade imbalance brought about by solids depositing. It is common therefore to find cleaning water nozzles installed on fans and these should be used from time to time to clean the fan blades. The impeller may be supported in plain or anti-friction bearings. Where the latter are used it is normal to mount them on resilient pads.

The fans discharge to the deck main via a seal which prevents the back flow of gases. The seals used can be classified as wet (Figure 6.13) or dry (Figure 6.14) seals. Both types involve feeding the inert gas through a flooded trough but in the dry type seal a venturi gas outlet is used which effectively pulls the water away from the end of the gas inlet at high flows allowing the inert gas to bypass the water trough. The reason for developing this type of seal was because early wet-type seals frequently caused water carry-over into the system. As with other components in the inert gas system the internal surfaces of the deck seal must be corrosion protected usually by a rubber lining.

Motor tankers and topping up system

Diesel engine exhaust gas has too high an oxygen content for use as an inert gas. In motor tankers, therefore, the exhaust gas from an auxiliary boiler may

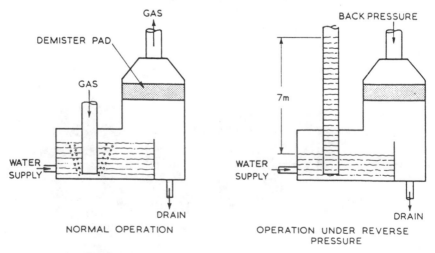

Figure 6.13 *Wet type deck seal*

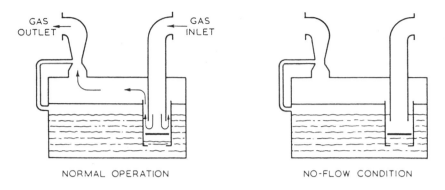

Figure 6.14   *Dry type deck seal in which the water-trough is by-passed at high gas glow rates*

be used or an inert gas generator. Flue gas is usually plentiful when discharging cargo since the auxiliary boiler will be supplying steam to cargo pumps and perhaps a turbo-alternator. At sea with the oil fired auxiliary boiler shut down, an alternative supply of inert gas must be found, to make good the inert gas losses. An oil-fired inert gas generator (Figure 6.15) may be installed for the purpose. (Inert gas generators have also been fitted in dry cargo ships for fire-fighting duties.) Units of the W. C. Holmes vertical chamber oil-fired design are capable of gas outputs at a pressure of 0.138 bar. Units have also been provided with output pressures up to 1 bar where required.

In this unit oil is drawn from a storage tank and is pumped by a motor-driven gear pump through a filter and pressure regulator to the pilot and main burners. The necessary air for combustion is delivered by a positive displacement Roots type air blower. Oil and air are mixed in the correct proportions in an air atomizing burner mounted on the top of a vertical, refractory lined combustion chamber. The burner fires downwards and the products of combustion leave the combustion chamber at the lower end. They then reverse direction and travel upwards through the cooling annulus surrounding the combustion chamber. The inert gas is cooled by direct contact with sea water in the cooling annulus to a temperature within 2°C of the temperature of the water. This water also keeps the shell of the combustion chamber cool, and in addition removes most of the sulphur oxides.

As the generator is of the fixed output type, a relief valve is fitted to exhaust excess inert gas to atmosphere should there be a reduction in demand. A single push button initiates the start up sequence. A programme timer subsequently controls the ignition of the pilot burner, ignition of the main burner and a timed warm-up period. The combustion chamber is then automatically brought up to the correct operating pressure. Operation is continuously monitored for flame or water failure and excessive cooling water level. Should emergency conditions arise, the generator will automatically shut down and an audible alarm sound.

In the Kvaerner Mult inert gas system (Figure 6.16) inert gas and electrical

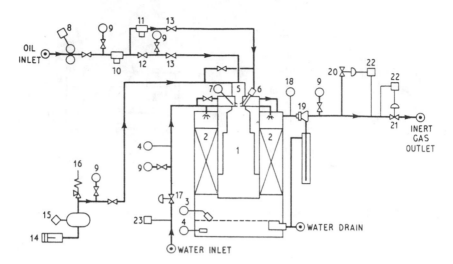

Figure 6.15 *Holmes vertical chamber oil-fired inert-gas generator*

1. Combustion chamber	9. Pressure gauge	17. Water inlet stop valve
2. Cooling annulus	10. Oil filter	18. Temperature switch
3. Float switch	11. Oil filter-pilot burner	19. Moisture separator
4. Cooling water thermometer	12. Pressure reducing valve	20. Inert gas relief valve
5. Main burner	13. Solenoid valve	21. Back pressure reguating
6. Pilot burner	14. Air filter	valve
7. Flame detector	15. Air blower	22. Pressure controllers
8. Motor driven oil pump	16. Air relief valve	23. Pressure switch

power are both produced from the one packaged unit. This consists of a generator driven by a diesel engine (formerly the system employed a radial flow gas turbine) the exhaust from which is delivered to a combustion chamber or after burner. The oxygen remaining in the diesel exhaust is reduced by further combustion with fuel to a very low level which makes it suitable for use as a tank inerting medium. The generator exhaust can be delivered to the after burner as required or directed to atmosphere.

Tank washing with sea water

Residues from crude oil accumulate in cargo tanks and must be regularly removed. The sludge blocks limber holes in frames and impairs or prevents final draining of tanks. Tank washing is a routine for crude oil carriers and also necessary when a vessel changes trade, from crude to clean products.

The accepted procedure for tank cleaning, before the introduction of crude oil washing, was to use rotating bronze nozzles, through which heated sea water was sprayed. Sea water is unsuitable as a solvent for oily sludges and static electricity generated during the washing process has caused numerous explosions. The nozzles, being suspended from hoses of non-conducting material, required an earth wire connected to the deck.

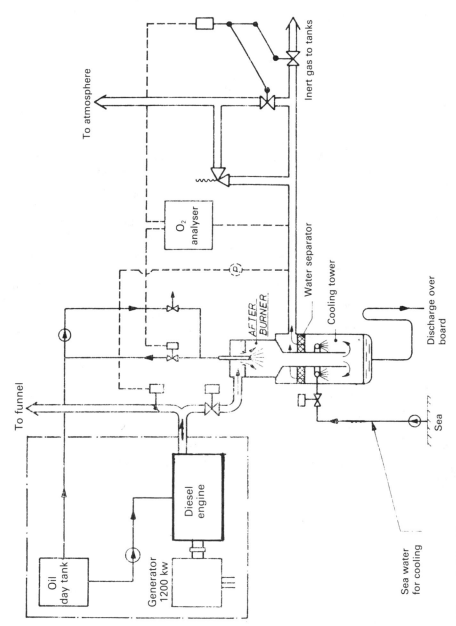

Figure 6.16 *Kvaerner inert gas system*

Crude oil washing (COW)

Crude oil washing of cargo tanks is carried out while the vessel is discharging, with the use of high-pressure jets of crude oil. For the process, a portion of the cargo is diverted through fixed piping to permanently positioned tank cleaning nozzles. Suspended nozzles are controlled to give a spray pattern on the upper areas and then progressively further down as surfaces are uncovered during the discharge. Washing is completed, with nozzles positioned on the tank bottoms. These are timed to coincide with the tank emptying so that oil below heating coil level will not solidify in cold weather. Effective washing can be carried out with crude, at the recommended tank heating temperature for discharge and even at temperatures as low as 5°C above the pour point. The waxy and asphaltic residues are readily dissolved in the crude oil of which they were previously a part and better results are obtained than with water washing. The oil residues are pumped ashore with the cargo. An inert gas system must be in use during tank cleaning.

Crude oil washing is necessary on a routine basis for preventing excessive accumulation of sludge. Tanks are washed at least every four months. Unless sludge is regularly removed drainage will be slow. If it is likely that ballast may have to be carried in cargo tanks (additional ballast to that in segregated ballast tanks, for example because of bad weather) then suitable tanks are crude oil washed and water rinsed. Water ballasted into a dirty cargo tank in emergency would be discharged in compliance with the anti-pollution regulations and a suitable entry would be made in the Oil Record Book. Crude oil washing must be completed before the vessel leaves port; a completely different routine from that of water washing which, when used, is carried out between ports.

Segregated ballast tanks

The segregated ballast tanks have lessened the pollution resulting from the discharge of contaminated ballast from dirty cargo tanks. These tanks are dedicated spaces and served by pumps and pipe systems completely separate from those for the cargo. The ballast arrangement shown (Figure 6.17) has a hydraulically operated submerged pump and an abbreviated ballast pipe for filling and emptying.

Chemical tankers

The enormous demand for crude oil and liquefied gas has meant that these cargoes are moved in very large quantities in ships that are dedicated to one cargo. Bulk chemicals are transported in smaller parcels and it is normal to find chemical tankers with almost as many different cargoes as there are tanks. With single cargo such as crude oil or one particular liquefied gas being carried continually, personnel become familiar with the risks. Dealing with a great variety of chemicals, each with different characteristics and properties, is much

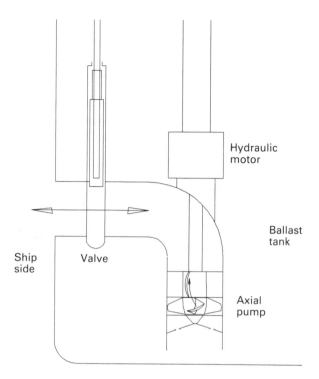

Figure 6.17 *Submerged ballast pump*

more of a challenge. Many chemicals are flammable and potentially explosive. Fires in a few are almost impossible to extinguish and there are restrictions on the type of fire fighting medium that may be used. Vapour from the cargo may be very toxic as well as flammable. The vapours from some cargoes react very strongly with oxygen so that there is violent combustion. To give oxygen to someone who has breathed in such fumes would produce unpleasant results. Many chemicals can poison due to absorption through the skin. The effects and remedies are so diverse that reference books are needed on chemical carriers, to summarize the dangers and to give the correct response.

Bulk liquid chemicals were initially transported in former oil tankers or vessels of much the same design as those for oil cargoes. These had pumprooms and wing tanks which extended to the shell plating. Corrosion of the steel plating in contact with chemicals over a long period inevitably caused the loss of some ships through final failure of the plating. The risk of pollution as the result of collision or grounding was also greater in single hulled vessels. Obviously there were some cargoes which corroded steel at too fast a rate to be considered as cargo but a number of acid and alkali substances were carried in the plain steel tanks of ships with single hulls.

As trading in chemicals increased, the scope of the problems became clear and improvements were made to the design of the ships. Guidelines and regulations relating to chemical tanker construction were also introduced (see the further reading section at the end of this chapter).

Cargo tanks

The tanks on a ship intended for type 1 chemical cargoes, are designed to provide maximum security in the event of collision, stranding or tank damage. Tanks for type 1 cargo must therefore be B/5 (one-fifth of the ship's breadth) or 11.5 m inboard from the ship's side (whichever is less), B/15 or 6 m from the bottom (whichever is less) and nowhere nearer to the shell plating than 760 mm. Tanks for type 2 cargoes must be at least 760 mm from the ship's side and B/15 from the bottom. The position of tanks in ships for type 3 cargoes, is not subject to special requirements.

Stainless steel is an ideal material for construction of cargo tanks, pipelines and pumps because it has the greatest overall resistance to corrosive attack by chemicals. However it is expensive and is vulnerable to attack by a few substances. The tanks of some chemical tankers are of plain steel but for greater resistance to corrosion, ease of cleaning and reduction of iron absorption by some chemicals and solvents tanks may be of steel with a protective coating of epoxy, polyurethane, zinc silicate or phenolic resins. Some tanks have been constructed with stainless steel cladding.

Epoxy coatings are suitable for alkalis, glycols, animal fats and vegetable oils but the acidity of the last two should be limited. Alcohols tend to soften the coating as do esters, ketones and chlorinated hydrocarbons.

Polyurethane coatings are suitable for the same types of cargoes as epoxies and some of the solvents compatible with zinc silicate.

Zinc silicate is used for aromatic hydrocarbon solvents, alcohols and ketones but not for acids or alkalis.

Phenolic resins have good resistance to strong solvents and most of the substances acceptable to the other coatings.

Prevention of pollution from bulk chemical carriers

Until the regulations to control and prevent pollution of the sea by chemicals came into force on 6 April 1987, there was no real restriction against discharge of cargo remains or tank washings from whatever cargo remained in the tanks of chemical tankers. The only factors limiting pollution, were goodwill and the fact that cargo remaining in the tanks after discharge constituted a loss to the shipper. For some tankers there were substantial remains because of the inability of the older type of cargo pumps to discharge completely. Later generations of cargo pumps were designed for more efficient discharge so that cargo tank remains were minimal. Improved clearing of tanks anticipated ideas put forward in draft regulations for more complete discharge of cargo, as a means to reduce pollution. Special draining and discharge methods have also been produced for fitting to existing vessels.

The anti-pollution regulations divide bulk liquid chemical cargoes into four categories (A – B – C – D) and give general direction for discharge and tank washing. There is a requirement in the rules for a Cargo Record Book and a Procedures and Arrangements Manual to be carried as a reference.

The list of Type A chemicals includes: Acetone cyanohydrin; carbon disulphide; cobalt naphthenate in solvent naphtha. The discharge into the sea of type A substances and any initial washings which carry them is prohibited. Chemicals in this category have to be totally discharged and delivered to the shore. Thus, when discharge is complete, any cargo remains must be removed and also discharged ashore by washing through. The washing process is continued until the content of the type A chemical falls below a certain value. After this, the discharge from the tank must continue until the tank is empty.

The washing through to clear the cargo is solely for that purpose and not intended as a complete cleaning operation. Traces of type 'A' cargo on the surfaces of tank bulkheads will remain until removed by a subsequent washing operation. These washings are considered as forming a residual mixture constituting a hazard if freely discharged. The rules are extended to include disposal of the subsequent tank washing operation residue.

Only wash water added after cargo discharge and completion of the 'in port' washing routine, can be pumped overboard at sea. The ship's speed may not be less than seven knots, with the vessel more than twelve nautical miles from land and in a water depth of 25 m minimum. The effluent may be pumped out through a discharge situated below the waterline and away from sea inlets.

A special low capacity pump which leaves the offending liquid mix in the film of water flowing over and adjacent to the hull, is used. It is intended that this flow shall carry traces of the chemical into the propeller where it will be broken up and dispersed in the wake. Presence of the chemical after this would not in theory exceed 1 ppm.

For most type A chemicals, the content in the pre-wash to the shore, must be reduced to less than 0.1% (weight) while in port if later washings are to be discharged outside special areas. If discharge of the later washings is to be in special areas (Baltic and Black Sea, etc.) then port washing must in general reduce the content of category 'A' chemicals to less than 0.05% by weight. Carbon disulphide is an exception for which the content must be less than 0.01 (not in special areas) and 0.005 (special areas).

Type B chemicals include: Acrylonitrile; some alcohols; calcium hypochlorite solution; carbon tetrachloride.

The cargo pumps for type B substances, in chemical tankers built after 1 July 1986, must be capable under test of clearing 'water' from the tank such that remains do not exceed 0.1 m^3 (0.3 m^3 for older vessels). Guidance for discharge ashore of category B cargoes, is obtained from the Procedures and Arrangements manual. Where difficulties prevent discharge according to the manual and for high residue substances, tanks are generally pre-washed with discharge of washings to reception facilities ashore.

Type C chemicals include: acetic acid; benzene; creosote (coal tar); ferric chloride solution.

The cargo pumps for type C substances, in chemical tankers built after 1 July 1986, must be capable of clearing 'water' from the tanks such that remains do not exceed 0.3 m^3 (0.9 m^3 for older ships). Guidance for the discharge of category C substances is obtained from the Procedures and Arrangements manual. These regulations are similar to those for type B substances.

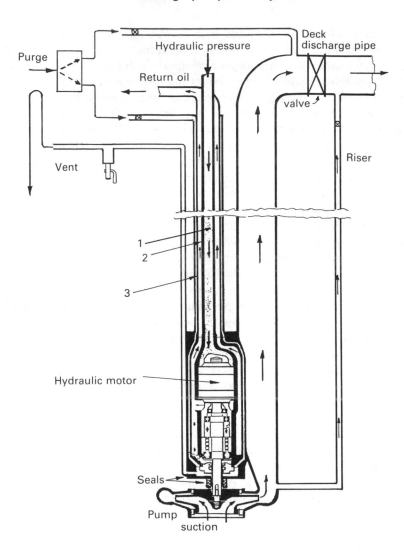

Figure 6.18 *Submerged cargo pump (Frank Mohn type)*

Type D chemicals include: calcium chloride solution; calcium hydroxide solution; castor oil; hydrochloric acid.

The discharge of type D chemicals into the sea is not permitted unless:

1 the ship is proceeding at not less than seven knots;
2 content of the discharge is made up of only one part of the substance with ten parts of water;
3 the vessel is more than twelve nautical miles from land.

There is a limit imposed on the quantity discharged if the vessel is in a special area.

Chemical tanker cargo discharge

Pumprooms in chemical tankers are very dangerous because of the risk of leakage from pump glands, of toxic/flammable vapour and corrosive or otherwise harmful liquids. The practice of positioning submersible or deepwell pumps within cargo tanks, eliminates pumproom dangers. Having individual pumps also reduces the risk of mixing cargoes and contamination. Deepwell pumps are described in the section on liquefied gas cargo pumping. To make them suitable for chemical pumping there will be a different gland arrangements and shaft bearings of Teflon.

Submerged cargo pump

Figure 6.18 shows a submerged pump based on a type produced by Frank Mohn AS. Pump elements, casing and pipework, are of stainless steel, which although expensive, will withstand the corrosive effects of most chemicals. Obviously, where a chemical tanker is to be engaged on a particular trade it may be possible to use cheaper materials. Working pressure for the hydraulic circuit is up to about 170 bar and return pressure about 3 bar. The high pressure oil supply pipe for the hydraulic motor, is placed central to and is surrounded by the return pipe. The return pipe and hydraulic motor casing are provided with a protective outer cofferdam to give complete separation of hydraulic oil and cargo. The arrangement of three concentric tubes used for this arrangement is shown in Figure 6.18.

The pump suction is positioned close to the bottom of the tank well for good tank drainage. Nevertheless, when pumping is completed, the vertical discharge pipe will be left full of liquid. Stopping the pump would allow the considerable amount of liquid in the discharge, to fall back into the tank so that clearing the tank of cargo remains or water used in tank cleaning, would be a constant problem. The remedy is provided by purging connections fitted to clear the discharge pipe. Purging is effected by closing the deck discharge valve as the tank clears of liquid, then with the pump left running to prevent cargo fallback, the purge connection is opened. Compressed air or inert gas at 7 bar will clear the vertical discharge pipe by pressurizing it from the top and forcing liquid cargo up through the small riser to the deck main. There is a small valve on the riser, to be opened before and closed after this operation.

The safety cofferdam around the hydraulic pipes and motor, is connected to the drainage chamber between the motor and pump. Seals above and below the chamber exclude ingress of low pressure hydraulic oil and liquid cargo, respectively. The bottom seal is subject only to pressure from the head of cargo in the tank, not to pump pressure.

The cofferdam can be pressurized while the pump is operating to check for leakage. Any liquid (oil or cargo) which collects in the chamber is forced up by the compressed air purge to the telltale, where it can be identified.

Stripping system

To enable older chemical tankers to comply with anti-pollution legislation, various stripping setups have been designed for retrofit. The drain tank for the retrofit system (Figure 6.19) can be fitted beneath the cargo tank, in the double bottom. Cargo remains drain through some type of automatically closing ball valve, to the drain tank below, and are then forced out through a discharge riser to deck level, by air or gas pressure. The procedure is repeated until the tank is empty. The discharge pipe is then blown dry. Complete cargo discharge is possible using this method, without the need to finish with a water wash to clear remains.

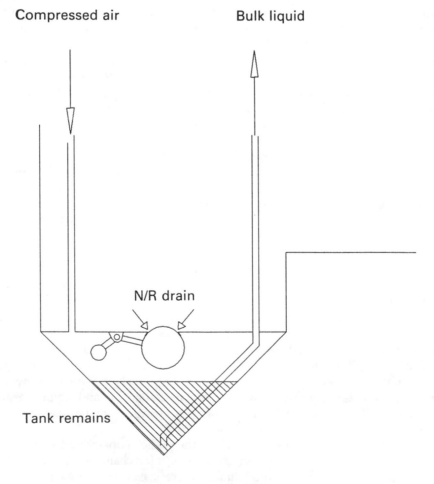

Figure 6.19   *Cargo tank stripping arrangement*

Fire risk

A dry powder installation on the deck of a chemical tanker, provides an extinguishing medium suitable for most chemicals carried. It may not be effective as a fire fighting agent for all chemicals, however. Guidance on the most appropriate action to be taken in the event of fire, is found from the reference books.

Liquefied gas carriers

The gas charge in a conventional refrigeration system, is first compressed to raise its pressure and relative boiling or saturation temperature, before being liquefied by cooling in a condenser. A gas which is to be transported by sea or land, can be liquefied in the same way; being first compressed and then cooled. Many gases, once liquefied, will remain in that state at atmospheric temperature provided that the necessary pressure is maintained. The storage pressures for these gases is lower if cooling is introduced.

There is a limiting feature with respect to pressure, in that there is an upper (critical) pressure for each gas, above which it cannot be liquefied.

Critical pressure is the ultimate for liquefaction; a gas at higher than critical pressure cannot be liquefied.

Whilst in general, gases can be liquefied by compressing and cooling, there are some such as methane, which cannot remain liquid at ordinary temperatures regardless of the pressure. Methane, must be cooled to less than its critical temperature of −82°C to become a liquid and the liquid state can only be maintained if the gas is held at a pressure of at least 47 bar. The pressure required to maintain liquidity, reduces with temperature however, and if methane is stored at −162°C it will remain liquid at atmospheric pressure.

Critical temperature is the limiting temperature for liquefaction of a gas. Every gas has a critical temperature above which it cannot be condensed regardless of the pressure.

Table 6.1 *Properties of some gases*

Gas	Critical pressure bar abs.	Critical temperature °C
methane	46.4	−82.1
ethylene	50.7	9.25
ethane	48.9	32.3
propylene	46.2	91.8
propane	42.6	96.8
butadiene	43.3	152
i-butane	36.7	135.2
n-butane	38	152
ammonia	112.8	132.9
vinyl chloride	53.4	158.4

The gases carried

Liquefied gas carriers may be categorized as suitable for the transport of LPG and ammonia or LNG or both if equipped appropriately.

LPG (liquefied petroleum gas) is the general term for gases such as propane, butane, propylene, butylene and C_4-isomers. These products can be liquefied at modest pressures.

LNG (liquefied natural gas) distinguishes methane and mixtures containing mainly methane (with small amounts of ethane or traces of other gases) from the gases mentioned above.

Liquefied chemical gases carried by gas tankers, include ammonia, vinyl chloride and chlorine.

Types of liquefied gas carrier

The carriage of gas in bulk, necessitates that it be maintained in a liquid condition by keeping it (1) under pressure; (2) at moderate pressure and moderately low temperature; or (3) at low temperature.

Fully pressurized ships used for the carriage of LPG and ammonia, have tanks in which the cargo is kept liquid, solely by pressure. At low ambient temperature the pressure is moderate but tanks are designed for pressures of 18 bar or more, relating to maximum possible temperatures which may be taken as plus 45°C. This pressure requires containment of great strength and fully pressurized tanks having great thickness, are consequently of small size and tend to be of cylindrical shape with rounded ends. Large tanks require disproportionately large wall thickness for the same design pressure, compared with small tanks. To avoid the excessive weight penalty, fully pressurized tankers have been built with a large number of small tanks. This makes poor use of hull space.

Part pressurized ships, suitable for LPG, have larger tanks, of cylindrical or near cylindrical shape with convex ends and insulation. Tanks are shaped to make better use of hull space with overall, a larger cargo quantity for the equivalent ship size, compared with the fully pressurized type. Tank strength must be adequate for pressures of perhaps 6 bar relating to a maximum tank temperature of 10°C. Tank material must be suitable for temperatures possibly down to −50°C.

Low temperature cargoes create difficulties because steel and other metals tend to become brittle at low temperature. Cracking (brittle fracture) can result if undue stress occurs from localized expansion or contraction, or from impact particularly if there is a flaw in the material. There is a range of nickel steels for very low temperatures. The nickel gives toughness and reduces the coefficient of expansion and thus the stresses due to expansion and contraction. Aluminium is used as an alternative to nickel steel where appropriate.

Non-pressurized ships, have the largest tanks for the transport of LPG, ammonia and vinyl chloride. The tanks are dimensioned for a maximum working pressure of only 0.25 bar above atmospheric, because boil-off enables carrying temperature to be maintained at the saturation temperature of the gases for atmospheric pressure. Self-cooling occurs as evaporation removes latent heat from the remaining cargo. Reliquefaction of the boil-off is usual with direct or indirect refrigeration equipment being fitted for the purpose. The saving in weight and cost plus the large size of tanks possible for pressures at just above ambient, makes refrigeration viable. The problem of metal brittleness due to low carrying temperatures, is overcome by using special nickel steels or aluminium alloys, for tank construction.

Liquefied gases are carried in tanks which are insulated but heat leakage is inevitable. The capacity of reliquefaction equipment is equated to heat leakage.

Insulated ships for carriage of LNG, are designed so that the boil-off, amounting to between 0.25 and 0.3% of cargo daily, can be used as fuel in boilers supplying steam to main propulsion turbines. In addition to providing fuel, the gas boil-off removes latent heat and keeps the rest of the liquefied gas cargo at a temperature of about − 162°C. The boil-off is delivered to the machinery space by compressor.

Diesel engines have also been designed for operation on boil-off, with ignition of the gas and air charge in the cylinder by a pilot injection of distillate fuel or a spark. Spark ignition was introduced to save liquid fuel in shore installation diesels operating on gas.

The use of liquefied natural gas boil-off to provide steam for main propulsion, continues to be preferred to reliquefaction for LNG carriers. Plant for reliquefaction and partial reliquefaction of LNG has been developed. However, operating costs are apparently still too high to achieve a saving by using conventional fuel. In the future, price fluctuations could make it more economic to reliquefy the gas boil-off and operate the ship on conventional fuel.

Liquefied gas cargo tanks

The non-pressurized tanks of larger LPG ships and those for LNG cargo, may be of spherical, prismatic or membrane construction, with heavy insulation.

Reliquefaction

Direct or indirect refrigeration is used for reliquefaction of the gases carried at low temperatures. Basically, the process involves removing boil-off from the cargo tank, reliquefying either by direct compression and cooling or by indirect cooling and returning the liquid to the cargo tank. The total heat quantity that is removed by the plant condenser from these gases comprises the heat which leaks into the cargo tank through the insulation plus heat picked up by the gas *en route* to the refrigeration plant plus any heat from a compression process.

Single stage direct system

In this arrangement (Figure 6.20) the cargo boil-off passes through a refrigeration circuit in the manner of a refrigerant. A separator for removal of

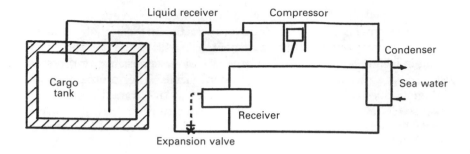

Figure 6.20   *Single stage direct reliquefaction*

any entrained liquid averts any risk of damage to the compressor. If the compressor piston has conventional cast iron rings, the necessary lubricating oil which also assists in forming a seal between the liner and rings, tends to be carried over to the tanks, despite the installation of an oil separator in the compressor discharge. To avoid the oil carry over problem, labyrinth seals are used or teflon piston rings. With LPG cargoes where a gas such as butane is to be used for domestic heating, slight oil contamination may not be a problem.

Compression of the gas is accompanied by an increase in temperature, due to the work done by the compressor. In the condenser, this heat and the heat associated with evaporation, is removed by sea water (but a lower temperature coolant could be used) so that after cooling down to the saturation temperature relating to the compression pressure, the gas is liquefied. The regulating valve controls pressure drop from that of the compressor discharge and condenser, down to that of the tank.

Two stage direct system

When the ratio of discharge to suction pressure for a compressor is excessive so that compressor capacity drops, two stages of compression are used. Ammonia, propane and propylene, carried at atmospheric pressure, require two stage compression.

The two stage system (Figure 6.21) is based on a Sulzer design installed on some combined LPG/LNG carriers for reliquefaction of commercial grades of butane and propane. (The system can be adapted for partial reliquefaction of LNG.) Vapour from cargo tanks reaches the boil-off compressor via the heat-exchanger (HE) where it is heated by sub-cooling the condensate. The compressed vapour is then delivered to the main (second) compressor for compression to about 26 bar depending on sea-water temperature and any ethane content of the boil-off. From the main compressor, the gas passes to the condenser and then the condensate is returned through the sub-cooling heat exchanger (HE) to the cargo tanks.

The final temperature of the boil-off after two stages of compression could be excessive (particularly with ammonia) and as with air compressors, a form of intermediate cooling (Figure 6.22) is available. The gas discharged from the

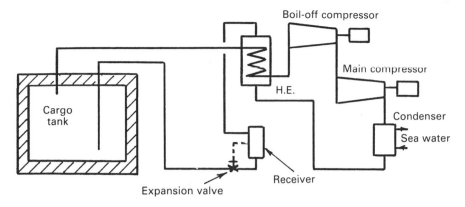

Figure 6.21 *Sulzer type 2 stage reliquefaction plant*

first stage (LP) compressor is delivered to an intermediate cooler in which heat is removed by letting the gas bubble through liquid. Gas from the second stage (HP) compressor is condensed and then returned to the cargo tank through the coil in the intermediate cooler. Sub-cooling in the intercooler is followed by expansion through a regulating valve. A regulating valve keeps the level in the intercooler constant.

Indirect system

Refrigerants such as R22, ammonia or propane are used in a closed cycle for the indirect system (Figure 6.23) which works in much the same way as a conventional refrigeration plant Heat exchange takes place in the evaporator which is cooled by evaporation of the refrigerant in the closed circuit. The vapour generated by heat leakage into the cargo tanks, condenses on the cold surfaces of the evaporator and returns by gravity to the tanks. The evaporator can be fitted inside the cargo tank if necessary due to the hazardous nature of the cargo.

The indirect system avoids any contact between boil-off and compressors, so oil contamination is impossible and ordinary refrigeration compressors with standard piston rings can be used. Oil contamination is not acceptable for cargoes such as vinyl chloride and butadiene.

The indirect system is also employed for cargoes for which the heat generated by compression could cause a change of state.

Cascade system

A combination arrangement consisting of a direct system with the cargo boil-off being compressed and condensed and the condenser for this circuit

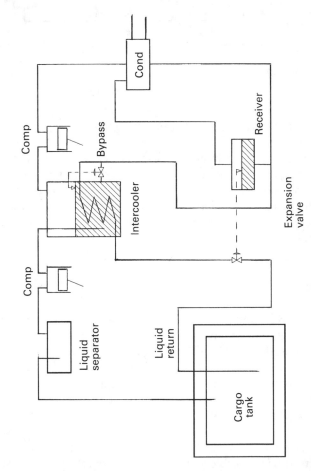

Figure 6.22 *Two stage reliquefaction with intercooler*

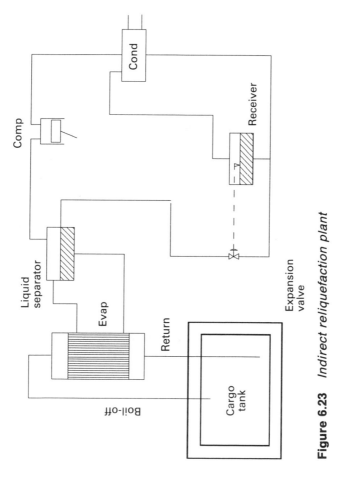

Figure 6.23 *Indirect reliquefaction plant*

being cooled by its use as the evaporator for a closed system is termed a cascade system. Single or two stage cycles could be used in a cascade system.

Cargo discharge

It is beneficial to site discharge pumps in the bottom of the cargo tanks, particularly for volatile cargoes, where NPSH becomes critical. The low temperature of liquefied gas prohibits the use of submerged hydraulically driven pumps. Thus deepwell pumps with motors at deck level are used, or submerged electrically driven pumps. The motors on deck may be induction motors constructed to comply with Ex e (enhanced safety requirements) or hydraulic motors.

Deepwell pumps

Deepwell pumps are centrifugal pumps driven, through a long vertical shaft, by a hydraulic or electric motor mounted at deck level above the cargo tank top. An electric motor for this duty would be a cage induction motor of the increased safety (Ex e) type. The multi-stage deepwell pump (Figure 6.24) is designed for liquefied gas cargoes and features an oil filled gas seal at deck level. The oil within the seal chamber is maintained constantly one bar above cargo tank pressure by an accumulator containing a plunger which is operated by air or nitrogen pressure.

The drive shaft is located within the pump discharge pipe and is supported in carbon bearings. The shaft is protected by way of the bearings by stainless steel sleeves. Positioning of the shaft within the discharge pipe allows the liquid cargo to lubricate and cool the bearings. It is essential that this type of pump is not allowed to run dry or with a throttled discharge valve as there is a risk of overheated bearings. A pressure cut-out or thermal switch may be arranged.

The weight of the pump shaft and impellers is considerable and one or more carrier bearings are fitted. Prevention of lift of the shaft due to ship movement or impeller surge makes a thrust bearing necessary.

The number of stages in the multi-stage pump shown is dictated by the discharge head required. The diffuser rings around each impeller serve to increase discharge pressure; this type of pump is often referred to as a turbine pump. As an alternative to the multi-stage unit and its high power drive through a long line shaft, a single stage pump can be installed to lift the liquid to deck level, where a booster pump is fitted to assist transfer of the cargo ashore. The inducer frequently fitted to the pump shown, is an archimedian screw, attached to the drive shaft just below the impeller at the pump suction. It improves the flow of volatile liquids into the pump.

Submerged electrically driven pumps

The pump and its electric motor are positioned at the bottom of the discharge pipe, within the cargo tank. The cage type induction motors used for this duty

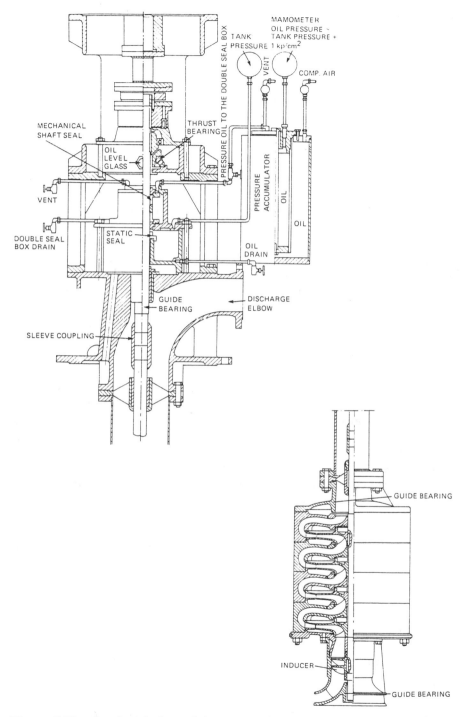

Figure 6.24 *Sectional view of deepwell pump showing tank top seal (above) and multi-stage pumping element (right) (Svanehoj, Denmark).*

are safe in an environment where no air is present and are also basically safe because there are no brush contacts or other sources of sparks. The arrangement can give problems with ammonia cargoes. Ammonia tends to corrode the electrical contacts.

Liquefied cargo

Liquefied natural gas, liquefied petroleum gas and ammonia all form flammable and potentially explosive mixtures with air. Internal tank pressure excludes air from the tank provided that it is maintained. Pressure in very low temperature cargo tankers is maintained just above atmospheric for this purpose. Leakage from these and from pressurized tanks is therefore always of gas to the outside; not of air to the inside. Flammable mixtures are formed externally when cargo leaks out and ignition can cause this to burn in the outside air (with a yellow flame). Gas boil-off passing through tank vents has been ignited by static electricity; to prevent this, the vents are earthed. Collision damage has also been responsible for external combustion and for increased safety, tanks for flammable cargo have a one fifth of the ship's breadth margin between them and the hull plating.

Potentially the most serious situation occurs when a tank is perhaps almost empty but open to the atmosphere and at atmospheric or lower pressure. A large amount of air entering and mixing with the small quantity of gas, would form a flammable and potentially explosive mixture. Ignition of such a mixture would result in a pressure rise and due to the confining effect a major explosion.

At the end of cargo discharge a residue of at least 2% liquid is left in tanks to maintain an atmosphere consisting totally of cargo vapour. With no air present and slight pressure above atmospheric the tank is safe. The heel of cargo also avoids the problem of draining and inerting. Only when the tank is to be emptied for repair or inspection is it drained and purged with inert gas and then finally purged with air.

The cargo remains are used to maintain tank temperature at carrying level during the ballast voyage. In this way, the probability of damage to tank structure due to uneven expansion and contraction is avoided, as is the problem of cooling down for the next cargo. A larger residue of LNG cargo is left if needed as fuel.

Further reading

Oil tankers

ICS (1988) *International Safety Guide for Oil Tankers and Terminals (ISGOTT)*. 3rd edn International Chamber of Shipping.
Day C.F., Platt E.H.W., Telfer I.E., Tetreau R.P. (1972) 'The Development and Operation of an Inert Gas System for Oil Tankers', Trans I Mar E, vol 84.

Chemical tankers

Statutory Instruments 1987 No. 551. Marine Pollution The Merchant Shipping (Control of Pollution by Noxious Liquid Substances in Bulk) Regulations 1987.
ICS (1971) *Tanker Safety Guide (Chemicals)*, International Chamber of Shipping.

Gas carriers

ICS (1978) *Tanker Safety Guide (Liquefied Gas)*, International Chamber of Shipping.
McGuire, G. and White, B. (1986) Liquefied Gas Handling Principles on Ships and in Terminals SIGTTO.

7

Auxiliary power

Auxiliary medium- or high-speed diesel engines, are mainly used as the prime movers for generators but they may also be coupled up to provide a direct drive for large pumps, bow thrusters or other machinery. The steam turbine, used as the alternative independent method of driving a generator, can also be employed for driving cargo or feed pumps.

Diesel driven generators can be expensive in terms of fuel cost and maintenance requirements. Continuous operation at sea and in port exacts a high price particularly if distillate fuel is used. This has led to the use of residual fuel and the introduction of fuel blenders so that heavy residual fuel can be added to the distillate as the load increases to give cost savings. Blenders improve fuel economy but running hours still mount up and routine maintenance is likely to be increased by the use of poor quality residual fuels. Fuel blending is described in Chapter 2 (pp. 73–4).

A better economy in the provision of electrical power is achieved with the use of a generator which can be operated at sea, on power derived from the main propulsion system.

The ship's electrical load can be provided by a turbo-alternator using steam from a waste heat boiler. The waste heat available from the exhaust of a very large slow-speed diesel engine is sufficient in many installations for the full electrical load at sea. Savings in fuel costs for such a system amount to about 10% with additional savings in maintenance. However, the continuing search for improvement in diesel engine and propeller efficiency has resulted in the development of very long stroke, very slow-speed engines with constant pressure turbo-charging and the waste heat available from the exhaust has gradually diminished. The quest for electrical power from the ever decreasing quantity of exhaust gas energy, has fostered progress in the area of harnessing otherwise wasted heat energy.

Mechanical generator drives from the main engine, gearbox or the propeller shaft temporarily lost popularity when alternating current power replaced direct current, because frequency demanded a constant speed. Constant speed drive schemes are now available for use with variable speed engines (and a number of electrical solutions to the problem – for further details see Marine Electrical Equipment and Practice).

Power turbines, driven by the exhaust gases from the engine in the same way as the turbo-charger, are used to convert waste heat into mechanical energy which is delivered to the main propulsion system through a fluid coupling, or used to provide an integrated drive with a diesel for a generator.

Medium speed auxiliary diesel engines

Over recent years, many owners have elected to burn low grade residual fuels in medium-speed auxiliary engines, sometimes with disastrous results. Fuels bunkered for slow-speed main engines may be of too poor a quality for use in auxiliaries even where the engines have been designed for heavy fuel operation. Major problems have been experienced on large slow-speed engines with some of the poor quality bunkers such as those containing catalytic fines. Fuel should conform to the specification given in the instruction book for the engine.

Typical of the medium-speed engines designed to be capable of running on heavy fuel are the Allen S12 series in-line engines having four, six, eight and nine cylinders respectively and the vee engines of similar cylinder size, designated VS12 which are produced with twelve or sixteen cylinders.

General construction (based on the Allen S12 engine)

The in-line S12 engine structure (Figure 7.1) is based on a deep section cast iron bedplate and a cast iron A-frame of monobloc construction which are flanged and bolted together. The bedplate carries thin wall, steel-backed, white-metal or aluminium-tin lined main bearings. An additional bearing is incorporated to carry the combined loads of the flywheel and part of the weight of the generator. Access doors are provided at the front and back. Those on the back of the engine are fitted with crankcase explosion relief valves.

In this style of construction, which is common to many medium-speed engines, it is necessary to lift the A-frame if the crankshaft is to be removed. Some designs incorporate a C-frame arrangement which permits side removal of the crankshaft.

The one-piece alloy steel crankshaft is slab-forged, oil-hardened and tempered. A solid half coupling forged integrally carries the flywheel to which the generator is coupled. The main coupling bolts pass through the crankshaft half coupling, the flywheel and the generator half coupling; two additional bolts are incorporated to retain the flywheel on the crankshaft when the generator is uncoupled. Additional machines such as an air compressor or a bilge pump may be driven from the free end of the crankshaft, through a clutch. Axial location of the crankshaft is maintained by renewable thrust rings. Drilled passages in the crankshaft feed oil from the main to the connecting rod bearings. In the S12-F engine these oil passages are arranged to provide a continuous supply of oil to the cooled pistons. Where necessary, balance weights are bolted to the crank-webs. In four-cylinder engines having cranks at 180°, secondary balancing gear is required (Figure 7.2).

The connecting rods are H-section steel forgings, bored to carry oil to the gudgeon pin bush, which is a light interference fit in the rod. Crankpin bearings are thin-walled steel, lined with aluminium-tin, split horizontally and secured

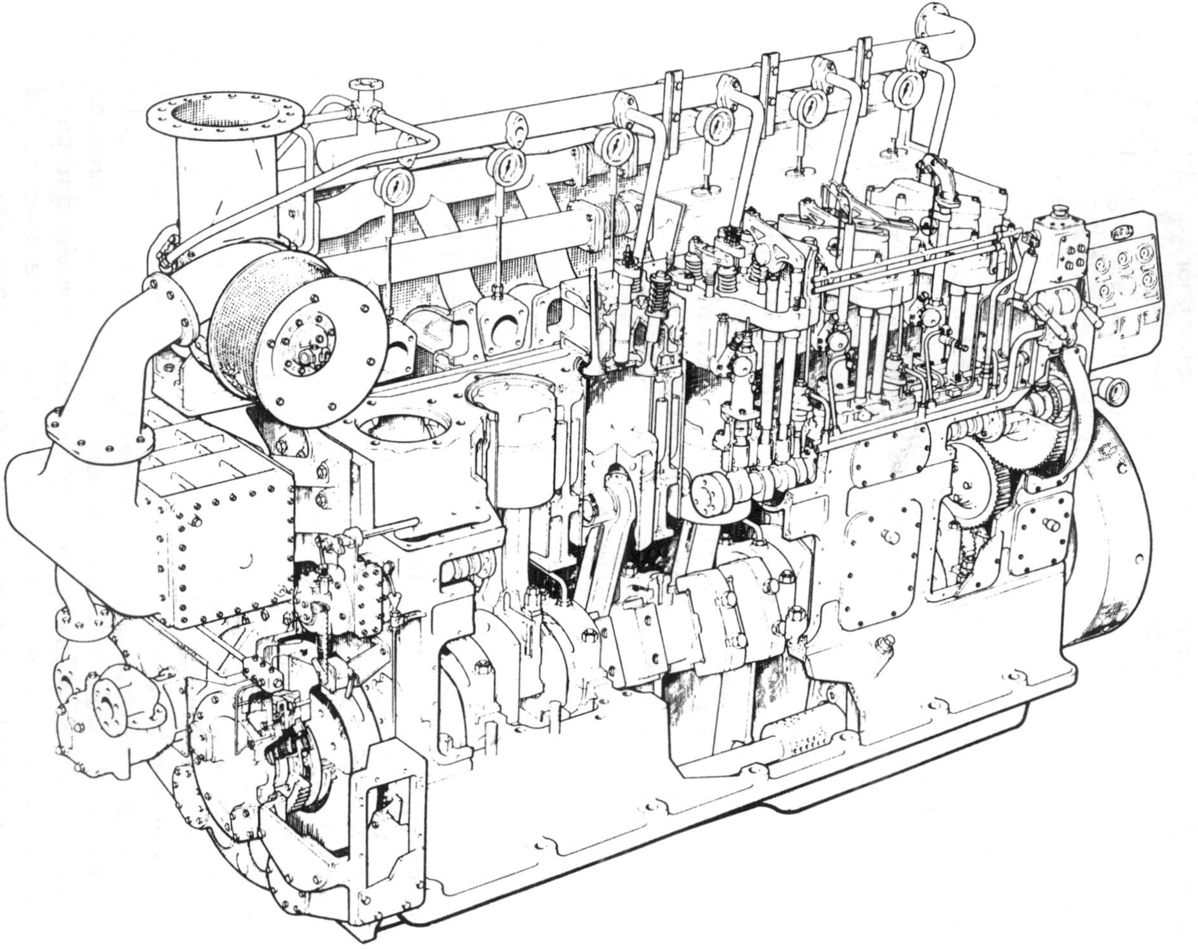

Figure 7.1   Six cylinder S12 F engine (ABE Allen Ltd)

Figure 7.2 *Secondary balancing gear (APE-Allen Ltd)*

by four fitted bolts. Connecting rods of differing designs will be found in some engines. Thus the bottom end assembly for the Allen VS12 (vee engine) series (Figure 7.3) is made simple by axially displacing both banks of cylinders and having two connecting rod bottom end bearings on each crankpin. The large end bearing is split diagonally to enable the connecting rod to be withdrawn through the cylinder. Bearing butts are serrated for location. Bronze bushes are used for the top ends; the large ends have steel backed, aluminium-tin bearings with an overlay which is deposited by plating.

The fully-floating gudgeon pin of the Allen S12 is steel, case-hardened and ground, retained in the aluminium alloy piston by circlips. The piston has an

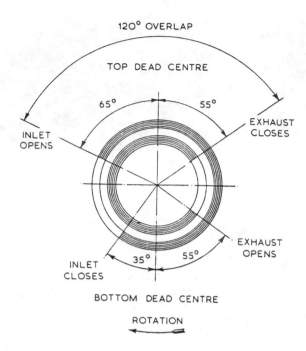

Figure 7.3   *Valve timing diagram for turbo-charged four-stroke engine (APE-Allen Limited)*

integrally cast alloy iron carrier for the top piston ring, two additional pressure rings and one slotted oil scraper ring, all above the gudgeon pin. Pistons for the S12-F are one-piece aluminium-alloy castings incorporating oil cooling cavities. The oil provides intensive cooling of the piston particularly in the region of the ring belt. The wet-type close grained, cast iron cylinder liners, are supported by a flange at the top which is sandwiched between the cylinder head and a spigot with ring gasket, on the engine frame. To permit vertical expansion of the liner it is free to move at its lower end, a seal being effected by two synthetic compound O rings carried in grooves in the liner wall.

Camshaft and cylinder head

The camshaft, is driven by a roller chain (some engines have a gear train). To allow accurate phasing of crankshaft and camshaft during initial set up and if timing has to be reset, elongated holes are provided at the coupling between the camshaft drive wheel and the camshaft. Adjustable packing pieces inserted into the elongated holes ensure that correct timing is maintained.

Lubrication of the camshaft bearings is by a forced feed system; an oilway bored through the full length of the camshaft conducting the oil to the bearings. An extension of the camshaft at the driving end, is provided with a flexible coupling for the hydraulic governor and tachometer drive.

The individual alloy cast iron cylinder heads have totally enclosed valve gear which is lubricated from the engine oil system. The S12-D engine has one inlet and one exhaust valve, the S12-F, because of its higher running speed, is fitted with heads having two inlet and two exhaust valves. The valve pairs are parallel and operated by rocking levers and guided bridges. Each valve has two springs and is fitted with a rotator. The valves seat in the cylinder heads on renewable inserts of iron alloy. The centrally placed fuel injector is situated between the valve covers so that fuel oil contamination of lubricating oil is avoided. This also enables the injectors to be withdrawn for servicing without disturbing the valve covers.

Fuel pump and timing diagrams

Separate camshaft-actuated helix-type fuel pumps are employed for each cylinder. These deliver fuel to the injectors which are set to lift at a pressure of 176 kg/cm² on the S12-D and 211 kg/cm² on the S12-F. Fuel pump delivery volume is controlled by a rack which alters the cut-off or spill point. The racks are linked through a control shaft to the engine governor which thus regulates the end of the fuel delivery period and hence the quantity of fuel delivered according to the power required. Fuel injection commences at approximately 15° before top-dead-centre and takes place over a period of about 35° of crank angle. Combustion should be completed within this period. A typical valve timing diagram (Figure 7.3) shows the large overlap between the opening of the air inlet and closure of the exhaust compared with the previous normally aspirated engine (Figure 7.4) which is no longer produced. The overlap allows a through flow of charge air which is essential for exhaust valve cooling, particularly for an engine which operates using residual fuel. Vanadium and sodium ash from the fuel tends to adhere to valves and seats if their temperature exceeds the melting point of the ash. The deposited ash itself causes surface damage in the form of pitting and also tends to prevent closure. Valve surface temperature should ideally not exceed 420°C if ash deposit is to be avoided. Localized high surface temperature can be prevented on that part of the valve adjacent to the fuel injector by the rotator. Valve surfaces can be protected by a stellite deposit or, alternatively, valves can be made from nimonic.

The turbo-blower, mounted at the free end of the engine, has a filter/silencer fitted on the air intake. The charge air cooler is similar to that described in Chapter 1.

Turbochargers

The turbocharger is driven by the exhaust gas leaving the cylinders of the diesel engine it serves. The gas has sufficient pressure and heat when released from the cylinder at exhaust opening, to drive the turbocharger. It is directed on to turbine blades by nozzles which are built into a nozzle ring in the axial flow type or into a radial turbine from the peripheral volute casing of smaller turbochargers.

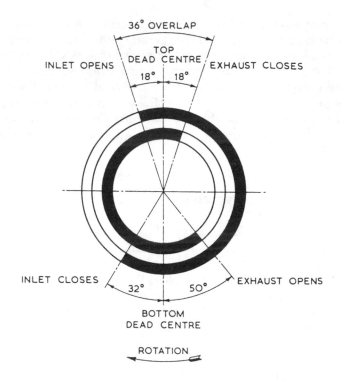

Figure 7.4   *Valve timing diagram for normally-aspirated four-stroke engine (APE-Allen Limited)*

A small turbocharger for a generator diesel prime mover may be driven by a radial flow gas turbine, which closely resembles the impeller it drives. This type of machine costs less to manufacture than the axial flow turbine, and has a simpler construction.

Turbocharger blades are rotated, partly by the impact of jets of gas from the nozzles and partly by the reaction, as gases leave the blades. Correct nozzle and blade shape is vital. Performance of turbocharger and engine can deteriorate seriously with sometimes very moderate surface marking due to erosion or corrosion.

Nozzle shape can be altered in service, by:

1   deposit build up;
2   corrosion of surfaces;
3   erosion, by solids entrained in the exhaust gas.

Oil refinery residuals are used in a blend with clean distillate fuels for economy in some engines. Deposits are common when heavy residual fuel is used. Regular cleaning is necessary to prevent nozzle blockage. Water washing rather than dismantling and cleaning is used for deposit removal. Fittings should be provided for water washing if an engine is to be operated using residual fuel.

Corrosion is a potential problem for the turbochargers of engines which operate using residual fuels containing vanadium, sodium and sulphur as impurities and in a marine environment where sodium chloride is present in the intake air. The impurities listed burn to form a number of different ash products which may adhere to surfaces at higher temperatures. Corrosion and surface damage follows breakdown of the protective film on the metal surface by ash compounds.

Remedies for these problems are based on designing for lower operating temperatures and regular water washing to remove the accumulated slag (ash). Corrosion problems are well known and documented in various technical papers such as those of the Institute of Marine Engineers.

Erosion by solids entrained in the exhaust gas is another potential problem. Catalytic fines (based on aluminium and silicon which are abrasive) will be present in some fuels and could cause surface damage to nozzles. Purification of fuels is necessary to remove solids and where catalytic fines are suspected, the use of centrifuges arranged as two purifiers in parallel or a purifier and a clarifier in series is recommended (see Chapter 2).

Serious damage by corrosion or erosion will finally require renewal of parts if the efficiency of the turbocharger and diesel is to be maintained.

Cooling system

A variety of cooling systems may be adopted for marine auxiliary engines but the most commonly used is the simple closed circuit system (Figure 7.5). Sea water is passed through the intercooler, the oil cooler and then the jacket water cooler in series flow. Engine driven fresh-water circulating pumps are normally fitted, but the sea-water pump may be either an independent unit or engine driven in tandem with the fresh-water pump. The cooling system may be arranged so that in an emergency sea water can be circulated through the engine jackets, after removal of certain blanks installed in the pipework.

In ships with diesel main propulsion engines cross-connections between the main and auxiliary engine jacket water systems are common. This enables the main engine to be kept warm in port from the heat in the auxiliary engine jacket water. To enable the auxiliary engine to be run in dry dock it is customary to arrange a connection from a double bottom or peak tank.

Bosch type fuel pump

The most common fuel pump used on auxiliary diesel engines is the Bosch type. This is a cam operated jerk pump with a helical groove on the plunger to control the fuel cut-off and therefore the quantity of fuel delivered to the cylinder for combustion. These pumps can be arranged singly along the camshaft, with one at each cylinder position or they may be housed in a single block. Each pump unit contains a pump plunger and guide together with a spring loaded delivery valve and its seat. Plungers and guides are not interchangeable – they should be treated as combined units or elements.

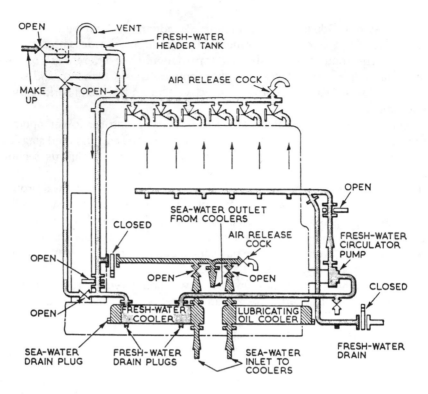

Figure 7.5 *Closed circuit cooling system*

Operation

The operation of helix type pumps is shown diagrammatically in Figure 7.6. With the plunger E at the lower limit of its travel (Figure 7.6a) fuel enters the barrel from the surrounding suction chamber, through the two ports. As the plunger rises, some fuel is displaced through the ports until they are just closed (Figure 7.6b) by the top edge of the plunger. Fuel trapped above the plunger is now forced out through the delivery valve above the top of the pump barrel.

The pressure exerted by the rising plunger causes fuel to lift the valve and to enter the pipe which connects the pump to the injector. As the pipe is already full, the extra fuel which is being forced in, causes a rise in the pressure throughout the line and lifts the needle valve of the injector. This causes fuel to be forced into the combustion chamber in the form of a fine spray.

As the plunger continues to move upwards, the lower edge of the control helix uncovers the spill port, allowing fuel to be bypassed from the barrel suction chamber through the vertical drilled hole in the plunger or via a machined slot or channel (Figure 7.7). This allows the delivery valve to shut under the action of its spring, and with the collapse of pressure in the pipeline, the injector also shuts. At the junction between the delivery valve and its guide, there is a short plane cylindrical piston which fills the seat aperture as the

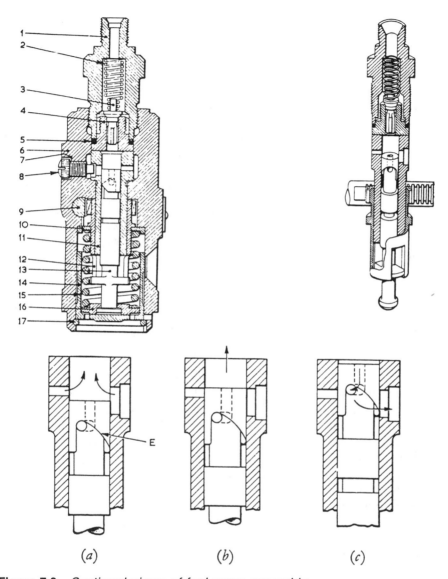

Figure 7.6 *Sectional views of fuel pump assembly*

1. Delivery pipe connection
2. Delivery valve spring
3. Delivery valve
4. Delivery valve seat
5. Delivery valve seat joint
6. Pump body
7. Locking screw joint
8. Locking screw
9. Control rack
10. Retaining collar top
11. Plunger guide
12. Control sleeve
13. Plunger
14. Tappet plunger
15. Pump spring
16. Retaining collar bottom
17. Circlips

valve closes. This emphasizes pressure drop in the delivery pipe so that the injector closes sharply.

The actual plunger stroke is constant, but an effective stroke depends on which part of the helix is moving up and down in line with the spill port. The effective stroke can be set between maximum fuel and no fuel. The latter

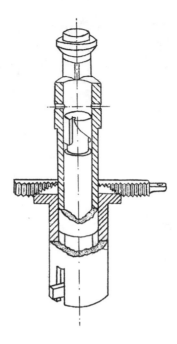

Figure 7.7   *Bosch type fuel pump*

setting, means that fuel spills for the full length of the plunger stroke. The plunger is moved around to the required position by the rack and quadrant (Figure 7.7). The quadrant collar is on a sleeve which has two vertical slots at the bottom. Two lugs projecting from the lower part of the plunger move up and down in these slots as the plunger reciprocates. Rotary movement of the sleeve (which has no vertical motion) moves the plunger. The rack which meshes with the toothed quadrant is externally connected to suitable linkage from the governor and the manual control lever.

Common fuel injector

The fuel injector nozzle contains a non-return valve labelled nozzle valve in the drawing (Figure 7.8), which is forced open against the spring loading by pressure of fuel from the fuel pump. The upper part of the nozzle (or needle) valve is lapped into the nozzle body within which it works freely. The lower part is of smaller diameter having at its extremity a valve face and at the junction of the two diameters a shoulder. The upper end has a small extension or spigot which engages with the valve spindle.

Fuel is delivered to an annular space in the nozzle via a hole, drilled through the nozzle body from the inlet. The nozzle valve is forced from its seat in the nozzle body by the pressure of fuel from the pump, acting on the shoulder of the needle valve. Pressure from the fuel pump also forces the fuel through the holes in the nozzle to form a fine spray in the engine combustion chamber.

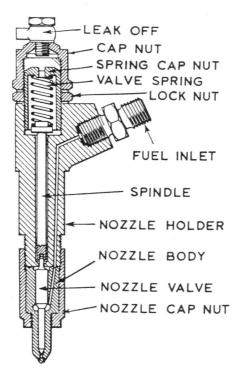

LEAK OFF
CAP NUT
SPRING CAP NUT
VALVE SPRING
LOCK NUT

FUEL INLET

SPINDLE

NOZZLE HOLDER

NOZZLE BODY

NOZZLE VALVE

NOZZLE CAP NUT

Figure 7.8   *Fuel injector assembly*

When fuel pump delivery ceases, there should be a sharp drop in pressure which means that the nozzle valve will close smartly with an instant interruption of flow.

Injector nozzles normally require cooling when residual fuel is used. Interruption of the fuel valve cooling supply, quickly results in vapour lock of the cooling passages within the nozzles and then blockage by scale. When fuel valve cooling is restored after a failure, it is essential to check flow from all returns, and to clear any with no return flow immediately. Loss of cooling will result in the formation of carbon deposits (trumpets) around the nozzles in a very short time.

Injectors should be taken out for examination, external cleaning and testing at regular intervals based on running hours as recommended by the makers handbook or as dictated by operational factors. Injector opening pressure, spray pattern and effective closing without leakage, are checked with a test pump. Defective injectors require overhaul in very clean conditions with special tools. On board overhaul may be carried out or alternatively nozzles may be sent to a specialist firm. After a replacement nozzle has been fitted in the nozzle holder, injector spring compression is set on the test rig and then the spray pattern and effective closing are checked.

Inexperienced personnel must be warned against contact with the nozzle tip when testing. Spray intensity is sufficient to puncture the skin and inject fuel.

For cleaning the fuel injectors, a special set of tools is available and contains the following main items: brass wire brush, nozzle body groove scraper, probing tool, nozzle body seat cleaner, nozzle body dome cavity cleaner.

Tracing faults

The failure of an engine to start or problems while running may be traced to faults with the fuel injection system or other possible causes. Instruction manual guidance on fault finding and remedies will include some of the typical problems listed below.

Difficult starting.
  Fuel system not primed through to the nozzle.
  Starting-air pressure too low.
  Water in the fuel.
  Air intake filter choked.
  Poor compression due to any of the following causes:
    Exhaust valve seats in bad condition.
    Exhaust valve spindle sticking.
    Starting-air valve sticking.
    Piston rings stuck or liners worn.

Engine uneven.
  Sticking fuel-pump control racks or linkage.
  Fuel pump delivery valves sticking.
  Overload or governor fault.

Smoky exhaust.
  Check fuel pump timing.
  Check maximum firing pressure and exhaust temperature.
  Injectors fault.
  Overload or unsuitable fuel.

Caterpillar engine fuel system

The range of larger Caterpillar engines use helix-type fuel pumps driven from a separate camshaft. Their principle of operation can be followed from the description given above, although the pump shown is different in detail. In the smaller range engine a sleeve metered fuel pump, unique to Caterpillar, is used.

A sleeve-metered fuel injection pump is shown in Figure 7.9a) and cross-sections of the plunger are shown in Figure 7.9b. The cams and plungers are completely immersed in diesel oil supplied under pressure. The diesel oil can enter the space above the plunger whenever either the fuel inlet port 5 or the fuel outlet port 9 is uncovered. The cam 12 moves the plunger 7 through a constant stroke. The fuel outlet port 9 in the plunger is uncovered earlier or later in the stroke by sleeve 8 controlled by sleeve control lever 10 which is positioned by the engine governor. This controls the spill point of the pump.

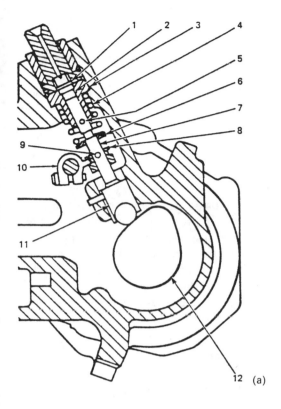

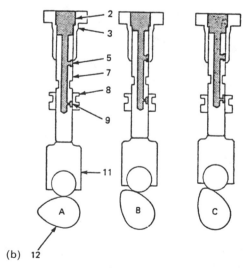

Figure 7.9 *Sleeve-metered fuel injection pump (Caterpillar Tractor Co.)*

1. Reverse flow check valve	6. Retainer	11. Lifter
2. Chamber	7. Plunger	12. Camshaft
3. Barrel	8. Sleeve	A Before injection
4. Spring	9. Fuel outlet (spill port)	B Start of injection
5. Fuel inlet (fill port)	10. Sleeve control lever	C End of injection

Hydraulic governor

When used for alternating current power generation, a diesel engine is normally fitted with a hydraulic governor. This incorporates a centrifugal speed sensing device (spring loaded flyweights) controlling a suitably damped oil operated servo-cylinder through a pilot valve. The governor has adjustable speed droop and load limit controls. A split field, electric motor-operated speeder gear to facilitate remote alteration of engine speed setting from the switchboard is incorporated. This alters the spring pressure usually through a screwed rod.

The Woodward hydraulic governor (Figure 7.10) is briefly described as an example of a commonly fitted type. This has a gear pump driven from the engine camshaft to supply the hydraulic oil first to accumulator pistons, under which is a bypass, to regulate maximum pressure. One branch supplies oil which acts on top of the power piston. The pressure from this supply tends to turn the terminal shaft to shut off fuel. The other branch supplies oil to the pilot valve which is operated by the linkage from the flyweights above.

Should the speed of the engine decrease due to increased load the flyweights (Figure 7.11) will move towards their centre of rotation and lower the position

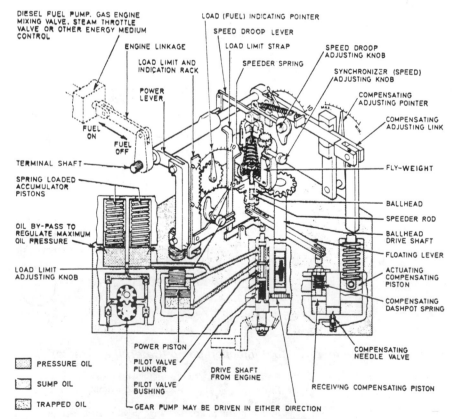

Figure 7.10 *Dial control governor without auxiliary equipment*

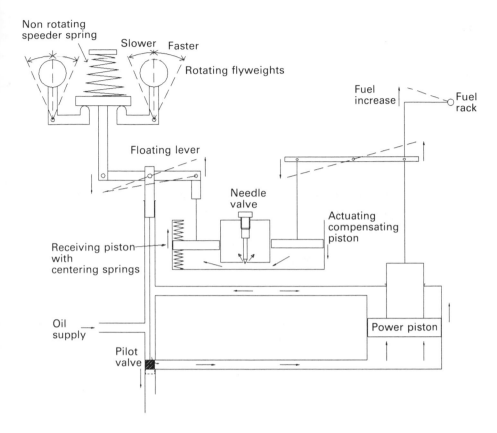

Figure 7.11 *Simplified diagrammatic sketch of Woodward governor*

of the pilot valve plunger, so that oil is admitted via the pilot to the underside of the power piston. The pressure is now equal on the top and bottom of the power piston. Because the area on the bottom is much greater than the top the net resultant force causes the piston to move upwards, so increasing fuel to the engine as the power lever turns the terminal shaft and fuel rack to give later fuel cut-off. As the power piston moves up, the actuating compensating piston moves down. Oil under this piston is now forced through to the receiving compensating piston, raising the outer end of the floating lever and closing the pilot valve. This stops excessive movement of the power piston and fuel rack. As the engine speeds up and the flyweights move out towards their former position, the oil holding the receiving compensating piston leaks through the needle valve. The two movements act on the floating lever without moving the closed pilot valve.

A decrease in the load causes the engine speed to rise so that the flyweights of the speed sensing assembly move outwards, raising the pilot valve through the floating lever. This allows oil to escape from beneath the power piston, so that the engine fuel setting is reduced. As the power piston moves down, the actuating compensating piston moves up causing the receiving compensating piston to move down, taking with it the outer end of the floating lever and closing the pilot valve. This action again stops excessive movement of the power piston and fuel rack. As the engine speed drops, the flyweights move back in towards their former position, while oil leaks through the needle valve allowing the receiving compensating piston to return towards its old position. Again, the two movements act on the floating lever without moving the closed pilot valve.

UMS operation

A survey conducted recently (1990) suggests that about 40% of ships in the world fleet operate with periodically unmanned machinery spaces (UMS). Of the 60% with watchkeepers, most are older ships, some are passenger vessels, or other vessel types where watchkeeping is justified for extra security. A few are ships with control equipment which is defective for various reasons.

UMS machinery spaces have automatic engine change-over in the event of a fault developing on the running machine. Some have programmed control of generators with automatic starting and stopping of stand-by engines as the demand for electrical power rises and falls. Synchronization, opening and closing of breakers, is automatic and load sharing is a function of speed sensing or load sensing governors.

The unattended installations require high dependability which demands intimate knowledge of the machines and strict attention to the maintenance schedule.

Generators driven from the main propulsion system

Generators can variously be driven from the propeller shaft, through a gearbox or by being mounted on the engine itself. Assuming that residual fuel is used in the main engine, then all electrical power at sea is provided at much lower cost, in terms of fuel price and auxiliary generator running hours. The diesel driven generator is needed only while manoeuvring and in port.

Shaft driven direct current generators

Direct current generators are not as sensitive to speed variation as are alternating current machines where frequency has to be maintained. If a direct current generator has an automatic voltage regulator, the output voltage can be maintained even with a 10 or 15 per cent speed reduction. Belt driven or shaft mounted direct current generators with automatic voltage regulators were therefore fitted in ships to save space and to reduce the workload. These machines could continue in operation with moderate speed reduction but auxiliary diesels were brought into use when manoeuvring.

Alternators driven from the main propulsion system

One answer to the frequency problem with alternators, is to supply direct current from a shaft driven direct current generator to a direct current motor and to use this to drive an alternator at constant speed. This arrangement permits moderate main engine speed reduction before a change over to auxiliary generators is necessary. Another solution to maintaining alternating current frequency, relies on the use of a controllable pitch propeller and constant speed engine, rather than one which has to be directly reversed. Manufacturers of electrical equipment have also developed various types of electronic circuits to maintain level frequency through main engine speed changes.

Mechanical constant speed drive from variable speed engine

The system shown (Figure 7.12) uses speed increasing gears to deliver drives from the main engine system to two parts of the installation. One gear train drives a variable delivery hydraulic pump (shown at the bottom). The other drives the planet carrier for the epicyclic gear train A. Rotation of the planet carrier A with the central sunwheel B fixed, causes the annulus C to drive

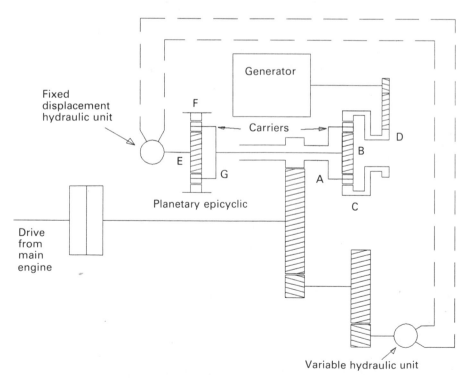

Figure 7.12 *Constant speed shaft generator drive (Vickers type)*

through its output shaft, the gear train for the generator. Any steady rotation of B will affect the generator speed and frequency.

When the speed of the main propulsion system is altered, this is sensed by an electronic device on the generator and the signal is used to control the swashplate for the variable displacement hydraulic pump unit. The output from the latter drives the fixed-displacement hydraulic unit which is connected to the sunwheel E. The annulus for this epicyclic gear is fixed so that rotation of the sunwheel E, drives the planet carrier G and through the shaft, sunwheel B. The speed and direction of B is used to maintain the speed of output shaft D and thus the speed and frequency of the alternating current generator.

Exhaust gas boilers

The original exhaust gas boilers or economizers were of simple construction and produced, from the low powered engines of the time, a very moderate amount of steam. As large slow speed engine powers increased, the larger quantity of steam that could be generated from otherwise wasted exhaust energy, was sufficient finally for provision of the ships entire electrical power requirement through a turbo-alternator, plus any necessary heating steam.

Slow-speed diesel power development has increased engine efficiency but actually reduced the waste heat available to an exhaust gas boiler. Waste heat systems have become more sophisticated (Figure 7.13) in order to continue to supply the electrical requirement and to obtain other economies.

Auxiliary steam turbines

Auxiliary steam turbines are used in turbo-generator sets and also for cargo pump and fan drives. Power outputs vary up to about 1.5 MW for generator sets. The single cylinder turbines can be arranged horizontally or vertically. Both condensing and back pressure turbines have been used, being designed for steam conditions ranging from about 6 bar to about 62 bar at 510°C.

The layout for a closed feed system featured in Chapter 1 shows how turbo-generators and turbine driven cargo pumps are incorporated into a steamship system.

Turbo-generators are also fitted in many motor ships in conjunction with waste heat recovery schemes, based on using the exhaust from very large and powerful slow-speed diesels. Diesel engine builders have developed engines with greater powers in response to the shipowners demand and also in competition with steam turbines, for propulsion. Diesels are now used almost exclusively for modern ships. Only for liquefied gas carriers where the gas boil-off can be burned in the boilers, are steam turbines still being installed.

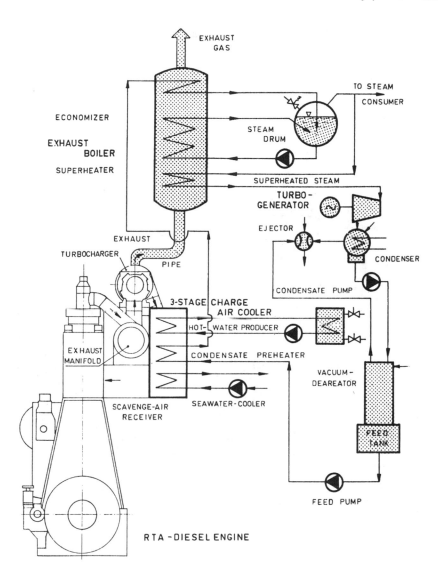

Figure 7.13 *Typical waste heat recovery system (courtesy of Sulzer)*

Turbo-generator construction

For electrical power generation, turbines are conventionally horizontal axial flow machines of the impulse reaction type. They may exhaust either to an integral condenser (invariably underslung) or to a separate central auxiliary condenser or the ship's main condenser. A typical auxiliary condensing turbine is shown in Figure 7.14. Alternatively the turbine may be a back-pressure unit in which the exhaust is used as a source of low pressure steam for other

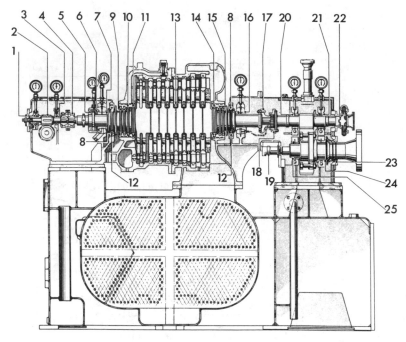

Figure 7.14 *Auxiliary condensing turbine (Peter Brotherhood Ltd)*

1. Pedestal end bearing
2. Oil pump and governor worm
3. Pedestal centre bearing
4. Internal tooth coupling
5. Thrust bearing oil seal
6. Michell thrust bearing
7. Rotor bearing
8. Oil seal labyrinths
9. Outer steam labyrinths
10. Centre steam labyrinth

11. Inner steam labyrinth
12. Oil seal housings
13. Interstage labyrinth packing
14. Inner steam labyrinth
15. Outer steam labyrinths
16. Rotor bearing
17. Gear half coupling turbine end
18. Gear half coupling pinion end

19. Tachometer generator
20. Pinion bearing turbine end
21. Pinion bearing outer end
22. Blower seal
23. Gear shaft oil seal
24. Gear shaft location bearing
25. Gear shaft bearing turbine end

services. The casings, split horizontally and supporting the rotors in plain journal bearings are cast mild steel or, for temperatures exceeding 460°C they are of 0.5% molybdenum steel, with cast or fabricated mild steel for parts not subject to high temperatures. Solid gashed rotors of chrome-molybdenum alloy steel are usual though some may be encountered having rotor spindles of this alloy, with shrunk and keyed bucket wheels. Blades may be of stainless iron, stainless steel or monel metal, with shrouded tips, fitted into the rotors in a number of root forms.

Depending on steam conditions and power the turbine will have a two row velocity compounded stage followed by a suitable number, probably five or more, single row pressure compounded stages, each separated by a cast steel nozzle. Steam enters the turbine at the free end via a cast steel nozzle box and flows towards the drive end which is connected to the pinion of the reduction gearing by a fine tooth or other flexible coupling designed to accommodate longitudinal expansion of the rotor. Typical rotating speed of the rotor is about 6500 rev/min.

The diaphragms separating each stage are split horizontally and fitted in grooves in the casing, to which they are securely fixed so as not to be disturbed when the top half casing is lifted. The diaphragms may be of steel or cast iron depending on the stage pressure.

Interstage leakage, where the rotor shaft passes through the diaphragm, is minimized by labyrinth glands of a suitable non-ferrous alloy such as nickel-bronze. Labyrinth packing may also be used for the turbine shaft/casing glands which are steam-packed. In some turbines contact seals utilizing spring-loaded carbon segments as the sealing media, are used instead of the labyrinth gland (Figure 7.15). A typical labyrinth gland arrangement is shown in Figure 7.16. The low pressure labyrinth is divided into three separate groups so as to form two pockets. The inner pocket serves as an introduction annulus for the gland sealing steam; this flows inwards into the turbine and some escapes through the centre labyrinth into the outer pocket. The supply of sealing steam is regulated to keep the pressure in the outer pocket just above atmospheric. Surplus steam in the outer pocket is usually led to a gland steam condenser. The gland at the high pressure end of the turbine is subject to a considerable pressure range from sub-atmospheric at low load to considerably above atmospheric at full load and is therefore arranged with three pockets. Gland steam is supplied to the centre pocket. The innermost pocket is

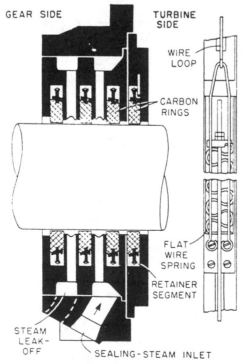

Figure 7.15 *Example of carbon ring shaft seal*

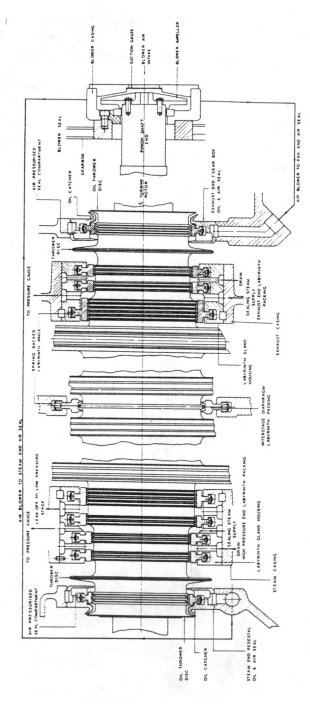

Figure 7.16 Typical labyrinth gland arrangement with air sealing system (Peter Brotherhood Ltd)

connected to a lower pressure stage further down the turbine, enabling the leakage steam to rejoin the main stream and do further work while the outermost pocket, connected to the gland condenser, prevents excessive leakage to atmosphere.

The labyrinth packings at both ends of the turbine and in the diaphragms are retained by T-heads on the outer peripheries which slot into matching grooves. Each gland segment is held in position by a leaf spring. The retaining lips of the T-head prevent inward movement and the arrangement permits temporary outward displacement of the segments. Rotating of the segments is prevented by stop plates or pegs fitted at the horizontal joint.

Although there is little residual end thrust on the rotor it is necessary to arrange a thrust bearing on the rotor shaft and it is normal to make this integral with the high-pressure end journal bearing. Sometimes the thrust is of multi-collar design but is more frequently a Michell-type tilting pad bearing.

Governing

Unlike propulsion turbines, generator turbines work at constant speed and must be governed accordingly. Classification Society rules require that there must be only a 10% momentary and a 6% permanent variation in speed when full load is suddenly taken off or put on. On an alternating current installation it is required that the permanent speed variations of machines intended for parallel operation must be equal within a tolerance of $\pm 0.5\%$. In addition to the constant speed governor an overspeed governor or emergency trip is also fitted.

Speed-governing system

Speed governing systems consist of three main elements:

1   A speed sensing device, usually a centrifugal flyweight type governor driven through worm and bevel gearing from the turbine shaft.
2   A linkage system from the governor to the steam and throttle valve; on larger turbines this is an oil operated relay consisting of a pressure balanced pilot valve controlling a supply of high pressure oil to a power piston.
3   A double-beat balanced steam throttle valve which regulates the amount of steam passing to the turbine nozzles, according to the speed and electrical load.

To ensure stability, that is freedom from wandering or hunting of the speed, the system is designed to give a small decrease in speed with increase in load. The usual amount of this decrease, called the 'speed droop' of the governor, is 3% between no load and full load. If the full load is suddenly removed, there will be a momentary speed increase to a value of 7–10% above normal before it returns to a value of 3% above normal above the full load speed (the droop value). Similarly if the full load is suddenly applied, a momentary fall in speed of 4–7% below normal will occur before recovery.

Figure 7.17 is a simplified schematic arrangement of a typical speed governing system from which the sequence of events during load changes may be more easily followed. In the diagram the throttle valve is operated via lever Y.

Overspeed trip

Overspeed occurs when the load is suddenly thrown off and while this is normally rectified by the speed governor, an emergency trip is always fitted. A common type is illustrated in Figure 7.18.

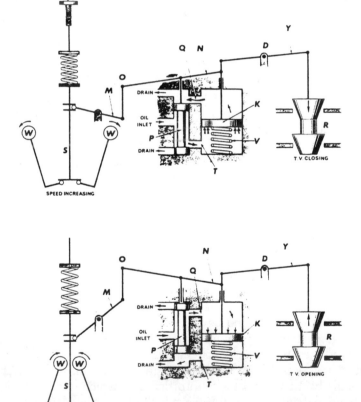

Figure 7.17   *Schematic arrangement of speed governing system (Peter Brotherhood Ltd.)*

D. Fulcrums	O. Fulcrum	S. Sleeve
C. Adjusting spring	P. Pilot valve	T. Port
G. Hand-operated wheel	Q. Port	V. Spring
K. Piston	R. throttle valve	W. Weights
M,N. Levers		

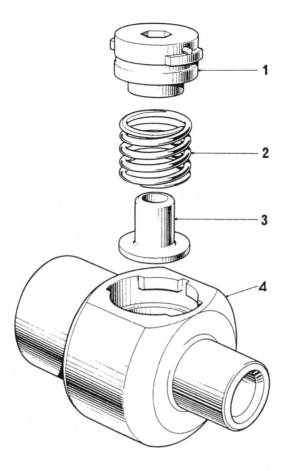

Figure 7.18   *Overspeed trip gear (Peter Brotherhood Ltd)*

1. Cap
2. Emergency valve spring

3. Emergency valve
4. Casing

An unbalanced steel valve 3, located in the pinion shaft extension, is held on to the valve seat by a helical spring 2, while the speed of pinion shaft remains below tripping speed. If the speed increases 10–15% above the turbine rated speed, the centrifugal effect on the trip valve, overcomes the spring force and the valve lifts rapidly from the valve seat. This allows lubricating oil, fed to the centre of the shaft extension through an orifice plate, to escape. Oil system pressure drops to zero downstream from the orifice and this causes the low pressure oil trip to operate and drain oil from the relay cylinder. The relay cylinder spring raising the relay piston and closing the throttle valve cuts off the steam supply to the inlet of the turbine. It is vital to maintain the trip gear in good working order and this can be greatly aided by testing at regular intervals.

In addition to an overspeed trip it is customary to fit a low pressure oil trip to steam turbines and frequently a back pressure trip (Figure 7.19) is fitted.

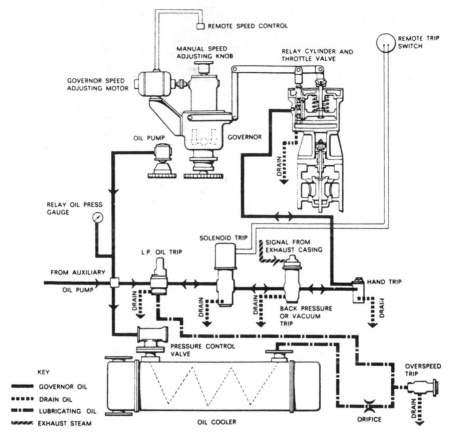

Figure 7.19 *Governor control oil system (Peter Brotherhood Ltd)*

Back-pressure trip

The purpose of the back-pressure trip is to protect the turbine and exhaust system in the event of over-pressure due to loss of condenser cooling water, extraction pump failure or accidental closure of the exhaust valve. The trip consists of a spring loaded bellows, connected to a hydraulic spool valve interposed in the governing oil circuit between the main oil pump and the governor relay.

When the turbine is running the valve is held upwards by the spring. In this position high pressure oil can pass freely across the upper ports of the valve to actuate the governor relay. If the back pressure increases to a predetermined level the load on the bellows unit is sufficient to overcome the adjusting spring and allow the operating spindle to move downwards and push the ball valve off its seat. In so doing, oil at relay pressure is admitted through the drilled passages in the trip body to the piston valve, so depressing the valve against its spring. The valve will, simultaneously, cut off and drain the high pressure supply to the governor relay. The throttle valve is consequently closed by the relay pistons under the action of the spring V (Figure 7.17) and the turbine is stopped.

Back-pressure turbines

Many ships have used an auxiliary steam turbine as a primary pressure reducing stage before passing the steam to other auxiliaries demanding steam at a substantially lower pressure than that available. Such an arrangement (Figure 7.20) gives a heat balance which is far more favourable than that obtained with a pressure reducing valve.

Back pressure turbines have most of the features of a condensing turbine but no condenser. The most important difference is in the governing. Designed to work against back pressures in the range 1–3.5 bar and with much lower available heat drops than with a condensing turbine, governing by the simple opening or closing of the throttle valve is inadequate. Instead the governor

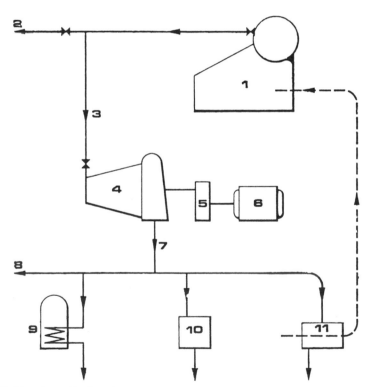

Figure 7.20 *Installation for back-pressure turbo-generator (Peter Brotherhood Ltd.)*

1. Main high pressure boilers	4. Auxiliary turbine	8. Steam to ship's
2. H.P. steam to main propulsion	5. Reduction gear	services
turbines	6. Generator	9. Evaporator
3. H.P. steam to back pressure	7. Auxiliary turbine exhaust	10. De-aerator
turbo-generator		11. Steam/air heater

(Figure 7.21) sequentially controls the opening of a number of nozzle control valves. The control system is arranged for a straight line regulation (load/speed) with a speed droop of about 4% between full load and no load. The turbine governor also incorporates speed droop control adjustment with a range of approximately 2.5% to 5.5% to enable load sharing between a group of generators to be readily adjusted.

Vertical steam turbines

Vertical steam turbines are extensively used for cargo, ballast and other pump drives. Like the horizontal machines used for power generation they can be condensing or back-pressure units. They are, however, invariably single stage machines having an overhung wheel (Figure 7.22). The steam casing of this turbine is a simple steel casting bolted to the top of the exhaust casing. The nozzles are fitted and seal welded into the underside of the steam casing inlet belt forming a ring which provides an uninterrupted arc of admission. The exhaust casing is split in the vertical plane, allowing removal of the front half for rotor inspection without disturbing the steam or exhaust piping.

The rotor shaft is bolted to a head flange on the pinion shaft of the single reduction gearing. A thrust bearing located below the pinion supports the weight of the rotor and absorbs any vertical thrust. This bearing is usually of the Michell multi-pad type.

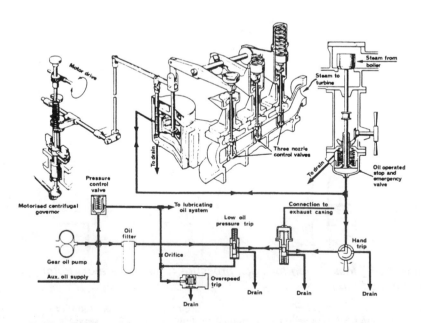

Figure 7.21 *Back-pressure governor and control system (Peter Brotherhood Ltd.)*

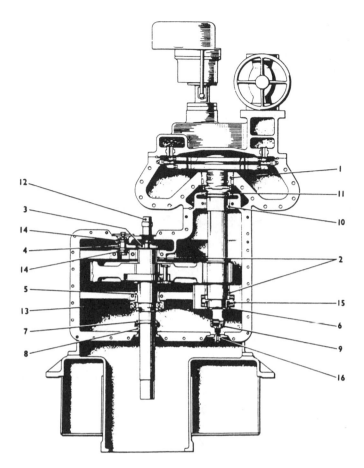

Figure 7.22 *Sectional arrangement of vertical turbine (Peter Brotherhood Ltd.)*

1. Labyrinth gland housing
2. Pinion and gear shaft bearing
3. Spur gear (govnr. and oil pump drives)
4. Idle gear (govnr. and oil pump drives)
5. Gear wheel shaft bearing
6. Thrust bearing oil seal
7. Oil thrower gear shaft
8. Oil seal gear shaft
9. Overspeed trip unit
10. Pinion shaft upper bearing
11. Pinion shaft upper oil seal
12. Tachometer generator
13. Gear shaft location bearing
14. Idler shaft bearings
15. Pinion thrust bearing
16. Trip oil inlet fitting

Further reading

Hensel, W. (1984) Energy saving in ships' power supplies, *Trans I Mar E,* **96,** paper 49.

McGeorge, H. D. (1993) *Marine Electrical Equipment and Practice,* 2n edn. Newnes (Butterworth-Heinemann).

Mikkelsen, G. (1984) Auxiliary power generation in today's ships', *Trans I Mar E,* **96,** paper 52.

Mitchell, R. W. S. and Kievits, F. J. (1974) Gas Turbine Corrosion in the Marine

Environment. Proceedings of a Joint Conference Corrosion in the Marine Environment, *Trans I Mar E* series B.

Murrell, P.W. and Barclay, L. (1984) Shaft driven generators for marine application, *Trans I Mar E*, **96** paper 50.

Pringle, G. G. (1982) Economic power generation at sea: the constant speed shaft driven generator, *Trans I Mar E*, **94**, paper 30.

Schneider, P. (1984) Production of auxiliary energy by the main engine, *Trans I Mar E*, **96**, paper 51.

8

The propeller shaft

The simplistic view of the main propulsion shaft installation is that the system is set up with initial straight alignment and remains in that state during the lifetime of the ship, unless affected by accident or wear. The reality is that there are many factors which can affect and alter alignment during building and throughout the working lifetime of a vessel.

Establishing the shaft centre line

Optical (or laser) equipment can be used to establish the centre line of the shafting system, to give a reference for cutting through bulkheads and machining of the aperture in the stern frame. One method employs a telescope with crosswires, set up on the shaft centre line at the forward end of the double bottom engine platform with a plain cross wire target on the same axis at the after end of the engine seating. With both in use, the centre of engine room and aft peak bulkheads can be located and marked prior to cutting holes for the shaft. The required centre of the aperture in the stern frame boss, can then be found by line of sight, using a crosswire in an adjustable spider. Replacement of the crosswire by a plug with a centre gives a location for the divider to be used when marking off the boss for boring. Importantly, the telescope and crosswire target method can also be used to check the accuracy of the boring operation, work on the installation of the stern tube and siting of shaft bearings. Some arrangements as for split stern tubes, involve the welding in of the boss and this operation can be guided by constant checking with optical equipment.

Deviation while building

With the ship under construction still firmly supported, faults causing shaft misalignment can and do occur. The stern tube aperture can be incorrectly machined due to flexure of the boring bar or human error. Any contraction or expansion of the hull as a result of temperature variation can conspire with changes caused by welding of the hull to effect change of hull shape. The welding in place of a fabricated stern tube requires constant checks to ensure alignment is maintained. Some stern tube bearing failures have been traced to alignment errors which should have been detected and remedied during installation.

After fitting the stern tube and propeller shaft, the propeller is mounted. The considerable weight of the propeller however, causes droop in the tailshaft and potential edge loading of the stern tube bearing. Arching tends to lift the inboard end of the propeller (or tail) shaft so that the next bearing forward whether in the stern tube or beyond, would tend to be negatively loaded. Deformation imposed by the propeller mass, remains even after installation of the rest of the shaft system.

The remedy for edge loading due to propeller shaft droop, is to arrange for the stern tube bearing to be slope bored or installed with a downward lie. Shaft weight is then fully supported along the bearing surface.

After the ship has been launched, the immersed section of the heavily framed stern with the propeller mass, being much less buoyant than the full hull further forward, flexes downward. This emphasizes the droop of the propeller shaft and resulting inherent misalignment. Downward flexing of the stern also deforms the hull, changing the line of the tank top. It was the normal practice to install the intermediate shafting after the launch, when the ship assumed its in-water shape. The shafting was installed from the tailshaft to the engine.

Optical equipment, as before, could be used to check the position of the propeller shaft inboard flange and to locate the centres of plumber bearings. Chocks for the shaft bearings are machined to the correct height.

Traditionally, the fairing of couplings has been used to align shafts and to check the alignment of adjacent shaft sections. The fairing of couplings involves the insertion of feelers between a pair of couplings to check that they are parallel and the use of a straight edge or dial gauge, to ensure that they are concentric. Incorrect alignment can result if it is assumed that shaft sections are rigid; particularly with the heavy shaft sections for engines of high power. Account must be taken of slight droop due to elasticity and overhanging weight at each shaft flange. The natural deformation of shaft sections is taken into account with rational alignment programmes and coupling conditions can be used to position shaft sections and to check alignment. For this procedure, pre-calculation is used to find the gap and sag that should exist between couplings, when shaft alignment is correct.

Alignment deviation in service

Shaft line is continually changed through the lifetime of a ship, as the hull is distorted by hog or sag due to different conditions of loading. The weight and distribution of cargo, ballast, fuel and fresh water are all subject to change and the changes are known from experiment to affect shaft alignment. (Incorrect cargo discharge procedures and resulting excessive hull stresses have resulted in ships actually breaking in two.)

High deck temperature in the tropics or low sea temperatures can cause differential expansion and hogging of the hull. These types of change can alter crankshaft deflection or shaft alignment readings which are taken even a few hours apart. Heavy weather produces cyclic change of hull shape so that the hull of a moderately sized ship can flex by as much as 150 mm. There are also local factors which alter shaft alignment. Thus the forward tilt of a loaded thrust block, and the lift of its after bearing, causes misalignment of the shaft and possible uneven loading of gear teeth. A build-up of fluid film pressure in bearings, as the shaft starts to rotate, lifts the shaft bodily. Sinkage of individual plumber blocks could be another problem.

Fair curve alignment

The method of fair curve alignment (using a computer programme developed at the Boston Navy Yard in 1954 and refined by others) accepts the changes of line endured by the shaft system and seeks a compromise installation to suit the varying conditions.

The initial calculation is to determine the load on each bearing, assuming all bearings to be in a straight line. The computer programme then simulates the raising of each bearing through a range and calculates, for each small change, the increase of its own load and alteration in load on each of the other bearings. The process is then repeated with a simulation of the lowering of each bearing in turn with the computer finding resultant load changes on the bearing in question and the others. Influence numbers, in terms of load change for each height variation, are calculated by this exercise for all bearings.

The data bank of influence numbers enables the effects of changes in alignment from hull flexure and local factors to be found. All of the variables described above can be assessed to determine the best compromise for shaft installation.

Shaft checks

The intention of good alignment is to ensure that bearings are correctly loaded and that the shaft is not severely stressed. Alignment can be checked with conventional methods, employing light and targets, laser or measurements from a taut wire. There is, however, a continuity problem because the line of sight or taut wire cannot extend over the full length of an installed shaft. There is no access to that part of the shaft within the stern tube and access is difficult in way of the propulsion machinery. Results are also uncertain unless the vessel is in the same condition with regard to loading and hull temperatures as when the shaft system was installed or previously checked.

The method of jacking (Figure 8.1) to assess correct bearing loads, is used as a realistic means of ensuring that statically, the shaft installation is satisfactory. In simple terms the load on each bearing can be stated as the total weight of the shaft divided by the number of bearings. The figure for designed load is normally given in a handbook with the usual permitted deviation of plus or minus 50%. The permitted variation may be less for some bearings.

The procedure involves the use of hydraulic jacks placed on each side of the bearing, to lift the shaft just clear. A dial gauge fixed to the bearing indicates lift. Hydraulic pressure, exerted by the jacks, registers the load on the bearing.

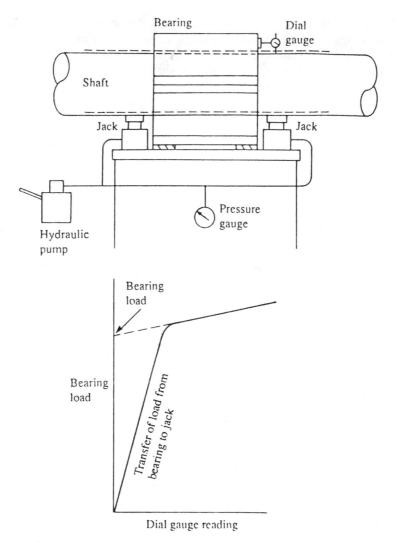

Figure 8.1   *Method for checking bearing load by jacking (R. C. Dean)*

A plot of lift and load made while hydraulically lifting the shaft, shows a distinctive pattern, due to the elasticity of the steel and removal of deformation from the bearing.

As the hydraulic jack pressures are raised from zero, the concentrated loading initially causes deformation of the shaft. Only after the journal section has been bowed up out of shape to some degree, and the bearing material resumes its relaxed primary shape, does the sagging centre part of the journal lift clear and out of contact with the bearing. The plot shows that the dial gauges register upward movement as soon as the shaft is pushed out of shape by increasing hydraulic pressure. The curve takes a different shape as the shaft lifts clear.

If the jacking is taken too far, then adjacent bearings gradually become unloaded and the plot is affected by a change in the elastic system. To guard against this, dial gauges are fixed on adjacent bearings to ensure that the lift is limited to the bearing that is being checked.

Strain gauges

Shaft stress is sometimes monitored in service, by fitting strain gauges on the shaft. These register alternating surface stretch and compression as the shaft rotates.

Change of engine position

The conventional midships position for the engines of older vessels, with the exception of tankers, was based on low engine power and strong hull construction. Shafts were long, but being of moderate diameter, were able to flex with the hull as loading or other conditions changed (and in heavy weather). A loading or ballast condition which changed hull shape and shaft alignment to an unusual degree, sometimes caused higher temperature in some bearings due to uneven load distribution. Shaft stress was the hidden factor.

The trend towards higher engine powers and the positioning of engines aft, gave rise to large diameter, short length shafts of increased stiffness. Excessive vibration and resulting damage in many dry cargo and container vessels is a common feature as a result. Hull detuners intended to reduce vibration have been fitted in steering gear compartments but the improvement to many ships seems to be marginal. Hull vibration seems to be less of a problem in ships with one cargo compartment aft of the machinery space.

Shaft bearings

The intermediate shafting (Figure 8.2) between the tailshaft and main engine, gearbox or thrustblock may be supported in plain, tilting pad or roller bearings. The two former types usually have individual oil sumps, the oil being circulated by a collar and scraper device; roller bearings are grease lubricated. The individual oil sumps usually have cooling water coils or a simple cooling water chamber fitted. Cooling water is provided from a service main connected to the sea-water circulating system. The cooling water passes directly overboard.

Usually for plain and tilting pad bearings, only a bottom bearing half is provided, the top acting purely as a cover. The aftermost plumber block however, always has a full bearing. This bearing and any bearing in the forward end of the stern tube, may be subject to negative loading.

Plain bearings

Any oil between a static shaft and plain journal bearing in which it rests, tends to be squeezed out so that there is metal to metal contact. At the start of the

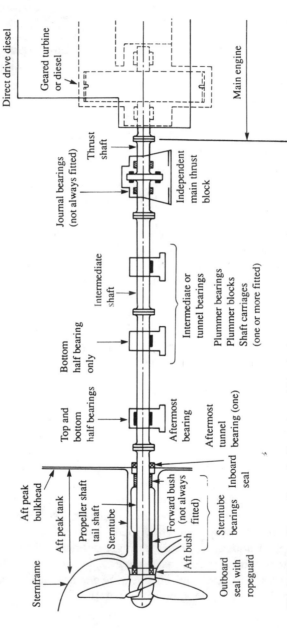

Figure 8.2 Propeller shaft system (Glacier)

rotation the journal is inclined to roll up the bearing surface against the direction of rotation until friction slip occurs. Then, provided there is oil in the clearance space, this will cling to the moving surface and be dragged between the shaft and the bearing. Shaft rotation, as it speeds up, continues to carry oil to the shaft underside so developing a film with sufficient pressure to hold the shaft clear of the bearing. The pressure build-up is related to speed of rotation. Thus oil delivered as the shaft turns at normal speed, will form a layer or film, separating shaft and bearing, and so prevent direct wear of metal to metal. Pressure generated in the oil film, is most effective over about one third of the bearing area (Figure 8.3) because of oil loss at the bearing ends and peripherally. Load is supported and transmitted to the journal, mainly by the area where the film is generated.

Tilting pad bearings

Replacement of the ineffective side portions of the journal by pads capable of carrying load will considerably increase its capacity. Tilting pads based on those developed by Michell for thrust blocks (Figure 8.4a) are used for the purpose. Each pad tilts as oil is delivered to it, so that a wedge of oil is formed. The three pressure wedges give a larger total support area than that obtained with a plain bearing. The arrangement of pads in a bearing is shown in Figure 8.4b.

The tilt of the pads automatically adjusts to suit load, speed and oil viscosity. The wedge of oil gives a greater separation between shaft and bearing than does the oil film in a plain journal. The enhanced load capacity of a tilting pad design, permits the use of shorter length bearings or fewer bearings.

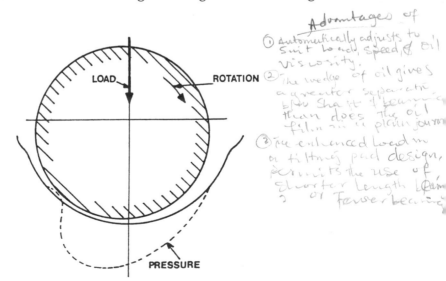

Figure 8.3 *Fluid film pressure in plain bearings*

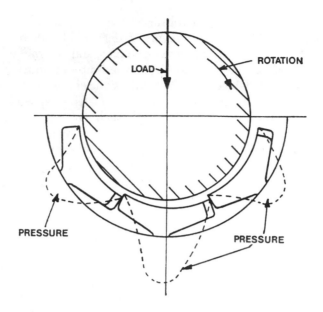

Figure 8.4a   *Tilting pad bearing*

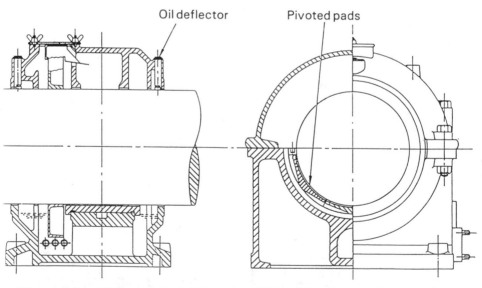

Figure 8.4b   *Tilting pad shaft bearing (Michell Bearings Ltd)*

Roller bearings

Roller bearings (Figure 8.5) are supplied in sizes to suit shafts up to the largest diameter. Flange couplings dictate that roller bearing races must be in two parts for fitting.

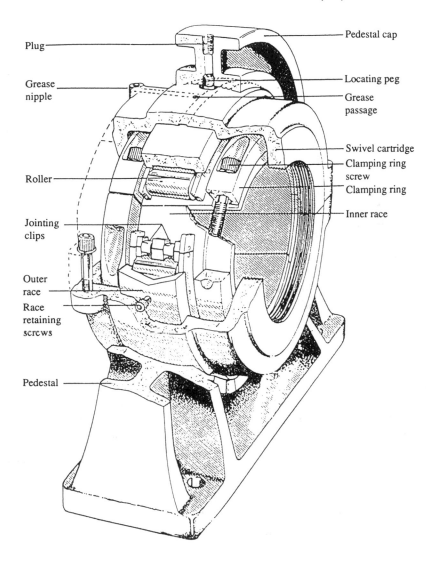

Plug

Grease nipple

Roller

Jointing clips

Outer race

Race retaining screws

Pedestal

Pedestal cap

Locating peg

Grease passage

Swivel cartridge

Clamping ring screw

Clamping ring

Inner race

Figure 8.5 *Roller bearings (courtesy of The Cooper Roller Bearing Company)*

The section of the shaft where the split roller bearing is to be fitted, must be machined very accurately and with good finish. The two halves of the inner and outer races are fitted and held with clamping rings.

Adequate speed for build-up of fluid film pressure is vital for journal bearings. At low speeds there may be metal to metal contact with wear and damage. Friction at low rotational speeds, is high. Roller bearings are not dependent on speed for effective lubrication. Friction is low at all speeds. This makes them suitable for steam turbine installations and in ships where slow steaming may be necessary. Roller bearings, where fitted, are grease-lubricated.

Thrust blocks

The main thrust block transfers forward or astern propeller thrust to the hull and limits axial movement of the shaft. Some axial clearance is essential to allow formation of an oil film in the wedge shape between the collar and the thrust pads (Figure 8.6) This clearance is also needed to allow for expansion as parts warm up to operating temperature. The actual clearance required, depends on dimensions of pads, speed, thrust load and the type of oil employed. High bearing temperature, power loss and failure can result if axial clearance is too small.

A larger than necessary clearance will not cause harm to the thrust bearing pads, but axial movement of the shaft must be limited for the protection of main machinery.

The accepted method of checking thrust clearance, involves jacking the shaft axially to the end of its travel in one direction and then back to the limit of travel in the other. Total movement of the thrust shaft (about 1 mm being typical) is registered on a dial gauge. Feelers can be used as an alternative, between thrust ring and casing. Use of feelers in the thrust pad/collar gap is likely to cause damage and may give a false reading.

Thrust block position

The siting of the main thrust block close to the propulsion machinery, reduces any problems due to differential expansion of the shaft and the hull. The low hull temperature of midship engined refrigerated cargo ships, caused a contraction relative to the shaft of perhaps 20 mm ($\frac{3}{4}''$). Variations can be caused by changes in water temperature or heating of fuel tanks. Other problems associated with the stern tube end of the shafting system include whirl of the tailshaft, relative movement of the hull and misalignment due to droop from propeller weight. Some thrusts are housed in the after end of large slow speed diesels or against gear boxes. Deformation produced by the thrust load, can cause misalignment problems, unless suitable stiffening is employed (particularly with an end of gearbox installation).

Thrust block support

The substantial double bottom structure under the main propulsion machinery, provides an ideal foundation for the thrust block and a further reason for siting it close to the engine. The upright thrust block and any supporting stool, must have adequate strength to withstand the effect of loading which tends to cause a forward tilt. This results in lift of the aft journal of the block (unless not fitted) and misalignment of the shaft.

Axial vibration of the shaft system, caused by slackening of the propeller blade load as it turns in the sternframe or by the splay of diesel engine crankwebs, is normally damped by the thrustblock. Serious vibration problems

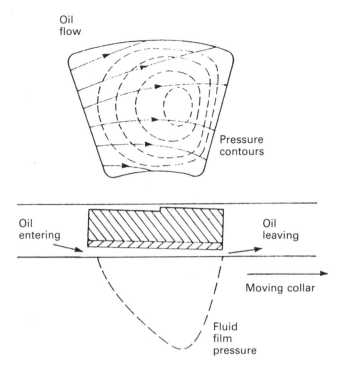

Figure 8.6 *Michell thrust pad*

have sometimes caused thrust block rock, panting of the tank top and structural damage.

Thrust pads

The pivot position of thrust pads may be central or offset. Offset pads are interchangeable in thrust blocks for direct reversing engines, where the direction of load and rotation changes. Offset pads for non-reversing engine and controllable pitch propeller installations are not interchangeable. Two sets are required. Pads with a central pivot position are interchangeable. Some modern thrust blocks are fitted with circular pads (Figure 8.7) instead of those with the familiar kidney shape. A comparison of the pressure contours on the conventional kidney shaped pads and the circular type shows why the latter are effective.

Shaft materials and couplings

The intermediate shafting and the propeller shaft for a fixed propeller are of solid forged ingot steel and usually with solid forged couplings. Shafts are

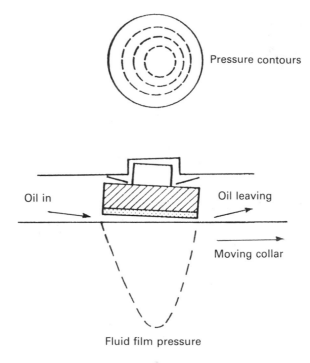

Pressure contours

Oil in

Oil leaving

Moving collar

Fluid film pressure

Figure 8.7 *Circular thrust pads*

machined all over but of a larger diameter and smooth turned in way of the bearings. The faces of flanged couplings (except where undercut in the centre area) are also smooth turned, with bolt holes carefully bored and reamered to give an accurate finish. Torque is transmitted by the friction between flanges and also through the shanks of the bolts. Each tightened bolt holds the flanges hard together in the area local to it. A circle of bolts is needed for a good all round grip. The design of flange couplings can be checked by formulae given in Lloyds or other classification society regulations.

Coupling bolts

The elongation of a bolt when tightened, causes a reduction in cross sectional area. The relationship between change in length and the change of cross sectional area is summarized by Poisson's Ratio. In a clearance bolt, this is not a problem, but with a normal fitted bolt, positive contact between the accurately machined bolt and the reamered hole is lost when the bolt is tightened. An oversize bolt could of course be used and cooling of the shank – probably with liquid nitrogen – would be necessary to cause contraction and reduction of cross sectional area before insertion. The effect of low temperature and the possibility of the steel becoming brittle as the result of the cooling must be considered.

Shaft coupling bolts are tightened to force the faces of the flanges together, so that friction between the faces will provide some proportion of the drive. However, fitted bolt shanks are also designed to take some load. A clearance bolt could provide the first requirement but not the second. A normal fitted bolt when tightened and subjected to a reduction in cross section, would also fail on the second count and probably be damaged by fretting. A tapered bolt (Figure 8.9) could be used instead of a conventional coupling bolt (Figure 8.8) to obtain a good fit and the required tightening.

The Pilgrim hydraulic bolt uses the principle embodied in Poisson's Ratio to provide a calculated and definite fitting force between bolt and hole. The bolt (Figure 8.10) is hollow and before being fitted is stretched with hydraulic

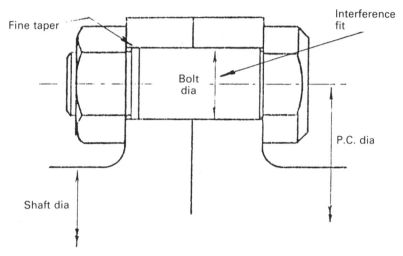

Figure 8.8 *Conventional coupling bolt*

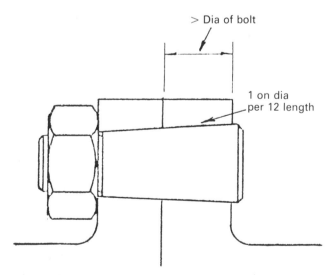

Figure 8.9 *Tapered coupling bolt*

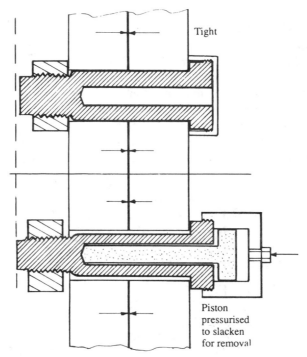

Tight

Piston
pressurised
to slacken
for removal

Figure 8.10   *Pilgrim type hydraulic coupling bolt*

pressure applied to an inserted rod from a pressure cylinder screwed to the bolt head. Stretching makes the bolt diameter small enough for insertion into the hole, after which the nut is nipped up. Release of hydraulic pressure allows the bolt to shorten, so that (1) predetermined bolt load is produced and (2) diametrical re-expansion gives a good fit of the shank in the hole. These bolts, when used in flange couplings and flange mounted propellers, have the advantage that they are easily removed for inspection and maintenance.

Muff coupling

An alternative to the conventional flange couplings for the tailshaft, the muff coupling allows the shaft to be withdrawn outboard. The SKF coupling (Figure 8.11) consists basically of two steel sleeves. The thin inner sleeve has a bore slightly larger than the shaft diameter and its outer surface is tapered to match the taper on the bore of the outer sleeve. The nut and sealing ring close the annular space at the end of the sleeves.

When the coupling is in position, the outer sleeve is hydraulically driven on to the tapered inner sleeve. At the same time, oil is injected between the contact surfaces to separate them and thus overcome the friction between them. Oil for the operation is supplied by hand pumps; two for the forced lubrication and another hand or power pump for the driving oil pressure.

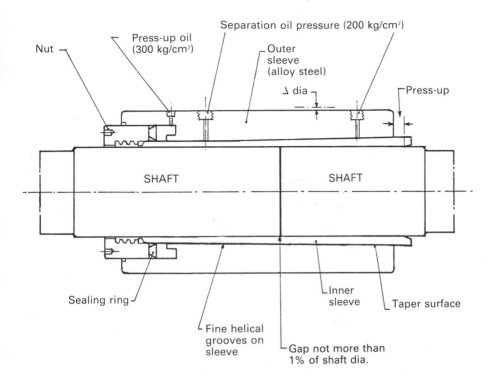

Figure 8.11 *SKF (muff) coupling*

When the outer sleeve has been driven on to a predetermined position, the forced lubrication pressure is released and drained. Oil pressure is maintained in the hydraulic space until the oil between the sleeves drains and normal friction is restored. After disconnecting hoses, plugs are fitted and rust preventive applied to protect exposed seatings. A sealing strip is pressed into the groove between the end of the sleeve and the nut.

The grip of the coupling is checked by measuring the diameter of the outer sleeve before and after tightening. The diameter increase should agree with the figure stamped on the sleeve.

To disconnect the coupling, oil pressure is brought to a set pressure in the hydraulic space. Then with the shafts supported, oil is forced between the sleeves. The outer sleeve slides off the inner at a rate controlled by release of the hydraulic oil pressure.

Stern tubes

The propeller shaft (or tailshaft) is supported in a stern tube bearing of one of a number of designs. The bearing, being at the end of the shaft, is affected by the overhanging weight of the propeller. The propeller mass pulls the outer end of the shaft down, so that there is a tendency for edge loading of the stern tube bearing to occur. The forward part of the propeller shaft is tilted upwards. Weardown of the bearing aggravates this misalignment and whirl due to weardown may give additional problems.

Sea-water lubricated stern tube bearings

The traditional stern bearing (Figure 8.12) is water-lubricated and consists of a number of lignum vitae staves held by bronze retaining strips, in a gunmetal bush. Lignum vitae is a hardwood with good wear characteristics and is compatible with water. The staves in the lower part of the bearing, are cut and fitted so that the end grain is vertical to give the longest possible life. Staves in the upper part are cut with grain in the axial direction for economy. The staves are shaped with V or U grooves between them at the surface, to allow access for water. The grooves also accommodate any debris. As an alternative to wood, reinforced rubber or Tufnol can be used. Bearing length is equal to four times shaft diameter.

Stern tubes (Figure 8.13) are supported at the after end by the stern frame boss and at the forward end in the aft peak bulkhead. Their cast iron construction requires strong support in way of the bearing, from the stern frame boss. A steel nut at the outboard end retains the tube in position, with its collar hard against the sternframe and the bearing section firm within the stern frame boss. Welded studs hold the forward flange against the aft peak bulkhead.

Sea water, which enters at the after end or from the circulation system to cool and lubricate, is an electrolyte which will support galvanic corrosion.

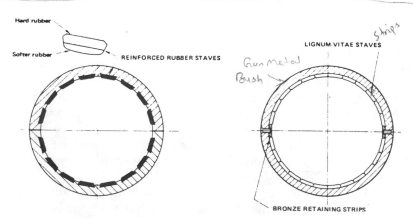

Figure 8.12   *(left) Rubber stave bearing (right) Lignum vitae bearing (Glacier Metal Co.)*

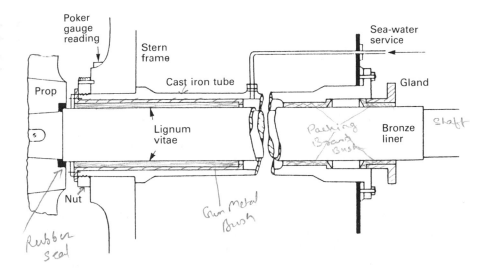

Figure 8.13 *Sea-water lubricated stern tube*

Wastage of the vulnerable steel shaft is prevented by a shrunk-on bronze liner and rubber seal sandwiched between the propeller hub and the liner end. It is essential that the rubber has freedom to flow when nipped between the hub and liner.

Excessive weardown of bearing materials due to vibration or whirl, poor quality of work when rewooding, inferior materials, presence of sand/sediment in the water or propeller damage, could necessitate early rewooding. The life of the bearing for vessels with engines aft, and particularly tankers and ore carriers which spend long periods in ballast, has been short with rewooding being needed in perhaps eighteen months.

The centre of the stern-tube is connected to a sea-water service line which, together with ingress of water between the shaft and bush, provides the cooling and lubrication. A packed gland seals the forward end of the bearing and is adjusted to permit a slight trickle of water along the shaft and into the tunnel well where it is regularly removed with the bilge pump. Bearing clearances are liberal both to accommodate the swelling which occurs when the staves are immersed in water and to permit the essential flow of water through the bearing.

A large number of vessels with water-lubricated bearings are still in service and they continue to be installed.

Inspection of sea-water lubricated stern tubes and tailshaft

During drydock inspection, bearing weardown is measured by poker gauge or by inserting a wedge between the shaft and bearing from the outside. The permissible wear is in the region of 9–12 mm on large diameter shafts.

The examination of the type of tailshaft described above requires removal of the propeller and inward withdrawal of the propeller shaft. The operation calls for the erection of staging, use of a large, suspended ram or tup for the spanner to slacken the nut and wedges to start the propeller. The nut remains on the thread after being slackened for safety reasons. Accidents have been caused by the sudden loosening of propellers with no nut in place to act as a stop. Timber between the aft peak bulkhead and the flange at the forward end of the tailshaft, supports the shaft against the action of the wedges. The examination when a tailshaft has operated in a sea-water lubricated bearing and where the propeller is keyed, may reveal (Figure 8.14) a number of defects. There is a potential for cracks in the keyway area but the likelihood of these occurring has been reduced by the employment of sled type keys, radiused corners within the keyway and spooning at the forward end.

A plain keyway milled in a shaft taper, is a weakening factor which allows deformation of the surface when push up is applied to the propeller and where there is any transmission of torque from the shaft via the key to the propeller hub. Torque causes a deformation which tends to open the keyway.

The rubber seal sandwiched by the propeller hub and protective bronze liner, prevents ingress of sea water which would act as an electrolyte to promote galvanic corrosion of the exposed part of the shaft. A defective seal permits corrosion and wastage. Fretting of the steel shaft tends to occur beneath the forward end of the propeller hub or under the after end of the liner. Any pitting or marking of the shaft surface in the area (or notch) between the propeller hub and the bronze liner can initiate a fatigue or corrosion fatigue crack in this vulnerable area. (Shaft droop from the overhanging weight of the propeller, stretches the upper surface and compresses the lower, to produce alternating stress when the shaft is rotating. The imposed alternating effect, likely to cause fatigue, is of a low frequency and high stress.)

The shrunk-on bronze liner, fitted to protect the steel shaft against black corrosion may itself be damaged by working conditions. Shaft whirl can lead to patches marked by cavitation erosion, scoring occurs due to the stern gland packing and liner cracking has sometimes penetrated through to cause corrosion cracking in the shaft.

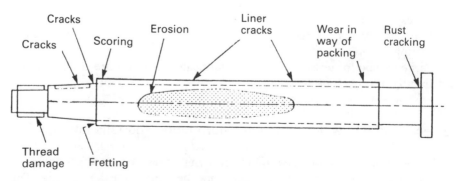

Figure 8.14   *Propeller shaft faults*

Oil-lubricated stern tubes

Progress from sea-water to early oil-lubricated stern tubes involved an exchange of the wooden bearing in its bronze sleeve for a white metal lined cast iron (or sometimes bronze) bush. Oil retention and exclusion of sea water necessitated the fitting of an external face type seal. The stuffing box was retained in many early oil-lubricated stern tubes, at the inboard end. In oil-lubricated bearings the shaft does not require a full length protective bronze sleeve.

Simplex type stern tube

The later designs of oil-lubricated stern tube (Figure 8.15) are fitted in a stern frame with an elongated boss to provide better support for the white metal lined bearing. A minimum bearing length of two times shaft diameter will ensure that bearing load does not exceed 0.8 N/mm² (116 lbf/in²). The bearing bush is normally grey or nodular cast iron centrifugally lined with white metal. A typical analysis of white metal would be 3% copper, 7.5% antimony and the remainder tin. White metal thicknesses vary according to the classification society. Figures of 3.8 mm for a shaft of 300 mm diameter to 7.4 mm for a 900 mm diameter shaft have been quoted, with bearing clearances of 0.51–0.63 mm and 1.53–1.89 mm respectively.

The forward part of the stern tube is fabricated and welded direct to the extension of the stern frame boss and into the aft peak bulkhead.

Oil pressure within the stern tube is maintained at approximately the same level as that of the surrounding sea water by a header tank. Oil is contained within the Simplex type stern tube by lip seals. The elastic lip of each nitrile rubber seal, grips a rubbing surface provided by short chrome steel liners at outboard and inboard ends of the steel propeller shaft. The outboard liner additionally protects the steel shaft from sea-water contact and corrosion.

Heat produced by the friction will result in hardening and loss of elasticity of the rubber, should temperature of the seal material exceed 110°C. Cooling at the outboard end is provided by the sea. Inboard seals, unlike those at the outboard end, cannot dissipate heat to the surrounding water. Oil circulation aided by convection, is arranged to maintain the low temperature of the seals at the inboard end. Connections for circulation, are fitted top and bottom between the two inboard seals and the small local header tank.

The chrome liners act as rubbing surfaces for the rubber lip seals but grooving from frictional wear has been a problem. The difficulty has been overcome by using a ceramic filler for the groove or alternatively a distance piece to displace axially the seal and ring assembly. Allowance must be made for the relative movement of shaft and stern tube due to differential expansion. New seals are fitted by cutting and vulcanizing in position.

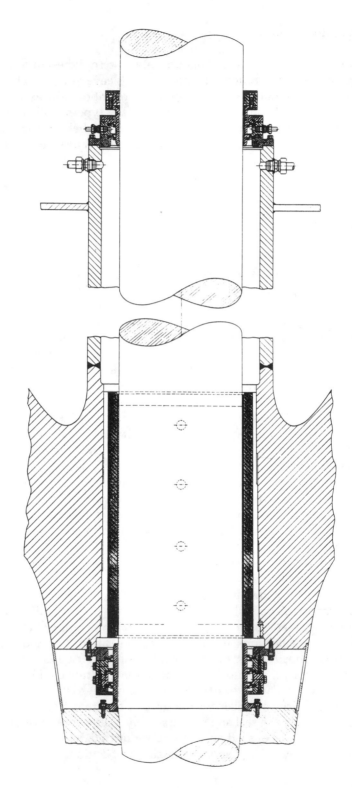

Figure 8.15 *Oil lubricated stern tube (Simplex type)*

Split stern bearings

To avoid the necessity for drydocking when an examination of stern bearings and tailshaft is needed, split stern bearings were developed. A suitable outboard sealing arrangement and design, permits the two halves of the bearing to be drawn into the ship, exposing the shaft and the white metal bearing.

Glacier-Herbert stern bearing

In the Glacier-Herbert system (Figure 8.16) the two completely symmetrical bearing halves are flanged along the horizontal centre line and held together by bolts. The after end of the bearing carries a spherical support ring to which is bolted the outboard seal housing. The spherical support ring rests in a carrier ring which is bolted to the after end of the stern frame boss. The forward end is supported by a circular diaphragm which is bolted to a flange provided in the stern frame casting. This diaphragm also acts as a carrier for the forward seal.

A series of axial bolts, fitted with Belleville washer packs to ensure virtually constant loading of these bolts and those securing the spherical seating ring, hold the diaphragm firmly in position. This arrangement permits sloped alignment of the bearing to give full support to the drooping tailshaft. Chocks are used to hold the bearing positively in its final position. The arrangement is such that it allows for the differential expansion of the bearing and its housing without detracting from the rigidity of support at the forward end of the bearing.

The propeller shaft is flanged at the after end and the hub of the propeller is bolted to the flange. On the inboard side there is a shroud around the spigot projecting aft from the carrier ring. Two inflatable seals with individual air

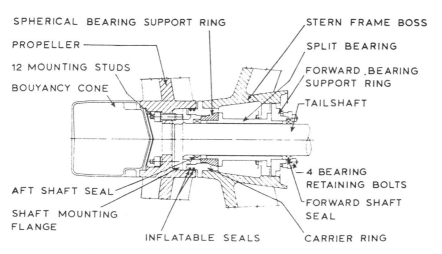

SPHERICAL BEARING SUPPORT RING   STERN FRAME BOSS

PROPELLER ———————   SPLIT BEARING

12 MOUNTING STUDS   FORWARD BEARING
                                         SUPPORT RING
BOUYANCY CONE
                                         TAILSHAFT

AFT SHAFT SEAL   4 BEARING
                                         RETAINING BOLTS
SHAFT MOUNTING   FORWARD SHAFT
FLANGE   SEAL

              INFLATABLE SEALS   CARRIER RING

Figure 8.16 *Solid propeller with hollow cylindrical boss and internal flanged mounting*

supplies are fitted in the periphery of the spigot. These can be inflated to provide a seal against inflow of water.

Sealing the stern bearing space permits work to be carried out on the stern bearing and seals, without the necessity of drydocking. An alternative to using the inflatable seals, is to apply a sealing bandage around the small gap at the carrier ring flange.

The propeller shaft has two short rotating liners of chrome steel. The liner at the after end is bolted to the propeller shaft flange. The inboard liner is fixed by a clamping ring. These liners act as rubbing surfaces for the rubber lip seals.

The coupling at the forward end of the tailshaft may be of the SKF oil injection type (i.e. muff coupling) described previously.

Ross-Turnbull split bearing

The Ross-Turnbull split stern bearing (Figure 8.17) has a bottom half bearing which is chocked on to two horizontal fore and aft machined surfaces within the stern frame boss. The whole bearing is held in position vertically by two 50 tonne Pilgrim type jacks, the chock thickness determining the bearing height. These jacks also hold the top half of the bearing in place. Lateral positioning of the bearing is by 30 tonne Pilgrim type jacks arranged on each side of the bearing. A running track is arranged above the bearing to allow easy transport of the top half. Skids are provided below the bearing to provide easy transport of the bottom half. When removing the bearing bottom half, a jack is first placed underneath it to lift it free of its chocks. The chocks are removed and skates are placed under the bottom half bearing. With the chocks out, the assembly is lowered until the propeller rests on the shroud or support cradle, which is part of the stern frame boss. Further lowering of the jacks, brings the bottom half away from the tailshaft until its weight is taken by the skates resting on the skids. The jacks are removed and the bottom half bearing is brought forward on the skates together with the seal face and the bellows section of the outboard seal.

Stern tube sealing arrangements

There are basically three sealing arrangements used for stern bearings. These are:

1 Simple stuffing boxes filled with proprietary packing material.
2 Lip seals, in which a number of flexible membranes in contact with the shaft, prevent the passage of fluid along the shaft.
3 Radial face seals, in which a wear-resistant face fitted radially around the shaft, is in contact with similar faces fitted to the after bulkhead and to the after end of the stern tube. A spring system is necessary to keep the two faces in contact.

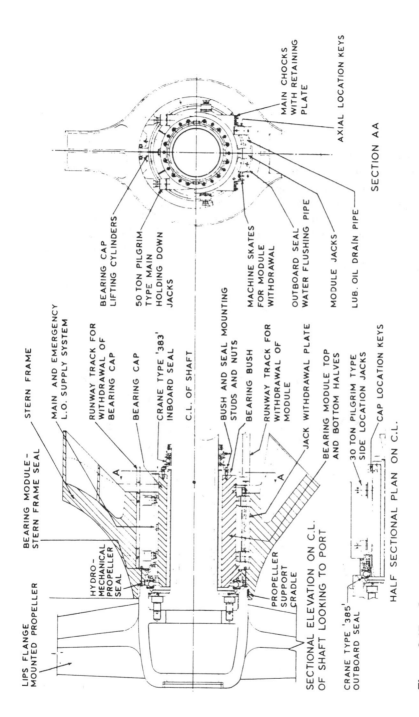

LIPS FLANGE
MOUNTED PROPELLER

BEARING MODULE –
STERN FRAME SEAL

STERN FRAME

MAIN AND EMERGENCY
L.O. SUPPLY SYSTEM

RUNWAY TRACK FOR
WITHDRAWAL OF
BEARING CAP

BEARING CAP

CRANE TYPE '383'
INBOARD SEAL

C.L. OF SHAFT

BUSH AND SEAL MOUNTING
STUDS AND NUTS

BEARING BUSH

RUNWAY TRACK FOR
WITHDRAWAL OF
MODULE

JACK WITHDRAWAL PLATE

BEARING MODULE TOP
AND BOTTOM HALVES

30 TON PILGRIM TYPE
SIDE LOCATION JACKS

CAP LOCATION KEYS

HYDRO-
MECHANICAL
PROPELLER
SEAL

PROPELLER
SUPPORT
CRADLE

SECTIONAL ELEVATION ON C.L.
OF SHAFT LOOKING TO PORT

CRANE TYPE '385'
OUTBOARD SEAL

HALF SECTIONAL PLAN ON C.L.

MAIN CHOCKS
WITH RETAINING
PLATE

AXIAL LOCATION KEYS

SECTION AA

BEARING CAP
LIFTING CYLINDERS

50 TON PILGRIM
TYPE MAIN
HOLDING DOWN
JACKS

MACHINE SKATES
FOR MODULE
WITHDRAWAL

OUTBOARD SEAL
WATER FLUSHING PIPE

MODULE JACKS

LUB. OIL DRAIN PIPE

Figure 8.17 *General arrangement of Ross–Turnbull Mark IV bearing (Ross Turnbull Ltd.)*

Lip seals

A lip seal assembly (Figure 8.18) consists of a number of nitrile rubber rings of special cross-section (Figure 8.19) sandwiched between bronze rings. Each individual rubber lip seal is held in contact with a renewable sleeve fitted to the shaft by its elasticity and a garter spring. The rings are renewed by cutting and then vulcanizing the ends *in situ.* The Simplex stern tube (Figure 8.15) has a forward seal with two rings and an after seal with three rings. A larger sketch of the after seal (Figure 8.18) shows how seals are built up from three basic assemblies namely the flange, intermediate, and cover rings and these parts can be used for either seal. It will be noticed that the garter spring holding the sealing ring against the shaft is located aft of the ring anchoring bulb in the case of both forward sealing rings. In the case of the after seal the two outboard sealing rings have their garter springs located aft of the ring anchoring bulbs while the inboard ring has its garter spring located inboard of the anchoring bulb. Lip seals will accept misalignment but a floating ring design was introduced by one manufacturer.

In some instances four or more sealing rings are installed. These are arranged so that one ring does not normally run on the shaft liner. In the event of leakage from the working seals, adjustment is made to bring the reserve ring into play.

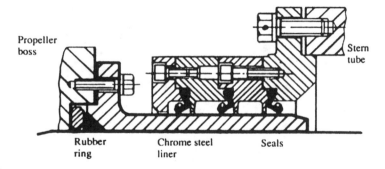

Figure 8.18   *Lip seal assembly*

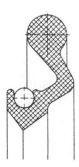

Figure 8.19   *Nitrile rubber ring cross-section*

Radial face seal

An example of a radial face seal, is shown in the general arrangement (Figure 8.20) and a detailed sketch (Figure 8.21) of a Crane seal. One of the principal features of the design and construction of this type of seal is the split construction of all component parts. This facilitates installation, and subsequent inspection and maintenance.

The function of sealing against leakage around the shaft is effected by sustaining perfect mating contact between the opposing faces of the seal's seat

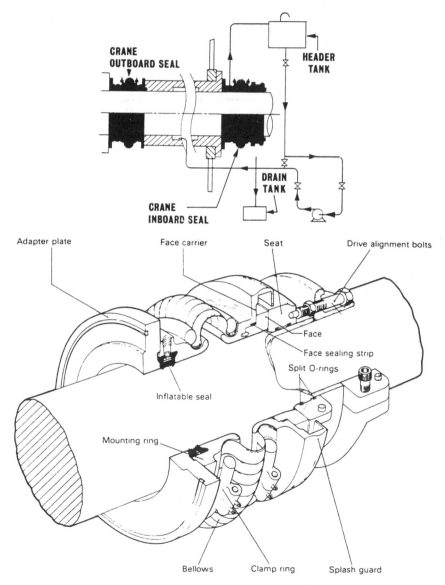

Figure 8.20 *Example of a radial face seal (Deep Sea Seals Ltd.)*

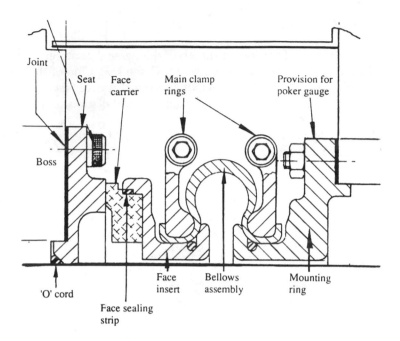

Figure 8.21   *Detail of a Crane seal*

which rotates with the shaft, and of the main seal unit which is stationary and clear of the shaft.

This mating contact of the seal faces, which are hydraulically balanced, is sustained by spring pressure and by the method of flexibly mounting the face of the main seal unit. The flexible member consists of a tough, but supple, reinforced bellows. Thus the main seal unit is able to accommodate the effects of hull deflection and vibration.

The bellows member is clear of the shaft, and its flexibility therefore cannot be impaired, as may happen when a flexible member is mounted on the shaft and hardens, seizes or becomes obstructed by a build-up of solids. The mechanical design principles also ensure continued sealing under fluctuating pressure conditions, i.e. changing draught.

An emergency sealing device can be incorporated into the design. The device, when inflated with air or liquid, forms a tight temporary seal around the shaft, enabling repairs to be made or a replacement seal fitted when the ship is afloat, without the shaft being drawn or drydocking being necessary.

Lubrication systems

The static lubrication system for vessels with moderate changes in draught, have header tanks placed 2–3 m above the maximum load waterline. The small differential pressure ensures that water is excluded. The cooling of simple stern tubes, necessitates keeping the aft peak water level at least 1 m above the stern tube.

Tankers and other ships with large changes in draught, may be fitted with two oil header tanks (Figure 8.22) for either the fully loaded or ballast condition.

Hydrodynamic or hydrostatic lubrication

The requirement for steaming at a slow, economical speed during periods of high fuel prices (or for other reasons) gives a lower fluid film or hydrodynamic pressure in stern tubes, due to the slower speed. The possibility of bearing damage occurring prompted the installation of forced lubrication systems to provide a hydrostatic pressure which is independent of shaft speed. The supplied oil pressure gives adequate lift to separate shaft and bearing and an adequate oil flow for cooling.

Fixed pitch propellers (Figure 8.23)

The normal method of manufacture for a fixed pitch propeller, is to cast the blades integral with the boss and after inspection and marking, to machine the

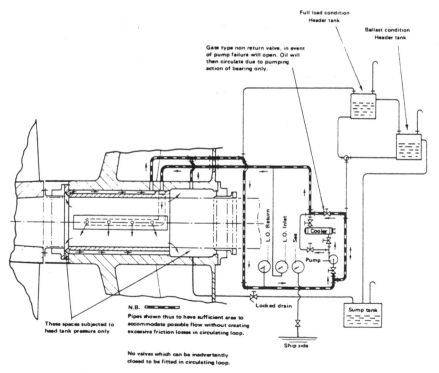

Figure 8.22 *Single-bush bearing showing also a forced lubrication system (Glacier Metal Co.)*

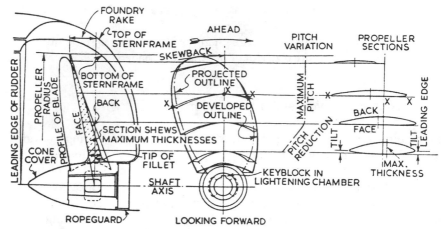

Figure 8.23   *Fixed pitch propeller terminology*

tapered bore and faces of the boss before the blades are profiled by hand with reference to datum grooves cut in the surfaces or with an electronically controlled profiling machine. Finally the blades are ground and polished to a smooth finish.

Built-up propellers, with blades cast separately and secured to the propeller boss by studs and nuts, were made obsolete as improvements permitted the production of larger one piece castings. The advantages of built-up propellers were the ease of replacing damaged blades and the ability to adjust the pitch, but these were outweighed by the loss of efficiency resulting from restricted width at the blade root, the greater thickness required to maintain strength and the larger hub diameter.

Methods of mounting propellers

Traditionally, fixed pitch propellers have been fitted to the tailshaft with a key and taper (Figure 8.24) being forced on to the taper by the tightening of a nut (see the section on sea-water lubricated stern tubes and inspection). The key was intended as a safeguard either against poor fitting, or against reduced grip due to higher sea-water temperature and differential expansion of bronze hub and steel shaft. Keyless fitting where reliance is placed entirely on a good interference fit, has proved effective, however, and this method removes problems associated with keyways and facilitates propeller mounting and removal. Many fixed propellers are of course flange mounted, being held by bolts as shown in the section on split stern bearings. For these, outward removal of the tailshaft is made possible with the use of a muff coupling.

Keyed propellers

For the conventional key and taper arrangement, keyways are milled in the shaft taper and the key accommodated in the bore of the hub, by slots

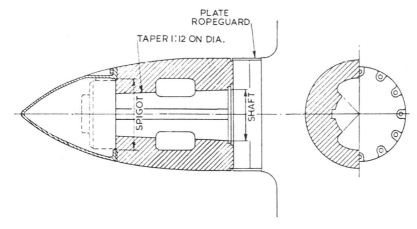

Figure 8.24   *Typical arrangement of solid propeller boss*

machined through. Ideally, the hub and shaft tapers would be accurately matched and the hub would be stretched by being forced past the point of fit on the shaft taper, by the propeller nut. The 'push up' of a few millimetres is calculated to give a good interference fit. Torque in the ideal condition is transmitted totally by the interference fit, with the key being merely a back up. If conditions are not as intended, fatigue cracks can occur at the forward end of the keyway and more serious fatigue cracks may result from fretting damage (or corrosion) particularly in high-powered single screw ships.

Keyless propellers

The success of a keyless propeller depends on the accuracy of the hub and shaft tapers and correct grip from the stretched propeller hub on the shaft. The degree of stretch (or strain) is controlled by push up. It must ensure adequate grip despite any temperature changes and consequent differential expansion of bronze hub and steel shaft. It must also avoid over stressing of the hub and in particular any permanent deformation.

Lloyds require that the degree of interference be such that the frictional force at the interface can transmit 2.7 times the nominal torque when the ambient temperature is 35°C. Lloyds also require that at 0°C the stress at the propeller bore, as given by the Von Mises stress criterion, shall not exceed 60% of the 0.2% proof stress of the propeller material as measured on a test bar.

Pilgrim nut method

The Pilgrim nut system used with the shaft and bore surfaces dry and degreased (except for cast steel propellers where wiping of the bore with an oil soaked rag is recommended) achieves the correct push up by a calculation based on the predictable friction of dry surfaces. The calculation gives the hydraulic pressure suitable for the prevailing ambient temperature to produce the required push up. The operation is of course checked by measuring the push up and the hub movement relative to the increase of jacking pressure is monitored by a dial gauge.

The Pilgrim nut (Figure 8.25a) employed for propeller mounting, has an internal nitrile rubber tube which when inflated hydraulically, forces a steel loading ring against the hub. Outward movement of the ring from the flush position must not exceed one third of the ring width, to avoid rupture of the rubber tube. Temperature of hub and shaft are recorded and also used to find the correct final push up pressure from the table provided in the instruction book.

The propeller, after a check with the blue marker of the mating surfaces, is positioned and initially jacked on to the shaft taper, before the Pilgrim nut is used to apply an initial loading of perhaps 67 bar pressure. A reference mark is made at this point about 25 mm from the forward end of the hub. The nut is then turned until the loading ring is again flush (venting hydraulic fluid) before full pressure is applied. During this stage, the dial gauge should show the movement. A second mark 25 mm from the forward face of the hub is then made. Push up, registered by the distance between the two reference marks, is measured and noted.

The nut is again vented and turned to bring the loading ring to the flush position and finally nipped up with a tommy bar. The Pilgrim nut can be reversed and used with a withdrawal plate and studs (Figure 8.25b) for removal of the propeller. To safeguard against any violent movement at release, wooden blocks are inserted as shown, and a gap of only a little more than the push up distance is left.

The Pilgrim keyless system owes its name to T. W. Bunyan.

The SKF system

The oil injection system of propeller mounting is associated with the name of SKF. With this method, instead of a dry push up, oil is injected (Figure 8.26) between the shaft taper and the bore of the propeller by means of high pressure pumps. Oil penetration is assisted by a system of small axial and circumferential grooves or a continuous helical groove, machined in the propeller bore. The oil reduces the coefficient of friction between the surfaces to about 0.015.

A hydraulic ring jack is arranged between the shaft nut and the aft face of the propeller boss, and with this it is a simple matter to push the propeller up the shaft taper by the required amount, overcoming the friction force and the axial component of the radial pressure. When the oil injection pressure is released, the oil is forced back from between the shaft/bore surfaces leaving an interference fit with a coefficient of friction of at least 0.12.

When it is required to remove the propeller, the process is equally simple and even quicker with the injection of oil between the surfaces obviating the need for any form of heating or mechanical withdrawal equipment. Precautions are necessary to prevent the propeller jumping at release.

A development of the keyless method involves a cast iron sleeve

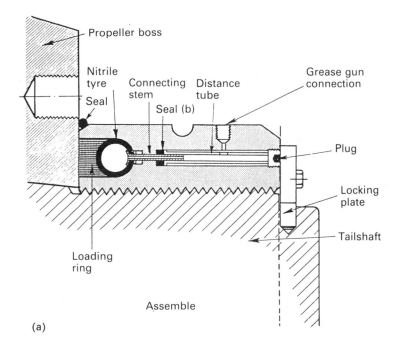

(a)

Assemble

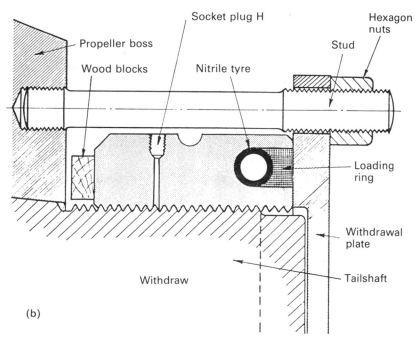

(b)

Figure 8.25 *The Pilgrim nut*

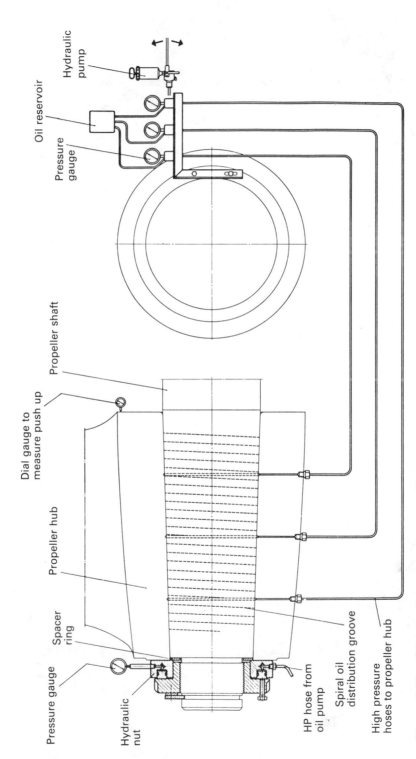

Figure 8.26 *Oil injection propeller mounting*

Hydraulic pump

Oil reservoir

Pressure gauge

Propeller shaft

Dial gauge to measure push up

Propeller hub

Spacer ring

Pressure gauge

Hydraulic nut

HP hose from oil pump

Spiral oil distribution groove

High pressure hoses to propeller hub

(Figure 8.27) which is bonded into the propeller boss with a special form of Araldite which is injected under pressure. The sleeve is machined and bedded to the shaft taper but can be used to adapt a general purpose spare propeller to a particular shaft taper. The sleeve is easier to handle when machining and bedding than a complete propeller. Another benefit is that cast iron has a coefficient of friction nearer to that of the shaft than to the propeller bronze.

Controllable pitch propellers

Controllable pitch propellers are normally fitted to a flanged tailshaft as the operating mechanism is housed in the propeller boss.

As its name implies, it is possible to alter the pitch of this type of propeller to change ship speed or to adjust to the prevailing resistance conditions. This change in pitch is effected by rotating the blades about their vertical axes, either by hydraulic or mechanical means. A shaft generator can be driven at constant speed while allowing at the same time a change of ship's speed through the propeller. Since it is normally possible to reverse the pitch completely, this type of propeller is used with a uni-directional engine to give full ahead or astern thrust, when manoeuvring. The most obvious application is for ferries or other vessels which regularly and frequently manoeuvre in and out of port. They are also used for double duty vessels, such as tugs or trawlers where the operating conditions for towing or for running free are entirely different.

One of the most widely used controllable pitch (c.p.) propellers is the KaMeWa, a hydraulically operated Swedish propeller first introduced in 1937. In this unit (Figures 8.28a and 8.28b) the blade pitch is altered by a servomotor piston housed within the hub body. The piston moves in response to the difference in oil pressure on its ends. Oil flow to and from the servomotor is controlled by a slide valve in the piston rod; the slide valve is part of a hollow rod which passes through a hole bored in the propeller shaft and is mechanically operated by operating levers located in an oil distribution box.

If the slide valve is moved aft, the valve ports are so aligned that oil under pressure flows along the hollow valve rod to the forward end of the piston, causing the piston to move in the same direction, until the ports are again in a neutral position. When the valve is moved forward, the piston will move in a forward direction.

When the piston moves, the crosshead with its sliding shoes moves with it. A pin on a crank pin ring, attached to each propeller blade, locates in each of the sliding shoes, so that any movement of the servomotor piston causes a pitch change simultaneously in the propeller blades.

Oil enters and leaves the hub mechanism via an oil distribution box mounted inside the ship on a section of intermediate shaft. Oil pressure of about 40 bar maximum in the single piston hub, is maintained by an electrically driven pump, which has a stand-by. A spring loaded, inlet pressure regulating valve on the oil distribution box, controls the pressure in the high pressure chamber from where the oil passes to the hub mechanism via the hollow rod in the propeller shaft. Oil passes from the hub mechanism to the low pressure

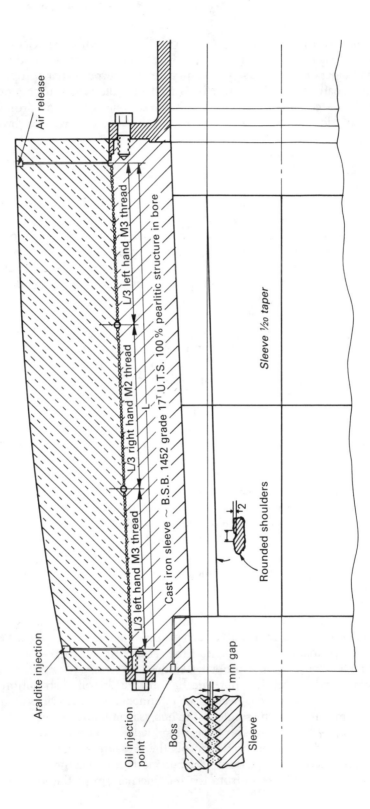

Figure 8.27 *Hub with cast iron sleeve*

chamber of the distribution box along the outside of the valve rod. A spring loaded back pressure regulating valve on the oil outlet, maintains a slight back pressure on the oil filled hub, when the vessel is underway. When in port this pressure is maintained by the static head of an oil tank mounted above the ship's waterline and connected to the oil distribution box. The oil pressure in the hub is needed to balance outside pressure from the sea and so make leakage in either direction unlikely.

The forward end of the valve rod connects to a T bar or key which is moved forward or aft by a sliding ring within the oil distribution box. (The T bar rotates with the shaft.) A servomotor mounted externally to the box is used to move the sliding ring through a yoke. In the event of a failure in the servomotor, an external lever can be used to shift the valve rod manually and so control blade pitch. A powerful spring may be fitted so that in the event of loss of hydraulic oil pressure, the blades will be moved towards the full ahead position. Forces on the blades tend to prevent the full ahead position from being attained and it may be necessary to slow the engine or even stop it, to allow the spring to act. The spring could be fitted to give a fail safe to astern pitch. Some controllable pitch propellers are arranged to remain at the current setting if hydraulic oil loss occurs.

Where a shaft alternator is installed, engine speed may remain constant as propeller pitch is altered for manoeuvring. Alternatively, propeller pitch and engine speed can be remotely controlled from a single lever known as a combinator. Any number of combinators may be installed in a ship. The combinator lever controls pitch and speed through cam-operated transmitters. These may be electrical or pneumatic devices.

Gears and clutches

For medium-speed engine installations in large ships (as opposed to coasters or intermediate sized vessels) reduction gears are needed to permit engines and propellers to run at their best respective speeds. Their use also permits more than one engine to be coupled to the same propeller. Gearboxes are available from manufacturers in standard sizes. Firms produce a standard range for different powers of single and multiple input, single reduction gearboxes for medium- (or high-speed engines) in a number of frame sizes. The input and output shafts for single input gears, may be either horizontally offset, vertically offset or coaxially positioned. From the appropriate selection chart, using figures for engine power, engine speed and reduction ratio (also Classification Society correction for ice if applicable), the size and weight of the appropriate gearbox can be found.

Ship manoeuvring is of course improved with twin screws and this is an added safeguard against total loss of power due to engine breakdown. The disposition of two engines and shafts can sometimes be improved with the use of offset gearboxes. Normally twin screw propellers turn outward when running ahead, i.e. when viewed from astern the port propeller turns anticlockwise and the starboard propeller turns clockwise. (Inward turning

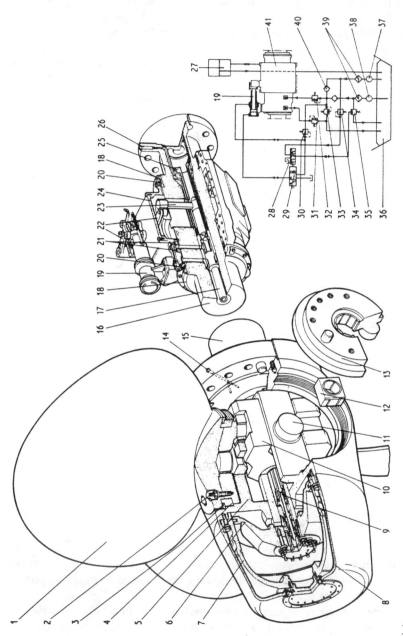

(a)

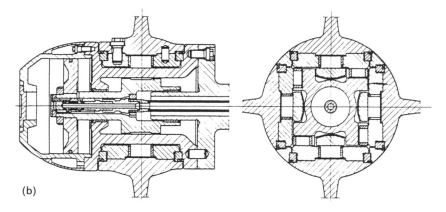

(b)

Figure 8.28 *(a) Single piston servomotor (KaMeWa) (opposite); (b) Detail of KaMeWa S1 propeller hub*

Key to Figure 8.28a
1. Blade with flange
2. Blade stud with nut and cover
3. Blade sealing ring
4. Bearing ring
5. Hub body
6. Servometer piston
7. Hub cylinder
8. Hub cone
9. Main regulating valve assembly
10. Piston rod with cross head
11. Centre post (integrated with hub body)
12. Sliding shoe with hole for crank pin
13. Crank pin ring
14. Safety valve for the low pressure part of the propeller hub

15. Propeller shaft with flange
16. Intermediate shaft
17. Valve rod
18. End cover
19. Pitch control auxiliary servomotor assembly
20. Low pressure seal assembly
21. High pressure seal assembly
22. Yoke lever
23. Valve rod key
24. Oil distribution box casing
25. Standy-by servo
26. Non-return and safety valve for stand-by servo
27. Oil tank e.g. Oil tank for static over-pressure in propeller hub

28. Regulating valve for unloaidng pump
29. Regulating valve for auxiliary servomotor
30. Reducing valve (auxiliary servomotor)
31. Back pressure maintaining valve
32. Sequence valve
33. Safety valve
34. Reducing valve (unloading)
35. Unloading valve
36. Main oil tank
37. Main pump
38. Unloaded pump
39. Main filter
40. Check valve
41. Oil distribution box

propellers, tend to make the movement of the stern unpredictable when manoeuvring and have given rise to other problems.)

Reverse reduction gearbox

Reversing with the use of a gearbox, after reducing engine speed as necessary, means that continually starting on cold air is avoided and less compressed air capacity is required. Reverse/reduction gearboxes, like straight reduction gears, are also obtainable in standard sizes, with manufacturers' charts for selection. Gear lubrication is by a self-contained system on many sets.

There are various arrangements possible for the shafts in a reverse/reduction gearbox to suit the required location of the engine input or drive shaft and the driven or output shaft. The sketch (Figure 8.29) shows a simplified, flat arrangement for ease of explanation.

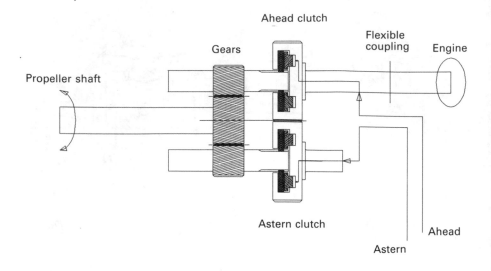

Figure 8.29   *Reverse/reduction gearbox arrangement*

The drive from the engine input shaft to the counter shaft, is through teeth on the outsides of both clutch housings, which are in continuous mesh. When the control lever is set for ahead running, the control valve supplies oil pressure to the ring piston of the ahead clutch. When the control lever is set for astern running, the control valve supplies oil pressure to the ring piston of the astern clutch. When either clutch is engaged, its pinion provides a drive to the large gear wheel of the driven shaft and the other pinion rotates freely. Oil pressure required for clutch operation is built up by a gear pump driven from the input shaft. Lubrication is by means of overflow oil.

The propeller thrust on the driven shaft is taken up by the thrust bearing. The driven (propeller) shaft, for ahead running, rotates in the opposite direction to the drive or input shaft. For astern running, the driven (propeller) shaft rotates in the same direction as the drive shaft. To stop the propeller shaft, the control is moved to the neutral position and both clutches are disengaged.

Flexible couplings

Where a gearbox is fitted, a torsionally flexible coupling (Figure 8.30) is necessary between the medium-speed diesel and the reduction gear. The coupling is necessary because the periodic application and reduction of torque as engine cylinders fire in turn, tends to result in alternate loading and unloading of the gear teeth. The torsional vibration effect is sufficient to cause serious gear tooth damage. Flexible couplings may be installed as separate entities or in conjunction with air or oil operated clutches. Flexible couplings may be built in a common casing with the clutch. Apart from protecting the gears, flexible couplings, are also able to withstand slight misalignment.

The Gieslinger coupling shown in Figure 8.30 has a housing and hub

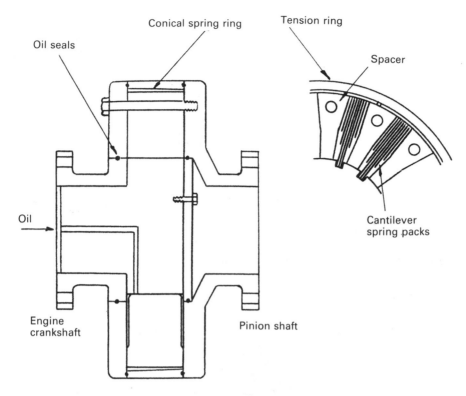

Figure 8.30 *Geislinger flexible coupling*

connected by leaf springs, which flex in service to absorb torsional effects from the engine.

Air operated clutches

Clutches which are not part of the gearbox, are usually air activated, with pads or linings which make either radial or axial contact. The application force for the friction pads or linings, is supplied by compressed air in a reinforced neoprene rubber tube. The compressed air is filtered and moisture is removed by drains provided in the system. Air pressure is monitored and the low pressure alarm is particularly important. Some form of rotary connection between the air supply pipe and the clutch is necessary, with the valve controlling the air supply to the clutch tube being operated by hand or remotely controlled by a solenoid or air pressure.

For a radial air operated clutch (Figure 8.31) the compressed air expands an actuating tube around the outside of the friction pads. Inward expansion of the tube forces the pads into contact with the friction drum. The transmission of torque relies on the air pressure and loss of pressure would allow slip.

The open construction of the clutch allows air access for pad cooling and the expanding tube compensates for wear. Springs (not shown) are incorporated

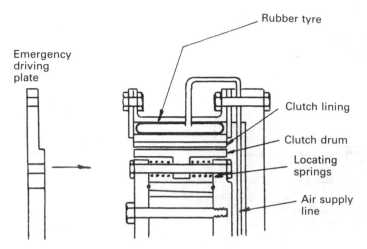

Figure 8.31   *Radial air operated clutch*

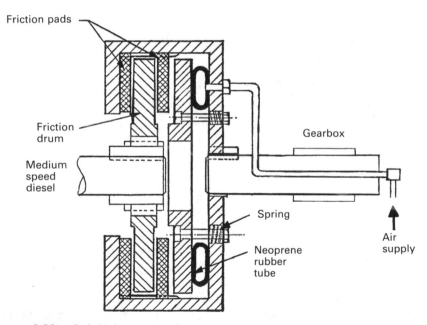

Figure 8.32   *Axial air operated clutch*

for disengagement of the clutch, which is also assisted by centrifugal effect. This type of clutch has been supplied in combination with a Geislinger coupling.

Axial air operated clutch

This type of clutch also uses a neoprene tube which is inflated by compressed air. Expansion of the tube (Figure 8.32) produces a sandwich action between friction pads and disc. The friction disc or drum is spline mounted and therefore has axial float. The friction pads are also free to float axially; being mated with teeth machined peripherally inside the casing. Springs cause disengagement of the clutch when the tube is deflated. Clutches produced by Wichita have a larger number of friction dics and pads than shown in Figure 8.32, which is intended to show the operating principle of axial air clutches.

Emergency operation

Failure of the air supply or other fault could render a clutch inoperative. To make provision for this eventuality, an emergency driving plate (Figure 8.31) or set of temporary coupling bolts is provided. The emergency arrangement shown, is for a combined clutch and coupling.

Prolonged use of the emergency solid coupling arrangement can result in serious damage to gear teeth. (The gear box of at least one medium-speed engined ship had to be expensively replaced after six months operation with the emergency coupling arrangement.)

Further reading

Sinclair, L. and Emerson, A. (1968) The design and development of propellers for high powered merchant vessels, *Trans I Mar E*, **80**, 5.

Bille, T. (1970) Experiences with controllable pitch propellers, *Trans I Mar E*, **80**, 8.

Crombie, G. and Clay, C. F. (1972) Design feature of and operating experience with turnbull split stern bearings, *Trans I Mar E*, **84**, 11.

Herbert, C. W. and Hill, A. (1972) Sterngear design for maximum reliability – the Glacier-Herbert system, *Trans I Mar E*, **84**, 11.

Rose, A. (1974) Hydrostatic stern gear, N.E. Coast 1. of Engineers and Shipbuilders.

Sterntube Bearings. The Glacier Metal Co. Ltd.

Wilkin, T. A. and Strassheim, W. (1973) Some theoretical and practical aspects of shaft alignment, I Mar E IMAS 73 Conference.

Pressicaud, J. P. Correlation between theory and reality in alignment of line shafting, Bureau Veritas.

9

Steering gears

Ships from at least the 1950s, have been fitted with automatic steering, making helmsmen redundant for deep sea passages. After the required course is set, the automatic steering maintains direction; correcting any deviations due to the weather. The automatic helm is consistent, with none of the fall off in performance that occurred in heavy weather with manual steering, when the human helmsman was changed. Some problems were experienced with the early versions of automatic steering systems when changing from automatic to manual steering and vice versa. An incorrect change over was blamed for at least one collision (the change over involved a hydraulic telemotor system, the by-pass for which had not been correctly set).

Automatic steering has improved and electrical control of the steering gear has now become the norm, with the hydraulic telemotor, if installed at all, being used only in an emergency. Hydraulic telemotors where fitted, should be regularly checked, however, with any leak being made good and the oil topped up.

Rudder carrier bearing

The rudder carrier bearing (Figure 9.1) takes the weight of the rudder on a grease lubricated thrust face. The rudder stock is located by the journal beneath, also grease lubricated.

Support for the bearing is provided by framing beneath the steering gear deck. There is thicker deck plating in the area beneath the carrier bearing and the latter may be supported on steel chocks. The base of the carrier bearing is located by side chocks welded to the deck. The carrier may be of meehanite with a gunmetal thrust ring and bush. Carrier bearing components are split as necessary for removal or replacement. Screw down (hand) lubricators may be fitted but automatic lubricators are common. The grease used for lubrication is of a water resistant type (calcium soap base with graphite).

The tiller (Figure 9.1) is keyed to the rudder stock and is of forged or cast steel with one (or two for a four ram gear) arms, machined smooth to slide in a swivel block arrangement designed to convert linear movement of the rams to the rotary movement of the tiller arms and rudder stock. This particular device, known as a Rapson slide, is used for many, but not all, ram type gears. The rams are one-piece steel forgings, with the working surface ground to a high finish. Each pair of Rapson slide rams, is bolted together, the joined ends being bored

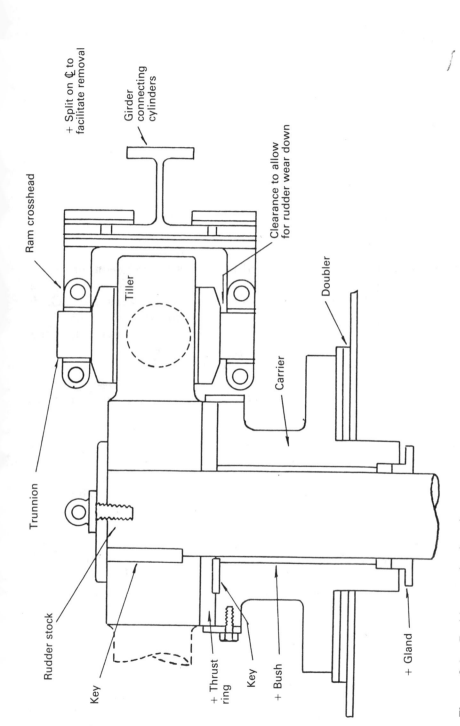

+ Split on ₵ to facilitate removal

Girder connecting cylinders

Ram crosshead

Clearance to allow for rudder wear down

Tiller

Trunnion

Doubler

Carrier

Rudder stock

Key

+ Thrust ring

Key

+ Bush

+ Gland

Figure 9.1 *Rudder carrier bearing*

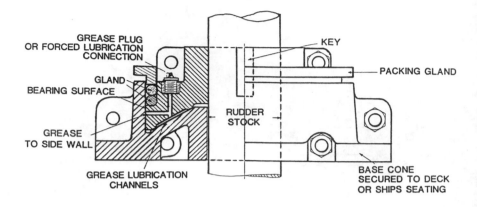

Figure 9.2   *Carrier with conical seat*

vertically and bushed to form top and bottom bearings for the projecting spigots on the swivel block. Crosshead slippers, bolted to the face of the central section of the rams, slide on the machined surfaces of the guide beam. Guide beams also serve to brace each pair of cylinders against the tendency for them to be pushed apart by the hydraulic pressure. The cylinders have substantial feet bolted to the stools on which the gear is mounted.

Weardown of the carrier bearing is monitored by periodically measuring the clearance marked. The original clearance is usually about 20 mm.

An alternative type of carrier bearing with a conical seat (Figure 9.2) has the advantage that the seat and side wall will locate the rudder stock. The angle of the conical seat is shallow to prevent binding.

Bearing weardown occurs over a period of time, and allowance is made in the construction of the steering gear (see Figure 9.1) for a small vertical drop of the rudder stock. This weardown allowance is checked periodically and restored as necessary. Lifting of the rudder and stock by heavy weather can be limited by jumping stops between the upper surface of the rudder and the stern frame.

The usual limit for movement of the rudder, is 35° each way from the mid position and this is controlled by the telemotor. External rudder stops if fitted, would limit movement to, say, 39° from the mid position. The steering gear itself will also impose a limit on rudder movement but with hydraulic oil loss and the ship stopped in heavy weather, there may be severe damage to the gear. The telemotor control imposes the usual 35° limit.

Ram type hydraulic steering gear

Figure 9.3 shows an arrangement of a two-ram steering gear with variable delivery pumps. Such gears may have a torque capacity of 120–650 kNm.

The cylinders for this gear are of cast steel but the rams comprise a one-piece steel forging with integral pins to transmit the movement through cod piece•

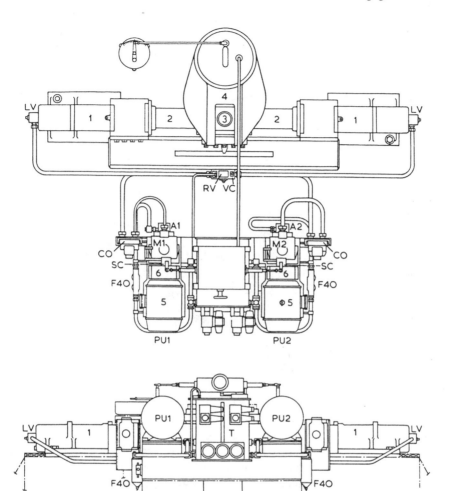

Figure 9.3 *Two-ram electro-hydraulic steering gear.*

1. Cylinders
2. Rams
3. Cod piece
4. Tiller
5. Motors
M1
M2 Variable delivery pumps
LV Locking valve

PU1, PU2 Power units
A1, A2 Auxiliary pumps
T Reservoir
F10 Filter
SC Servo-controls
CO Changeover valves
PC20 Pressure limiting valves
CV Check valves

P1, P2 Isolating valves
LV Locking valves
BP By-pass valve
RV Relief valve
HP Hand-pump shut-off valves
WP Non-return valves

which slide in the jaws of a forked tiller end. The rams are machined and ground to slide in the gunmetal neck bushes and chevron type seals of the cylinders.

Hydraulic pressure is supplied to one cylinder or the other, by uni-directional, variable delivery pumps, with electric drive, running at constant speed. The pumps may be Hele-Shaw radial piston type or a development of the axial piston V.S.G. pump. The strokes of the pump pistons in both types of pump can be varied and the flow of oil to and from the pump can be reversed. When the operating rod of the pump is in mid position, there is no flow of oil.

Variable delivery pumps

Variable delivery pumps can run continuously in one direction but have the capability of an infinitely changeable discharge from zero to a maximum either way. The principle of operation is based on altering the stroke of the pump pistons in radial or axial cylinders, by means of a floating ring or swash plate respectively, to change the amount of oil displaced. There is very little shock to the hydraulic system as the pump commences delivery, because the piston stroke increases from zero gradually. For a small rudder movement, piston stroke is small; the stroke becomes full only for larger rudder movements. At the end of the rudder movement, pump discharge tapers off; it does not cease abruptly as with constant delivery pumps which have valve control.

The radial cylinder (or Hele-Shaw) pump

The pump (Figure 9.4a) consists of case A, to which are attached two covers, the shaft cover B and the pipe connection cover C. This latter cover carries the D tube (or central valve), which has ports E and F forming the connections between the cylinders and branches G and H. The cylinder body J is driven by shaft K, and revolves on the D tube, being supported at either end by ball bearings T.

The pistons L are fitted in radial cylinders, and through the outer end of each piston there is a gudgeon pin M, which attaches the slippers N to the piston. The slippers are free to oscillate on their gudgeon pins and fit into tracks in the circular floating ring O. This ring is free to rotate, being mounted on ball bearings P, which are housed in guide blocks R. The latter bear on tracks formed on the covers B and C and are controlled by spindles S, which pass through the pump case A. The maximum pump stroke is restricted by the guide block ends coming in contact with the casing. Further restriction of the pump stroke is effected externally.

Figure 9.4b shows sections through the D tube, cylinder body, pistons and slippers at right angles to the axis. XY is the line along which stroke variations take place. The arrow indicates the direction of rotation.

With the floating ring central, i.e. concentric with the D tube, (1) the slippers move round in a circle concentric with the D tube, and consequently no pumping action takes place. With the floating ring moved to the left, (2) the slippers rotate in a path eccentric the D tube and cylinders, consequently the pistons, as they pass above the line XY, recede from the D tube and draw oil through the ports, E, whilst the pistons below XY approach the D tube and discharge oil through ports F.

With the floating ring moved to the right (3) the reverse action takes place, the lower pistons moving outwards drawing oil through ports F and the upper

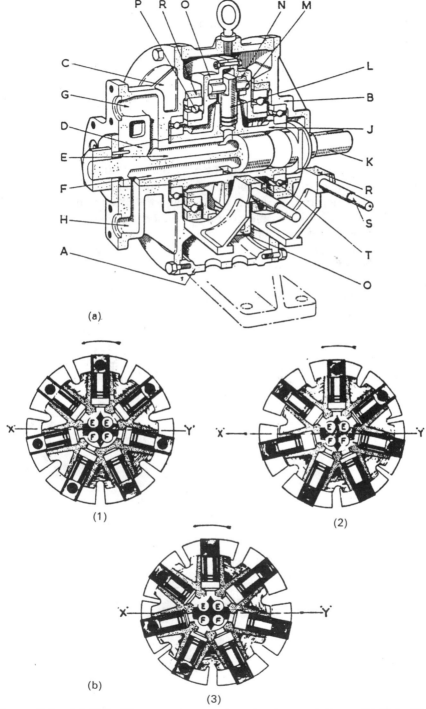

Figure 9.4 *(a) Hele-Shaw pump, see text for key to letters; (b) Hele-Shaw pump (1) Floating ring central (2) Floating ring moved to left (3) Floating ring moved to right. (John Hastie & Co Ltd)*

pistons moving into the cylinders and discharging oil through ports E.

The direction of flow depends on the location of the floating ring, left or right of the centre. The floating ring can be moved to any intermediate position between the central and maximum positions; the quantity of oil discharged varies according to the amount of displacement of the floating ring from its mid-position.

Non-reverse locking gear

When two pumping units are fitted and only one is running, the idle pump might be driven in the reverse direction by fluid under pressure from the running pump, if non-reverse locking gear (Figure 9.5) were not fitted. This gear is integral with the flexible coupling connecting motor and pump. It consists of a number of steel pawls so mounted on the motor coupling that, when pumping units are running, they fly outward due to the centrifugal effect and remain clear of the stationary steel ratchet secured to the motor supporting structure.

The limit of this outward movement is reached when the pawls contact the surrounding casing, which revolves with the coupling.

When the pumps stop, the pawls return to their normal, inward position and engage the ratchet teeth, so providing a positive lock against reverse rotation. This action is automatic and permits instant selection and commissioning of either unit without needing to use the pump isolating valves, which are normally open – and are only closed in an emergency.

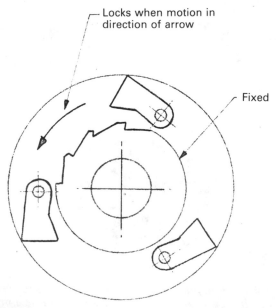

Figure 9.5 *Non-reverse locking gear to stop pump idling (Dean)*

The swash plate axial cylinder pump

This pump (Figure 9.6) has a circular cylinder block with axial cylinders disposed on a pitch circle around a central bore which is machined with splines to suit the input shaft with which it revolves. The individual cylinders are parallel with the shaft, with one end of each terminating in a drilled port at the end face of the block. This face bears against a stationary valve plate and is maintained in contact by spring pressure. The spring compensates automatically for wear. Semi-circular ports in the valve plate, in line with those from the cylinders, are connected by external pipes to the steering cylinders. The connection is often direct to the cylinders for two ram gears, but by way of a change over valve chest in four ram gears.

In the Mark III design, the cylinder barrel is driven by the input shaft through a universal joint and the valve plate contact springs are supplemented by hydraulic pressure. Each cylinder contains a piston, connected by a double ball-ended rod to a socket ring driven by the input shaft through another

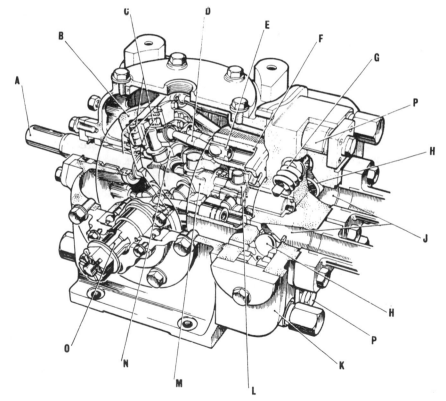

Figure 9.6   *Arrangement of 'VSG' Mark III pump (Vickers Ltd)*

A. Input shaft	F. Cylinder barrel	L. Barrel joint
B. Tilting box	G. Relief valve	M. University joint
C. Roller bearings	H. Replenishing valve	N. Socket ring
D. Connecting rod	J. Ports	O. Control trunnion
E. Piston	K. Valve plate	P. Control cylinder

universal joint and rotating on roller thrust bearings (in some cases on Michell pads) within a tilt box. This is carried on trunnions and can be tilted on either side of the vertical by an external control; the telemotor for a steering gear.

When the tilt box is vertical, the socket ring, cylinder barrel and pistons all revolve in the same plane and the pistons have no stroke. As the box is tilted, and with it the socket ring, stroke is given to the pistons at each half revolution, the length of stroke being determined by the angle of tilt.

The slipper pad axial cylinder pump

This is another development of the pump described above, suitable for the higher pressures demanded as steering gear and fin stabilizer systems were developed. The socket ring and connecting rods are replaced by slipper pads in the tilt box, the spherical ends of the pistons being carried in the pads. Inclination is given to the tilt box by a servo piston, which is operated by hydraulic pressure. Figure 9.7 shows a cut-away section.

Another variant is the 'Sunstrand' pump (Figure 9.8) in which a reversible swashplate, vertically disposed and given the desired angular rotation by an integral servo-piston, is used to vary the quantity and direction of the hydraulic fluid.

Hydraulic telemotor

The telemotor has become, on many vessels, the stand-by steering control mechanism, used only when the electric or automatic steering fails. It comprises a transmitter on the bridge and a receiver connected to the steering gear variable delivery pump, through the hunting gear. Transmitter and receiver are connected by solid drawn copper pipes. Liquid displaced in the transmitter causes a corresponding displacement in the receiver and movement of the pump control through the hunting gear.

The transmitter (Figure 9.9) consists of a cylinder with a pedestal base which contains a piston operated by a rack and pinion from the steering wheel. The make-up tank functions automatically through spring loaded relief and make-up valves. Excess pressure in the telemotor system causes oil to be released through the relief valve to the make-up tank and loss of oil is made up through the lightly loaded make-up valve. The two valves are connected through a shut-off valve, which is normally left open, and the bypass which connects both sides of the pressure system, when the piston is in mid position. There is also a hand operated bypass.

The tank must be kept topped up. The non-freezing working fluid is normally a mineral oil of low viscosity and pour point, which gives some protection against rusting. Before the general use of mineral oil, it was common to employ a mixture of glycerol and water as the low pour point (non-freezing) working medium.

The section through the receiver, shows two receiving cylinders in one

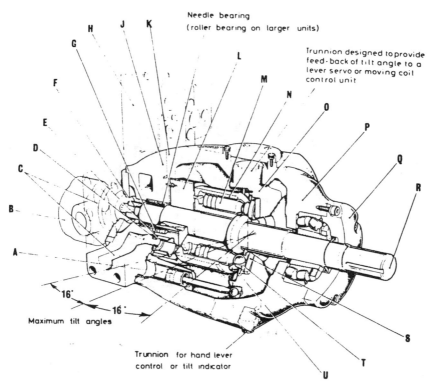

Needle bearing
(roller bearing on larger units)

Trunnion designed to provide feed-back of tilt angle to a lever servo or moving coil control unit

16°

16°

Maximum tilt angles

Trunnion for hand lever control or tilt indicator

Figure 9.7 *Arrangement of 'VSG' slipper pad (Vickers Ltd)*

A. Control piston	H. Valve block	P. Casing, end section
B. Auxiliary pump	J. Casing, centre	Q. End cover
C. Ports	K. Casing, centre	R. Shaft
D. Floating seal	L. Floating film face ring	S. Thrust plate
E. Floating piston	M. Working piston	T. Slipper pad
F. Splined coupling	N. Cylinder barrel	U. Retaining plate
G. Compensating piston	O. Tilt box	

casting, with circuit pipes connected to the outer end of each ram. Any fluid displaced in the transmitter cylinder by the piston will therefore be forced through the pipes and circuit valve to the receiving cylinder. The receiver rams are fixed in the arrangement shown and any displacement of the fluid causes the cylinder body to move along the rams against the compression of one of the springs. The compressed spring serves to return the receiver, and with it the transmitter, to midship position when the helmsman releases the steering-wheel. The linkage to the pump control through the hunting lever, is fitted to one end of the cylinder body.

Variations of the hunting gear floating lever

Figure 9.10 shows diagrammatically three variants of the floating lever arranged as a control and cut-off gear. In each diagram, the control movement

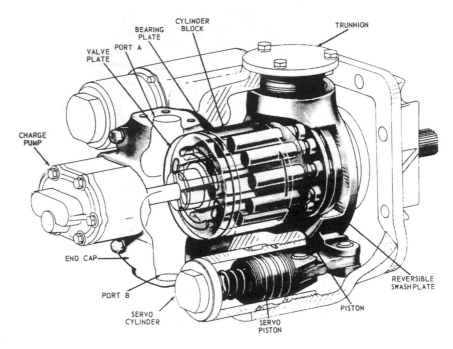

Figure 9.8   *Sunstrand pump*

is applied at point A, point B is linked to the body whose movement is to be controlled and point C is linked to the control mechanism of the power source i.e. in a ship's steering gear, movement of point A is controlled by the helmsman, point B, known as the hunting point, is linked to a point on the tiller and point C is linked to the control lever(s) on the pump(s). The floating lever pivots alternately about points B and A while executing the control and cut-off functions respectively so that the lever 'floats' in space and, as can be seen in the diagrams, the distance between one or more pairs of attachment points on the lever varies as movement takes place. In consequence, the motion of the lever is extremely complex and is difficult to visualize as a continuous process. However, if the geometry of the system is examined step by step at each of its rest positions, the principles of operation become clear.

To render the geometry of the system determinate and to ensure that it functions correctly, it is prudent to impose the following limitations on the design:

(a)   Either point A or point C, but not both, should occupy a fixed location in the length of the lever: the other must be left free to move longitudinally in the lever to accommodate the variation that occurs in the distance between them.

(b)   Points A and C should be constrained to move on known loci to ensure the accuracy of the system,

(c)   Point B should be free to move longitudinally in the lever.

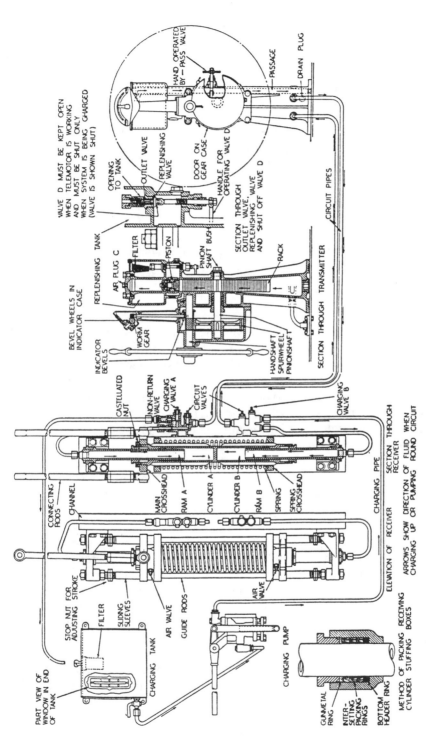

Figure 9.9 Patent hydraulic steering Telemotor (Brown Bros & Co Ltd)

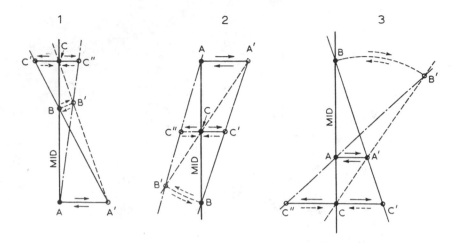

Figure 9.10 *The floating lever (Brown Bros & Co Ltd)*

A. Control point   B. Hunting point   C. Pump control point

In the three diagrams (Figure 9.10) the control point A has been selected as the datum point and constrained to move only in a straight line such as AA', the hunting point B is free to slide in a slot in the floating lever and the pump control point C is arranged to slide in another slot and constrained to move along a straight line such as CC'.

Now, if we assume a ship to be proceeding on a straight course, points A, B and C in each of the diagrams lie on the mid line since the steering controls and the rudder are centralized and the pump is in the no-stroke or neutral condition. Movement of point A to A', which corresponds with the rudder angle required, causes the floating lever to pivot about point B and point C moves to C' placing the pump(s) on-stroke in the correct sense. As the rudder moves over towards the angle 'ordered' by the movement of A to A', point B, which is linked to the tiller, moves towards B' with the floating lever now pivoting about A', being held there by the helmsman, causing point C' to return to C. Points B'and C' arrive at B and C simultaneously, placing the pumps in the no-stroke condition and bringing the rudder to rest at precisely the angle ordered. If, now, the control point is moved back from A' to the mid position A, the floating lever pivots about B', point C moves to a new position C'' on the opposite side of the mid line, placing the pump(s) on-stroke to drive the steering gear back towards the mid position. The lever again pivots about A as the gear returns, point B' moves back to B, causing C'' to move back to C thus placing the pump(s) in the no-stroke condition and bringing the rudder to rest in the mid position. Movement of point A to a position A'' (not shown) on the opposite side of the mid line and then back to A would have a similar effect except, of course, that the rudder movements and all the hunting movements would also occur on the opposite side of the mid line to that shown.

The mechanical arrangement of a floating lever system may not always be in one of the three forms illustrated but, if the principle is understood and it is kept in mind that its basic functions are to initiate movement of the steering gear and to stop movement when the rudder arrives at the angle ordered by the helmsman, the reader should have no difficulty in recognizing any floating lever arrangement he may encounter.

Detailed description of two ram gear hydraulic system

With reference to the sketch (Figure 9.11), the duplicate power units PU1 and PU2, each have a continuously running electric motor driving, through a flexible coupling a variable delivery axial-cylinder pump and auxiliary pumps, A1, A2. The latter draw filtered oil from the reservoir T and discharge through a 10 micron filter F10 to supply oil at constant pressure to the servo-controls SC, to the automatic change-over valves CO, to maintain a flow of cool oil through the main pump casings and to make up any oil loss from the main

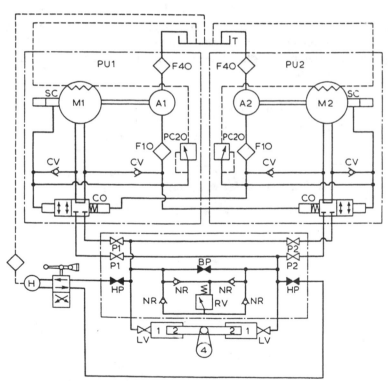

Figure 9.11 *Hydraulic diagram for typical two-ram electro-hydraulic steering gear*

PU1, PU2,	Power units	PC20	Pressure limiting	BP	By-pass valve
A1, A2,	Auxiliary pumps		valves	RV	Relief valve
T	Reservoir	CV	Check valves	HP	Hand-pump shut-off
F10	Filters	P1, P2	Isolating valves		valves
SC	Servo-controls	LV	Locking valves	WP	Non-return valves
CO	Changeover valves				

system. When the main pumps are at no-stroke, the auxiliary pumps discharge to the reservoir via a pressure-limiting valve PC20, set at 20 bar, and to the pump casings. When the main pumps are on-stroke, the auxiliary pumps discharge to the main pump suction.

A main pump may be brought into operation at any position of the gear, at any time, by starting the motor. The servo-operated automatic change-over valves (Figure 9.12) are held in the bypass condition by a spring, while the associated pump is at rest. When a pump is started, the auxiliary pump pressure builds up, overcomes the spring, closes the bypass and connects the main pump to the hydraulic system. Thus, the main pump starts in the unloaded condition; it cannot be motored when idle by cross pressure flow and load is held off until the electric motor high starting current has reduced to running level. When a pump is stopped, the spring returns the valve to the bypass condition. The spring end of the valve is connected to the constant pressure line and, to obviate hydraulic locking, the spring chamber has a bleed line. From the automatic change-over valves CO the main pump discharge passes to the pump isolating valves, P1 and P2, and to the cylinders, through the locking valves LV. These valves are incorporated in a group valve chest so arranged as to provide cross-connections with the bypass valve BP, relief valve RV and the emergency hand pump shut-off valves, HP, with appropriate non-return valves NR.

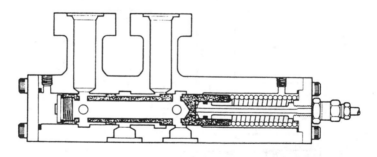

VALVE IN STRAIGHT THROUGH CONDITION

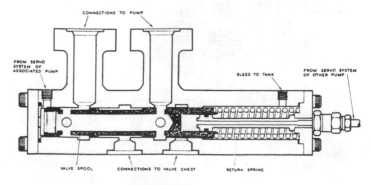

VALVE IN BY-PASS CONDITION

Figure 9.12   *Servo-operated automatic change-over valve*

In open water it is usual to have one power unit in use. If quicker response is required from the gear, two units may be run simultaneously to double oil flow and increase speed of operation.

Normally the gear is controlled from the bridge through an electric telemotor and local control box, but a local control handwheel is also provided as is a means of communication with the bridge.

Enclosed hunting gear

The handwheel (Figure 9.13) for local control has a bevel gear which engages with a similar wheel on the screwed shaft, when a catch (detent) is lifted and the handwheel is pushed in. The screw shaft is normally turned to the required position, by a split field electric motor.

The light construction of the combined control and hunting gears is possible because the forces concerned are moderate. The self-contained unit is self-lubricating, and contained in an oil-tight case. Steering gear pump output and therefore rudder movement, is controlled by a floating lever, one end of which is moved by the control motor (or telemotor), the other end by the movement of the tiller. A rod attached to its mid-point and to the pump control lever, puts the pump on stroke, in response to movement of the floating lever by the control motor or telemotor. As the tiller moves, the cut-off linkage acts to counteract the movement and brings the gear to rest by restoring the pump

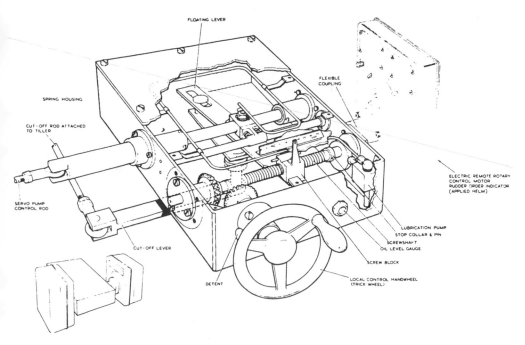

Figure 9.13 *Control box, Electrical remote input and local mechanical input version (Brown Bros & Co Ltd)*

control lever to the no-stroke position. Spring links, suitably disposed, obviate over-stressing of the mechanism.

The end of the floating lever connected to the remote control mechanism, is attached to a block which moves along the screw shaft when the latter is rotated by the control motor or by the local control handwheel. Stroke is restricted by stop collars.

Detailed description of four ram gear hydraulic system

The drawing of a four ram gear (Figure 9.14) shows the Rapson slides, the combined guide and brace arrangement previously described and the control box with its connecting link from the rudder stock. A second link from the stock to the rudder angle indicator transmitter, is also shown.

The hydraulic circuit (Figure 9.15) incorporates an arrangement of stop and bypass valves in the chest VC, which enable the gear to be operated on all four or on any two adjacent cylinders but not with two diagonally disposed cylinders. Inactive cylinders are isolated from the pumps by valves while the bypass valves connecting them, are opened to permit free flow of idle fluid. Either or both duplicate independent power units may be employed with any

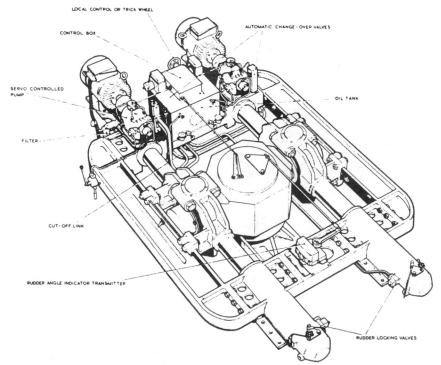

Figure 9.14 *Four-ram electro-hydraulic steering gear with electric control (Brown Bros & Co Ltd)*

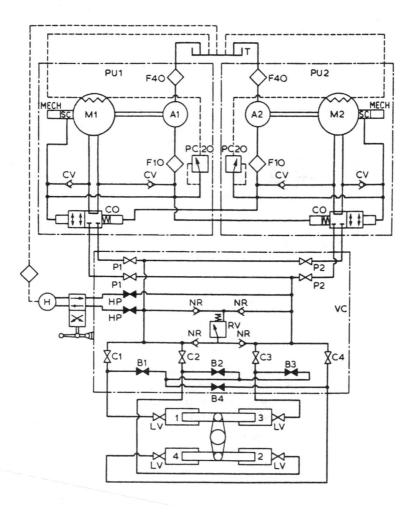

Figure 9.15 *Hydraulic circuit for four-ram electro-hydraulic steering gear (Brown Bros & Co Ltd)*

B1-B4.   By-pass valves   C1-C4.   Cylinder isolating valves
For other key letters see Figure 9.11

usable combination of cylinders. It will be seen that the torque available from two cylinders is only one-half of that from four, even when both power units are working, though the speed of operation will be increased if both are used.

The mechanical arrangement of the control gear and the basic hydraulic system, in all but their layout (Figure 9.15) are identical with the two ram gear already described. The valve chest however, must cater for four cylinders in all useful combinations. This demands four cylinder isolating valves, C1–C4, and four bypass valves, B1–B4. The emergency hand pump arrangement, its directional control valve, the main system relief and the locking valves remain unchanged, as do the remote, local and emergency control arrangements.

Normally, the pump and the four cylinder isolating valves, P1, P2, C1–C4, and the rudder locking valves LV, are open. The bypass valves B1–B4 and the emergency hand pump isolating valves HP are closed. Power units may be brought into action or shut down, by starting or stopping the associated motors.

To change from four ram to two ram working, it is only necessary to make two cylinders inoperative by closing their isolating valves C1–C4 and opening the bypass valves between them. For example to steer on cylinders 1 and 3, valves C2 and C4 are closed; B2 and B4 are opened so that cylinders numbers 2 and 4 are isolated from the main hydraulic system and the oil in them is free to flow from one to the other. The cylinder isolating valves and the bypass valves are shown as separate items in diagram but each pair may be combined as a double seating valve so that, as any cylinder is isolated from the main hydraulic system it is automatically opened to a bypass manifold and to the other inoperative cylinder.

Four ram gear with servo-controlled axial cylinder pumps

Variants of the servo-controlled swash plate axial cylinder pump (Figure 9.16) are capable of working at 210 bar. Each pump is complete with its own torque motor, servo-valve, cut-off mechanism, shut-off valve and oil cooler. These pumps are brought into operation as described earlier and an idle pump is prevented from motoring.

The rotating assembly of the pump, which consists of cylinder, nine pistons, valve plate, slippers, slipper plate and retaining ring, is manufactured from EN8 steel, which is finally machined, heat treated and then hardened for long wear. The nine pistons are fitted with return springs. The casing and covers are of nodular cast iron. The main valve block is of EN8 steel and houses five check valves, main pump relief valve, boost and servo-relief valves, boost gear pump and servo-gear pumps. Piping from the valve block supplies the servo pistons via the servo valve. The main drive is through a splined shaft to the cylinder

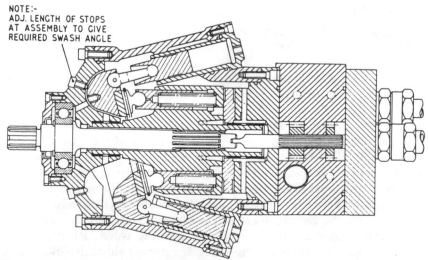

NOTE:-
ADJ. LENGTH OF STOPS
AT ASSEMBLY TO GIVE
REQUIRED SWASH ANGLE

Figure 9.16 *Hastie axial-cylinder pump (John Hastie and Co Ltd)*

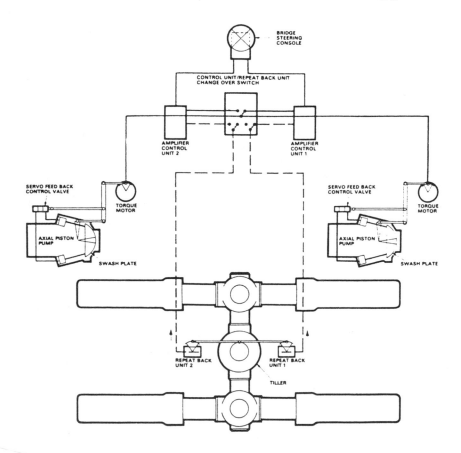

Figure 9.17 *The control system for the Hastie pump (John Hastie & Co Ltd)*

body of the pump. The pump swash plate is actuated by two servo cylinders which receive oil at the desired pressure through a directional servo-valve. The servo-valve is displaced initially by the torque motor acting on the input signal demand and is returned to the neutral position by the hunting linkage connected to the swash plate. The hunting action is achieved through the application of a simple lever system connecting input and displacement servo-valve and the hunting action from the pump swash plate angle. This allows for a very fast response.

The control system is shown in Figure 9.17. As will be seen, a torque motor, receiving the appropriate signal from the bridge through an amplifier unit actuates the floating lever, putting the pumps on stroke in response. The hunting action of the floating lever is no longer required as the normal control of the steering gear from the bridge is by electric signal. The signal is directed to the torque motor which operates the servo valve that in turn controls the pump. When the steering gear has attained the required rudder angle, the electric feed back unit connected directly to the rudder-stock cancels the input signal to the control amplifier, and the steering gear is held at that angle until another rudder movement is required.

This form of control eliminates the need for mechanical linkage and hunting gear on the steering gear.

Later modifications

The *Amoco Cadiz* disaster focused attention on the fact that failure of the common hydraulic pipe system of a four ram steering gear with duplicated power units, could result in rapid discharge of oil from the circuit and loss of steering capability. Four ram or double vane type gears with duplicated hydraulic circuits, as well as duplication of pumps, were developed. This arrangement cannot, however, be operated with both pumps running and the duplicated hydraulic circuits isolated from each other. The systems have to be connected in common for operation with both pumps. Either pump can provide hydraulic power for the combined circuit or for an isolated half, with the bypass open on the other part.

Figure 9.18 illustrates a four ram gear which complies with the International Convention for the Safety of Life at Sea (SOLAS) regulations (1974, amended 1981) relating to tankers, chemical tankers or gas carriers of 10 000 gt and upwards (see especially Chapter 11-1 Regulation 29, Paragraph 16).

Two main power and servo-power units draw from a two-compartment tank fitted with oil level switches arranged at three levels. Level I gives an initial alarm following loss of oil from either system. In normal operation one or

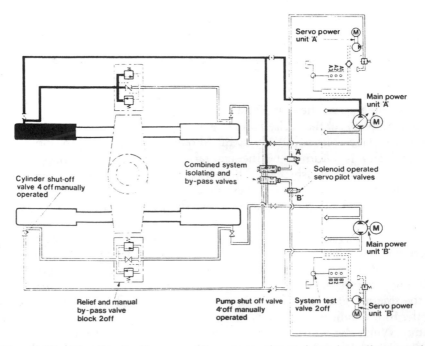

Figure 9.18   *The Hastie-Brown split system, shown here in split operation, is arranged to give two ram operation automatically in the event of loss of fluid from one system*

both power units provide hydraulic power to all four rams. Continued loss of oil initiates one or both of the level 2 switches. These energize their respective solenoid operated servo valves, causing the combined isolating and bypass valves to operate, splitting the system such that each power unit supplies two rams only. At the same time if one power unit is stopped it is automatically started.

Further loss of oil, and the system on which it is occurring will operate one of the level 3 switches. This will close down the power units on the faulty side. Steering then continues, uninterrupted but at half the designed maximum torque on the sound system. The defective system is out of action and isolated.

Vane type gear

These may be regarded as equivalent to a two-ram gear, with torque capacities depending on size. An assembly of two rotary vane gears, one above the other, provides the security of a four ram gear. Figure 9.19 illustrates the principle.

The rotor C is fitted and keyed to a tapered rudder stock, the stator B is secured to the ship's structure. Fixed vanes, secured equidistantly in the stator bore and rotating vanes secured equidistantly in the rotor, form two sets of pressure chambers in the annular space between the rotor and stator. They are interconnected by a manifold. Fluid supplied at pressure to one set of these chambers will rotate C clockwise and the rudder will turn to port, or to starboard if the alternate set is put under pressure.

Three fixed and three moving vanes are usual and permit a total rudder angle of 70°, i.e. 35° in each direction. The movement obtainable from a gear with two fixed and two moving vanes, may be 130°. Figure 9.19 (b) shows the detail of a typical unit.

The fixed and rotating vanes may be of spheroidal graphite cast iron. They are securely fixed to the cast steel rotor and stator by high tensile steel dowel pins and cap screws. Keys are also fitted along the length of the rotary vanes, for mechanical strength. Assembly of the gear would not be possible if the fixed vanes were keyed; they rely on the dowels to provide equivalent strength. The vanes fixing is considered to be of sufficient strength to make them suitable to act as rudder stops. Steel sealing strips, backed by synthetic rubber, are fitted in grooves along the working faces of the fixed and rotary vanes, thus ensuring a high volumetric efficiency, of 96–98% even at the relief valve pressure of 100 bar or over. Rotation of B is prevented by means of two anchor brackets, and two anchor pins. The anchor brackets are securely bolted to the ship. Vertical clearance is arranged between the inside of the stator flanges and the top and bottom of the anchor brackets to allow for vertical movement of the rudderstock. This clearance varies with each size of the rotary vane unit, but is approximately 38 mm in total and it is necessary that the rudder carrier should be capable of restricting the vertical movements of the rudderstock to less than this amount.

The method of control for these gears and also for the hydraulic supply system is as described for electro-hydraulic gears.

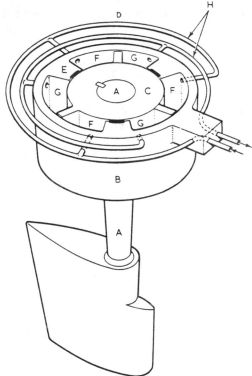

Figure 9.19 (a) *Rotary vane operation*

A. Rudderstock
B. Stator

C. Rotor
E. Fixed vanes

F. G. Pressure chambers
H. Manifolds

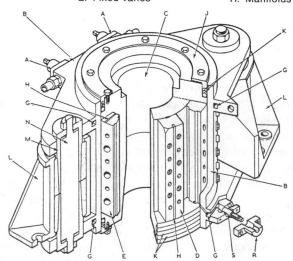

Figure 9.19 (b) *Rotary vane unit*

A. Stop valves
B. Stator
C. Rotor
D. Rotary vanes
E. Fixed vanes

G. Upper (starboard) and
lower (port) manifolds
H. Steel sealing strips
J. Gland
K. Gland packing

L. Anchor bracket
M. Span bush
N. Anchor bolt
R. Stop valve casting
S. Stop valve spindle

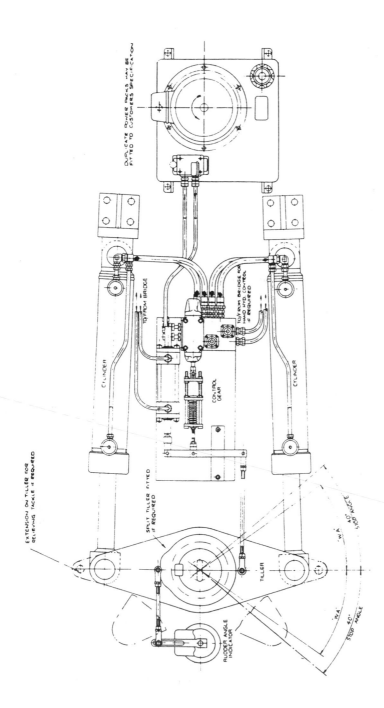

DUPLICATE POWER PACKS MAY BE
FITTED TO CUSTOMERS SPECIFICATION

EXTENSION ON TILLER FOR
RELIEVING TACKLE IF REQUIRED

CYLINDER

TO/FROM BRIDGE

CONTROL
GEAR

TO/FROM BRIDGE FOR
HAND HYD. CONTROL
IF REQUIRED

CYLINDER

SPLIT TILLER FITTED
IF REQUIRED

RUDDER ANGLE
INDICATOR

TILLER

'WA'
40°
STOP ANGLE

'WA'
40°
STOP ANGLE

Figure 9.20 *Electro-hydraulic steering gear (hydraulic control). General arrangement of actuator type steering gears (Brown Bros & Co Ltd)*

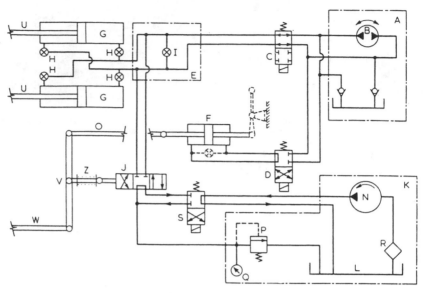

Figure 9.21 *Hydraulic circuit diagram for hand and power steering. For key to letters, see text*

Steering gear with constant output pump

A steering gear as with any hydraulic system, can be operated by a constant delivery pump as an alternative to the conventional variable delivery type. Output from the pump, which runs continuously, is circulated through a bypass until required for steering gear movement.

Small hand and power gears

A simpler variant of the electro-hydraulic gear, for small ships requiring rudder torques below say, 150 kNm is shown in Figure 9.20. The hydraulic circuit is shown diagrammatically in Figure 9.21.

The rams (U) in the double-acting steering cylinders G, which are free to oscillate on chocked trunnions, are linked directly to the tiller. F is a double-acting control cylinder, linked to the floating lever V by rod O. J is a directional control valve linked to the mid point of the floating lever by a spring link Z. W is the cut-off link from tiller to floating lever. H is a locking valve and I a bypass valve.

Valves C, D and S are solenoid controlled. When not energized, C is open and D is closed to through flow but acts as a bypass between the ends of control cylinder F when solenoid S is closed. When the solenoids are energized, C is closed, D and S are open. J is only operative when steering by power.

When steering by hand, C, D and S are not energized. A pump B in the steering pedestal, coupled to the wheel, deliver fluid under pressure direct to the steering cylinders G, so moving the tiller in the sense and to the extent appropriate to the movement of the wheel. There is no hunting action.

To change to power steering, the power pump is started and C, D and S are energized, i.e. C is closed, D ceases to be a bypass and it connects the steering pedestal pump to the control cylinder F. S, now open, allows fluid to pass from the power pump to J, which now comes under the influence of the floating lever. Steering-wheel movement now moves the piston in F, the floating lever pivots on its attachment to W, and J opens to allow the power pump to discharge to the appropriate ends of the steering cylinders G. As the tiller moves, a hunting movement occurs, the cut-off link W acting on the floating lever which, pivoting on its attachment to O, closes J and brings the gear to rest with the rudder at the angle required.

Rudder angle indicators, either mechanically linked or electrically powered, are fitted as required. For local control, the ends of the control cylinder are made common by opening a bypass valve and the control piston is moved by a hand lever (shown dotted) or, for example, by wheel, rack and pinion. It will be seen that in the off-loaded condition, the pump discharge circulates through J.

Emergency steering is by relieving tackles, fitted when the rudder is locked by closure of the valves H.

If hand steering only is required, the gear is reduced to tiller, cylinders and rams, locking and bypass valves, rudder indicator and steering pedestal with pump. A simple form of this gear for torques below 11 kNm is shown in Figure 9.22.

Figure 9.22 *Hydraulic steering gear*

Steering gear failures and safeguards

The vital importance of the steering gear is reflected in the regulations of the government department with responsibility for shipping (usually Department of Transport or Coastguard) the requirements of the classification societies (Lloyd's Register, American Bureau of Shipping, Bureau Veritas and others) and the recommendations of the International Maritime Organization (IMO).

Some general requirements for steering gears, based on the various regulations and SOLAS 1974, are given below:

(1)   Ships must have a main and an auxiliary steering gear, arranged so that the failure of one does not render the other inoperative. An auxiliary steering gear need not be fitted, however, when the main steering gear has two or more identical power units and is arranged such that after a single failure in its piping system or one of its power units, steering capability can be maintained. To meet this latter alternative the steering gear has to comply with the operating conditions of paragraph 2 – in the case of passenger ships while any one of the power units is out of operation. In the case of large tankers, chemical tankers and gas carriers the provision of two or more identical power units for the main steering gear is mandatory.

(2)   The main steering gear must be able to steer the ship at maximum ahead service speed and be capable at this speed, and at the ship's deepest service draught, of putting the rudder from 35° on one side to 30° on the other side in not more than 28 secs. (The apparent anomaly in the degree of movement is to allow for difficulty in judging when the final position is reached due to feedback from the hunting gear which shortens the variable delivery pump stroke.) Where the rudder stock, excluding ice strengthening allowance, is required to be 120 mm diameter at the tiller, the steering gear has to be power operated.

(3)   The auxiliary steering gear must be capable of being brought speedily into operation and be able to put the rudder over from 15° on one side to 15° on the other side in not more than 60 secs with the ship at its deepest service draught and running ahead at the greater of one half of the maximum service speed or 7 knots. Where the rudder stock (excluding ice strengthening allowance) is over 230 mm diameter at the tiller, then the gear has to be power operated.

(4)   It must be possible to bring into operation main and auxiliary steering gear power units from the navigating bridge. A power failure to any one of the steering gear power units or to its control system must result in an audible and visual alarm on the navigating bridge and the power units must be arranged to restart automatically when power is restored.

(5)   Steering gear control must be provided both on the bridge and in the steering gear room for the main steering gear and, where the main steering gear comprises two or more identical power units there must be two independent control systems both operable from the bridge (this does not mean that two steering-wheels are required). When a hydraulic telemotor is used for the control system, a second independent system need not be fitted except in the case of a tanker, chemical carrier or gas carrier of 10 000 gt and over. Auxiliary steering gear control must be arranged in the steering gear

room and where the auxiliary gear is power operated, control must also be arranged from the bridge and be independent of the main steering gear control system. It must be possible, from within the steering gear room, to disconnect any control system operable from the bridge from the steering gear it serves. It must be possible to bring the system into operation from the bridge.

(6) Hydraulic power systems must be provided with arrangements to maintain the cleanliness of the hydraulic fluid. A low level alarm must be fitted on each hydraulic fluid reservoir to give an early audible and visual indication on the bridge and in the engine room of any hydraulic fluid leakage. Power operated steering gears require a storage tank arranged so that the hydraulic systems can be readily re-charged from a position within the steering gear compartment. The tank must be of sufficient capacity to recharge at least one power actuating system.

(7) Where the rudder stock is required to be over 230 mm diameter at the tiller (excluding ice strengthening) an alternative power supply capable of providing power to operate the rudder, as described in paragraph 3 above, is to be provided automatically within 45 seconds. This must supply the power unit, its control system and the rudder angle indicator and can be provided from the ships emergency power supply or from an independent source of power located within the steering compartment and dedicated for this purpose. Its capacity shall be at least 30 minutes for ships of 10 000 gt and over and 10 minutes for other ships.

Steering gear testing

Except in the case of ships regularly engaged on short voyages, the steering gear should be thoroughly checked and tested within 12 hours before departure. These tests should include testing of power unit and control system failure alarms, the emergency power supply (when relevant) and automatic isolating arrangements.

Every three months an emergency steering drill should be held and should include direct control from within the steering compartment at which time the use of the communications procedure with the navigating bridge should be practised.

Further reading

Cowley, J. (1982) 'Steering Gear: New Concepts and Requirements', *Trans I Mar E,* **94**, paper 23.

The International Convention for the Safety of Life at Sea (1974) as amended in November 1981, Chapter 11–1: Regulation 29.

The Merchant Shipping (Passenger Ship Construction and Survey) Regulations 1984, HMSO.

The Merchant Shipping (Cargo Ships Construction and Survey) Regulations 1984, HMSO.

'Steering a Safer Course', *MER*, October 1988 p. 31.

10

Bow thrusters, stabilizers and stabilizing systems

Bow thrusters

The transverse thruster, installed in the bow and/or the stern, has become an essential item of equipment on many vessels. It enables the normal process of docking to be managed without tug assistance because the vessel is made more manoeuvrable at low speeds. Safety is increased when berthing in adverse weather conditions provided that the required thruster capacity has been correctly estimated. Transverse thrusters are installed to facilitate the positioning of some types of workboats. Some craft have thrust units for main propulsion and azimuth thrusters with computer control for position holding.

Thrust calculations must be based on the above water profile of a ship as well as the under water area. For passenger ships and ferries, the above water area may be three times that of the under water lateral area. For loaded tankers and bulk carriers, the situation is reversed but the unloaded profile must also be considered.

The regular and frequent use of electrically driven bow thrust units on ferries and other vessels operating on short sea routes means that motor windings are kept dry by the heating effect of the current. This helps to maintain insulation resistance.

There are potential problems with the electric motors and starters of infrequently used units, particularly where installed in cold, forward bow thrust compartments. They are subject to dampness through low temperature and condensation. Insulation resistance is likely to suffer unless heaters are fitted in the motor and starter casings. Space heaters may be fitted also. A fan is beneficial for ventilation before entry by personnel, but continuous delivery of salt laden air could aggravate the difficulties with insulation resistance.

Bow thrust compartments below the waterline should be checked frequently for water accumulation and pumped out as necessary to keep them dry. Vertical ducts for drive shafts should also be examined for water and/or oil accumulation. Flexible couplings with rubber elements quickly deteriorate if operating in oily water. Thruster shaft seals must be inspected carefully during preliminary filling of a drydock. Failure to detect and rectify leakage at this stage can be expensive later.

Bow thrusters with diesel drive

By installing diesel drives various problems are avoided, for example the very large power demand of electrically driven bow thrusters, the insulation problems associated with the windings and the complications involved with starting, speed control and reversing. For a conventional thruster in an athwartship tunnel, the diesel engine may be mounted at the same level as the propeller to provide a direct drive through a reverse/reduction gear. An alternative diesel arrangement (Figure 10.1) where space is limited, has the diesel mounted above the thruster. Both of the units shown, have horizontally mounted diesel engines with simple speed control through the fuel rack, and a reverse/reduction gearbox. The second arrangement requires an extra gearbox with bevel gears to accommodate change of shaft line. Flexible couplings are also fitted.

The reversing gearbox has ahead and astern clutches, with one casing coupled to the diesel engine shaft and a drive to the other clutch casing, through external gear teeth. The clutch casings rotate in opposite directions and whichever is selected, will apply drive, ahead or astern, to the output shaft. The engine idles when both clutches are disengaged.

Alternating current electric motor drives with pitch control

An alternating current (a.c.) induction motor of the (squirrel) cage type is used for many bow thrust units, with the motor being mounted above the

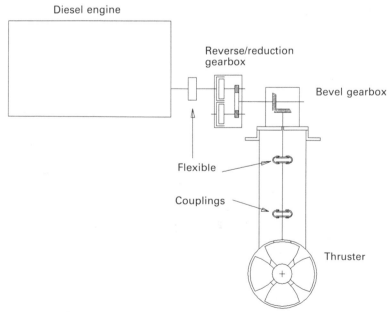

Figure 10.1 *Diesel bow thruster drive*

athwartships tunnel (Figure 10.2). Thrust is varied in direction and strength through a controllable pitch propeller.

This arrangement permits the use of a simple and robust induction motor, which operates at one speed. Starting current for a large induction motor tends to rise to about eight times the normal full load figure and to reduce this a star-delta or other low current starter is used. Low current starting implies low starting torque as well. It is important that the hydraulic system is operative and holding the propeller blades at neutral pitch when starting.

Pitch control for a thruster, is very similar to that for a controllable pitch propeller. The shaft of the lips arrangement shown, is hollow and has a flange to which the one-piece hub casting is held by bolts. The hub is filled with lubricating oil and there is free flow from the hub to the pod through the hollow shaft. The four blades are bolted to the blade carrier and have seals to prevent oil leakage. The pitch of the blades is altered by means of a sliding block, fitted between a slot in the blade carrier and a pin on the moving cylinder yoke. A piping insert in the hollow shaft connects the cylinder yoke to the oil

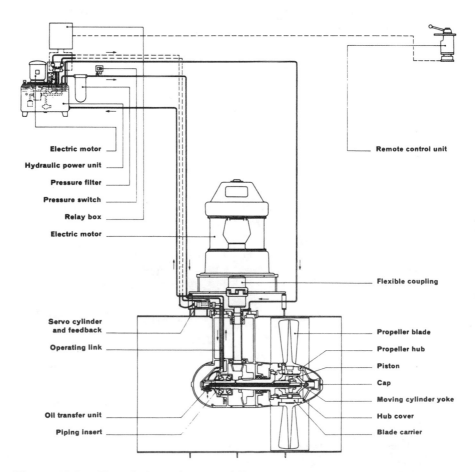

Figure 10.2   *Electric bow thruster drive*

transfer unit which contains a servo valve for follow-up pitch control. A mechanical connection between the oil transfer unit and the inboard servo cylinder facilitates accurate pitch settings and provides feedback for remote control. The hydraulic power unit is supplied with two safety valves, suction and pressure filters, a pressure gauge and pressure switch, as well as an electrically driven pump with a starter. To complete the equipment an electric switch is supplied which, in combination with the pressure switch, prevents the prime mover from starting when the pitch is in an off-zero position and/or no hydraulic pressure is available.

Hydraulic thruster

An external hydraulic drive motor can be used as the alterative to an electric motor but a design with the hydraulic unit within the bow thruster pod, was produced by Stone Manganese Marine. The variable displacement hydraulic pump (Figure 10.3) is powered by a constant speed, uni-directional electric motor or diesel prime mover connected through a flexible coupling. Pump output is controlled by means of a servo-control operated direct from the bridge (or locally) to give the required speed and direction to the hydraulic motor inside the thruster. The pod and propeller are suspended in a conventional athwartship tunnel below the waterline.

Other considerations and designs

The customary transverse thruster has a limited application because it is based in an athwartships tunnel. It cannot contribute to forward or reverse motion of the ship and ship speed must be less than four knots for it to be effective. Some schemes to improve performance have variously used double entry tunnels, shallow vee or curved tunnels and different flap arrangements.

The White Gill type thruster which is fitted on a number of existing ships, can provide thrust in any direction and is also used as the propulsion unit for some small craft.

The White Gill type thruster

This type of thruster (Figure 10.4) is positioned at the bottom of the hull so that the suction and discharge are at bottom shell plate level. Water is drawn in and discharged by a propeller through static guide vanes, much as with an axial pump. The guide vanes remove swirl and the water passes out as a jet through a rotatable deflector. The latter can be turned through 360°. The deflector has curved vanes, resembling in section a turbine nozzle, which produces a near horizontal jet of water. The deflector is rotated by a steering shaft which passes through a gland in the casing. This in turn is controlled from the bridge. No

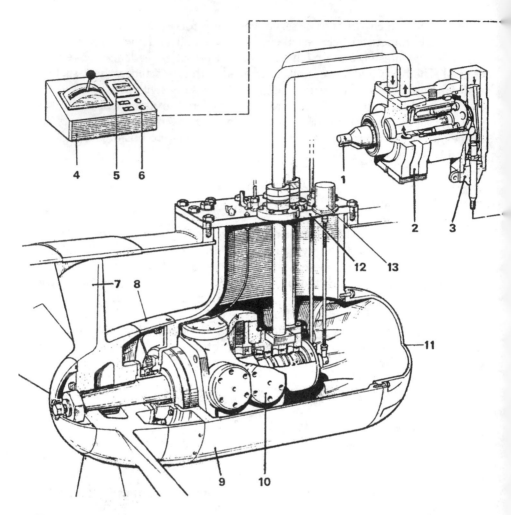

Figure 10.3   *Hydraulic drive bow thruster*

1. Prime mover output shaft	6. Running lights	10. Hydraulic motor
2. Variable delivery pump	7. Propeller	11. End cover
3. Servo valve assembly	8. Fairing cover	12. Mounting plate
4. Bridge control unit	9. Main casing	13. Tacho generator
5. Thrust indicator		

reverse arrangements are needed because thrust is available in any horizontal direction. The drive for the propeller may be applied vertically (Figure 10.4a) or horizontally (Figure 10.4b) depending on the design of unit installed.

Stabilizers and stabilizing systems

A ship at sea has six degrees of freedom, i.e. roll, heave, pitch, yaw, sway and surge (Figure 10.5). Of these, only roll can effectively be reduced in practice by fitting bilge keels, anti-rolling tanks or fin stabilizers. A combination of fins and

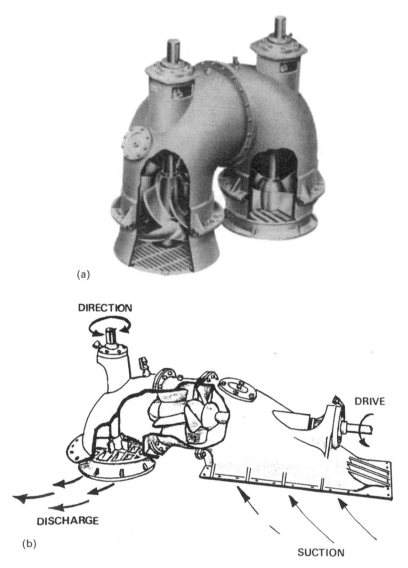

Figure 10.4 *(a) Vertically driven White Gill bow thruster; (b) Horizontally driven White Gill bow thruster*

tanks has potential advantages in prime cost and effective stabilization at both high and low speeds.

Since a ship is a damped mass elastic system, it has a natural rolling period and large rolling motions may be induced by resonance with relatively small wave forces. Large resonant rolls can be avoided by generating forces equal and opposite to the impressed sea force. Figure 10.6 shows that the roll amplitude at resonance is much greater than that at long wave periods. The ratio of these amplitudes is the dynamic amplification factor which is limited by

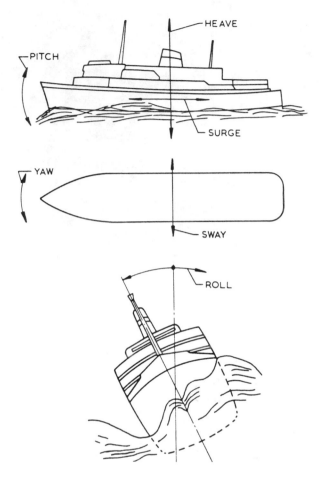

Figure 10.5   *Ship motions*

the inherent damping of the ship, i.e. viscous damping and the action of bilge keels.

Natural roll period

In Figure 10.7 G is the centre of gravity, B is the centre of buoyancy, W is the weight or displacement of the ship, θ the roll angle. M, the point of intersection of a vertical line through B with the centre line of the ship, is known as the metacentre and the distance GM as the metacentric height which is of great importance in relation to the rolling characteristic and the stability of the ship. When a stable vessel is heeled over, the couple W.GM.sin θ tends to restore it to the vertical whereas, in the unstable condition, the couple tends to turn the vessel further over. In the latter case, G is above M and GM is said to be negative, an unacceptable condition.

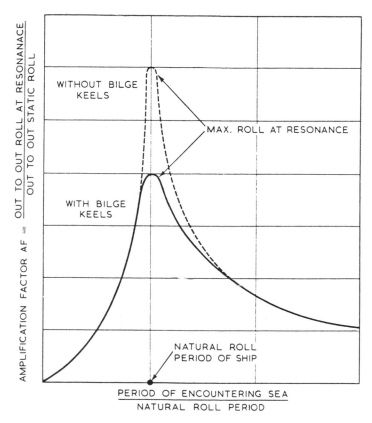

Figure 10.6 *Roll amplification factor*

The natural rolling period, i.e. the time taken to roll from out-to-out and back again is

$$T = \frac{C \times \text{beam}}{\sqrt{GM}}$$

where C is a constant determined by experience on ships of similar hull form and usually falls within the range 0.72 to 0.80 for merchant vessels.

For example, where maximum beam is 16 m GM 0.49 m and C is 0.80 then

$$T = \frac{0.80 \times 16}{\sqrt{0.49}} = 18.29 \text{ s.}$$

If the GM changes to, say, 0.64 m

$$T = \frac{0.80 \times 16}{\sqrt{0.64}} = 16.0 \text{ s.}$$

A 'stiff' vessel has a large GM and the natural period is short: a 'tender' vessel

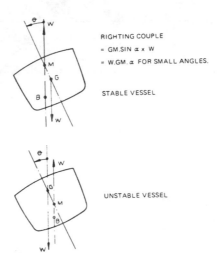

Figure 10.7   *Stable and unstable vessels*

has a small GM and the period is long. It is important to note that, particularly in cargo vessels, the draught and dead weight may change appreciably during a voyage hence the natural rolling period may also change, varying inversely as the square root of the GM.

A ship in a beam sea will roll to an angle $\theta°$ if the rolling moment applied by wave action is equal to that required to heel the ship $\theta°$ in calm water. When θ is small, the heeling moment is W.GM. θ (θ now expressed in radians), which provides a basis for calculations of the stabilizing force required.

Fin stabilizers

The stabilizing power of fins is generated by their movement through the sea and 'lift' created by the flow of water above and below the 'aerofoil' or hydrofoil shape. When the front edge of the fin is tilted up, water flow across the top of the profile produces lift due to a drop in pressure while a lifting pressure is provided by flow along the underside. Downward tilt of the forward edge of the fin, inverts the effect, so that a drop in pressure occurs at the underside and increased pressure at the top to give a downward force. Without a reasonable rate of forward movement of the ship, the small size of the fins makes them ineffective. Thus active fin stabilizers are fitted to the faster types of ship, operating at perhaps fifteen or more knots.

The hydrofoil section may be all-movable, with or without flaps or partly fixed, partly movable, (Figure 10.8). These fins are tilted, usually hydraulically, in phase with the roll at long wave periods, 90° out of phase at resonance and in phase with roll acceleration at short periods.

Non-retractable fins are commonly used where space within the hull is limited. They are usually fitted at the turn of the bilge and do not project beyond the vertical line from the ship's side or below the horizontal line of the

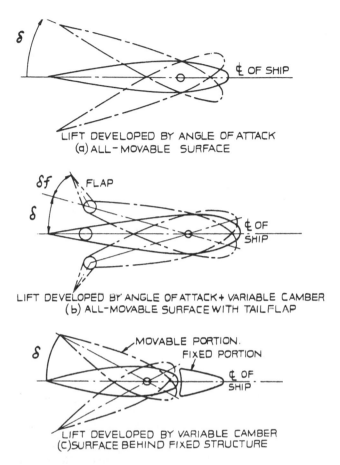

Figure 10.8 *Fin arrangements*

ship's bottom, to minimize the risk of contact with a quay wall or the dock bottom.

The fin shaft, to which the fin is rigidly attached, passes through a sea gland in a mounting plate welded or bolted to the hull and is supported by two substantial bearings. A double-ended lever keyed to the inner end of the finshaft is actuated by two hydraulic rams supplied from an electrically driven variable delivery pump (Figure 10.9).

Control of fin movement is automatic and is usually derived from gyroscopic sensing gear which, in its simplest form — velocity control — is based on one small, electrically driven gyroscope mounted horizontally with its axis athwartships. The angular velocity of roll of the ship causes the gyroscope to precess against centralizing springs to an amount proportional to the velocity and it generates a small force which is hydraulically amplified by a hydraulic relay unit to provide power sufficient to operate the controls of the variable delivery pump via suitable linkage. Part of the linkage is coupled to the fin shaft to transmit a cancelling signal to the pump control and to bring the fin to rest at the angle of tilt demanded by the sensing unit.

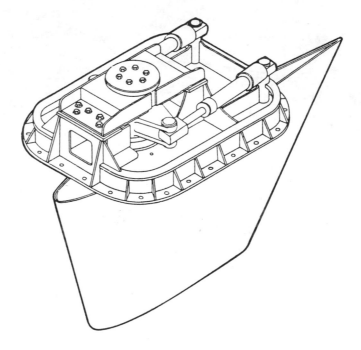

Figure 10.9   *Non-retractable fin assembly*

This type of control is often fitted in small installations, usually for economic reasons, and is most effective against resonant rolling.

Ships seldom roll in a purely resonant mode; the sea state is often highly confused. More elaborate, and more expensive, control systems are required to deal with suddenly applied roll, rolling at periods off resonance and rolling in conditions arising from the combination of several wave frequencies. A sensing unit based on a vertical-keeping gyroscope and a velocity gyroscope coupled into differentiating and summation units enables fin movement to be controlled by a composite function derived from roll angle, roll velocity and roll acceleration. By adding a 'natural list' unit, stabilization is achieved about the mean point of roll and so reduces both propulsion and stabilizing power demand. This is known as a compensated control system (Figure 10.10) and is generally used in large installations.

Roll reduction in excess of 90%, typically 30° out-to-out reduced to less than 3° out-to-out, can be achieved at resonance and low residual rolls can be maintained over a wide range of frequencies. However, since the stabilizing power varies as the square of the ship's speed, fins are least effective at low or zero speed where they function only as additional bilge keels.

Retractable fin stabilizers

Activated fin stabilizers which extend and stow athwartships, require spaces for fin boxes and guide bars of a length more than twice that of the fins, within

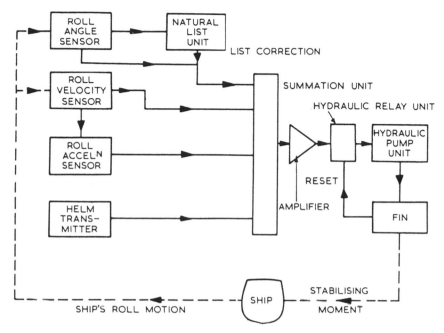

Figure 10.10 *Principles of Multra control system*

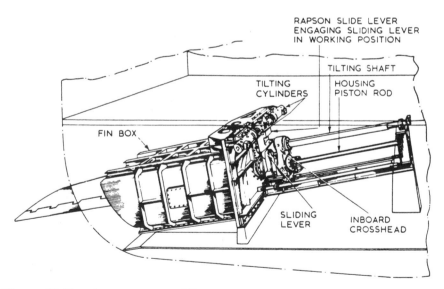

Figure 10.11 *Arrangement of Denny-Brown athwartships retracting fin*

the ship. The sketch of the general arrangement of a retractable fin (Figure 10.11) shows the amount of space required and the support seatings.

The finshaft (Figure 10.12) has a tapered outboard end to which the fin is keyed. The parallel inboard end passes through a sea gland on the inboard face of the fin box and is supported by two bearings. One, close to the inboard end

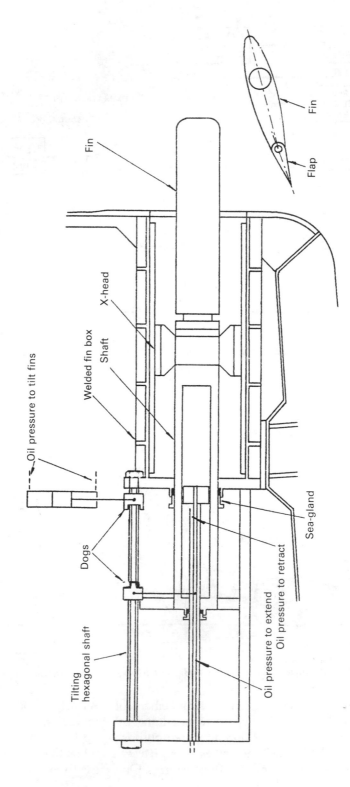

Figure 10.12 *Detail of retractable fin (R. C. Dean)*

of the fin, is carried in a heavy crosshead, arranged to slide in top and bottom guides within the fin box and the other in a crosshead slidably mounted on the extension guides, within the hull. The hollow bore of the parallel section of the finshaft houses a double acting piston to act as housing and extending gear. Tilting of the fin is by two or four hydraulic cylinders which may be of the simple oscillating type or arranged on the Rapson-slide principle as described for steering gears. Power units, control and sensing equipment are as for non-retractable fins.

Folding fin stabilizer

The Denny-Brown-AEG hinged or folding fin stabilizer (Figure 10.13) rotates to a fore and aft stowed position with the fins folded forward into the hull. The finshaft (Figure 10.14) is rigidly fixed into the crux which has a heavy vertical trunion that is housed in bearings top and bottom. The fin is free to oscillate on the finshaft and the tilting force is provided by a vane type servo-motor the stator of which is secured to the crux with the rotor being keyed to the fin through a flexible coupling. The vane servo-motor is housed in an oil-tight casing which is secured to the fin. There is a stainless steel sleeve fitted to the crux, for the casing lip seal. The whole of the casing and the interior of the fin is full of oil under pressure, to prevent the ingress of sea water.

Housing and extending the fin is achieved by a double acting hydraulic

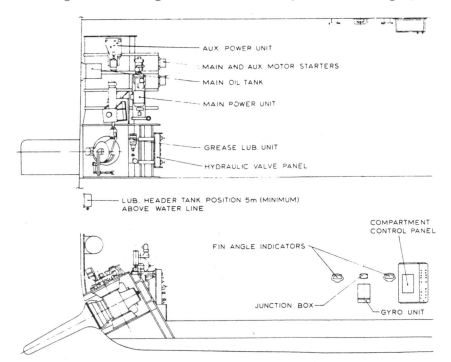

Figure 10.13 *Arrangement of Denny-Brown-AEG forward folding fin*

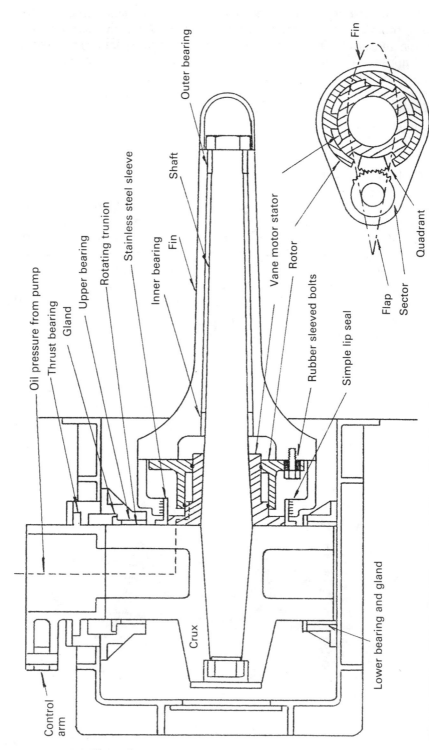

Figure 10.14 *Detail of folding fin stabilizer (R.C. Dean)*

Oil pressure from pump

Thrust bearing

Gland

Upper bearing

Rotating trunion

Stainless steel sleeve

Inner bearing

Fin

Shaft

Outer bearing

Vane motor stator

Rotor

Rubber sleeved bolts

Simple lip seal

Control arm

Crux

Lower bearing and gland

Fin

Quadrant

Flap

Sector

cylinder connected to the upper part of the trunnion. Power units, control and sensing equipment are generally similar to other types of stabilizer except that feed-back of fin angle is accomplished electrically by synchros.

Details of the variable delivery pumps also used for stabilizer operation, are given in Chapter 9 'Steering Gears'.

Tank stabilizers

Tank stabilizers (like bilge keels) are virtually independent of the forward speed of the vessel. They generate anti-rolling forces by phased flow of appropriate masses of fluid, usually water, in tanks installed at suitable heights and distances from the ship's centre line. Fluid transfer may be by open flume or from and to wing tanks connected by cross ducts. The tank/fluid combination constitutes a damped mass elastic system having its own natural period and capable of developing large forces at resonance with the impressed wave motion.

Since the fluid can only flow downhill and has inertia, it cannot start to move until the ship has rolled a few degrees (Figure 10.15). The natural restoring forces limit the maximum roll angle and initiate roll in the opposite sense. In the meantime the fluid continuing to flow downhill, piles up on the still low side and provides a moment opposing the ship motion. As the ship returns and passes its upright position, fluid again flows downhill to repeat the process.

The fluid flow tends to lag a quarter of a cycle behind the ship motion, a phase lag of approximately 90°, to generate a continuing stabilizing moment. This is due, mainly, to the transfer of the centre of gravity of the fluid mass away from the centre line of the ship. The transverse acceleration of the fluid generates an inertia force and thereby a moment, about the roll centre, which reduces the gravity moment when the tanks are below the roll centre and increases it when they are above. In practice, tanks may be placed 20% off the beam below the roll centre without serious loss of performance. Above the roll centre, other factors associated with the phase of fluid motion prevent augmentation of the gravity stabilizing power being realized. The phase lag may be increased, within limits, by placing orifice plates or grillages in the fluid flow path, to increase the damping.

In the wing tank system the mode of operation is similar to the simple flume but the tank geometry combined with the dynamic amplification of the flow tends to make fluid pile up to a greater height at a greater distance from the ship's centre line to give more effective stabilization. The wing tanks must be of sufficient depth to accommodate the maximum rise of the fluid without completely filling them. For purely passive action the tank tops are vented to atmosphere but in a controlled passive system, such as the Muirhead-Brown (Figure 10.16) they are connected by an air duct fitted with valves, controlled by a roll sensing device, which regulate the differential air pressure in the tanks to modify the natural fluid flow rate. This system will generate its full stabilizing power from a residual roll of about 7° out-to-out at resonance, due to the fact that dynamic amplification of the fluid motion may be from twice to

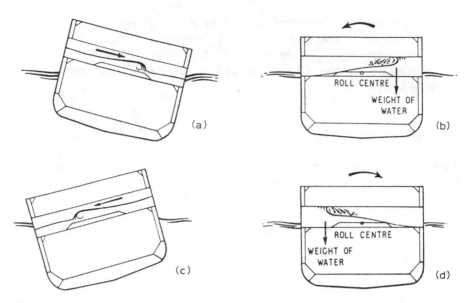

Figure 10.15   *Brown-NPL passive tank stabiliser: (a) Stern view of ship with passive tank rolled to starboard. The water is moving in the direction shown; (b) Ship rolling to port. The water in the tank on the starboard side provides a moment opposing the roll velocity; (c) Ship at the end of its roll to port. The water is providing no moment to the ship; (d) Ship rolling to starboard. The water in the tank on the port side provides a moment opposing the roll velocity.*

six times the long period effect. The natural period of the fluid is a function of tank geometry and the volume of fluid contained. It is arranged to be equal to or slightly less than the lowest natural roll frequency of the ship. Provided the system has little damping, maximum roll reduction is achieved at resonance and the roll amplitude/roll period characteristic is virtually a straight line at about the optimum residual roll characteristic (Figure 10.17).

Anti-heeling tanks

Container ships and RO-RO (roll on–roll off) vessels are usually fitted with anti-heel tanks which enable the ship to be kept upright during uneven loading of cargo. Transfer of liquid from one tank to the other is by pump or compressed air.

Controlled passive system stabilizing tanks are used in the anti-heeling arrangements of some ships. The simplest method in this scheme, for transferring the liquid, makes use of compressed air admitted to the top of either tank, to force the liquid from one to the other.

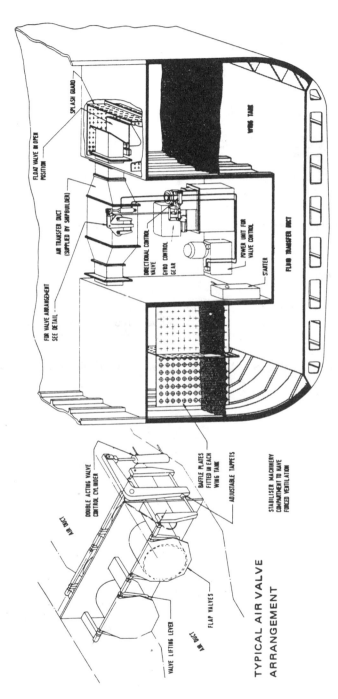

SPLASH GUARD

FLOAT VALVE IN OPEN
POSITION

WING TANK

FOR VALVE ARRANGEMENT
SEE DETAIL

AIR TRANSFER DUCT
(SUPPLIED BY SHIPBUILDER)

DIRECTIONAL CONTROL
VALVE

GYRO CONTROL
GEAR

POWER UNIT FOR
VALVE CONTROL

STARTER

FLUID TRANSFER DUCT

MUIRHEAD–BROWN TANK STABILISER
(AIR CONTROLLED)

DOUBLE ACTING VALVE
CONTROL CYLINDER

BAFFLE PLATES
FITTED IN EACH
WING TANK

ADJUSTABLE TAPPETS

STABILISER MACHINERY
COMPARTMENT TO HAVE
FORCED VENTILATION

TYPICAL AIR VALVE
ARRANGEMENT

AIR DUCT

VALVE LIFTING LEVER

FLAP VALVES

AIR DUCT

Figure 10.16   Muirhead-Brown controlled passive tank system (R.C. Dean)

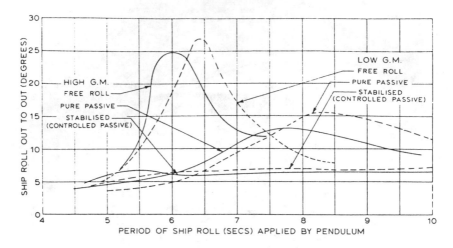

Figure 10.17   *Typical performance curves for Muirhead-Brown tank stabilizer*

Further reading

Bell, J. (1957) Ship stabilization controls and computation, *Trans RINA*, July.
Conolly J. E. (1969) Rolling and its stabilisation by active fins, *Trans RINA*.
Mitchell, C. C. and Stewart, D. (1966) Hydraulic power applied to ship stabilizers, *Proc Inst Mech Engrs*, **180**, pt 3L 1965–66.
Rorke, J. and Volpich, H. (1970) Selection of a ship's stabiliser, *Trans I Mar E*, February.

11

Refrigeration

Refrigeration is used in the carriage of some liquefied gases and bulk chemicals (Chapter 6), in air conditioning systems (Chapter 12) to cool bulk CO_2 for fire fighting systems (Chapter 14) and to preserve perishable foodstuffs during transport or storage, as described in this chapter. The principles of refrigeration, briefly stated below, are the same for each of the applications. Details of the different types of plant are described in the relevant chapters.

The main consideration with a ship built solely for the carriage of refrigerated cargo, is the high value of the produce which could be lost in the event of serious failure of the refrigerating machinery. A second very important consideration is that the produce should reach the consumer in good condition.

The purpose of refrigeration in the carriage of perishable foodstuffs, is to prevent or check spoilage, the more important causes of which are:

1 excessive growth of micro-organisms, bacterial and fungal;
2 changes due to oxidation, giving poor appearance and flavours;
3 enzymatic or fermentive processes, causing rancidity;
4 drying out (dessication);
5 The metabolism and ripening processes of fruit and vegetables.

The perishable foodstuffs carried as refrigerated cargo or as stores on ships can be categorized as dead produce such as meat and fish or as live produce such as fruit and vegetables. The dead cargoes tend to be carried frozen, an exception being made for meat on voyages of moderate length which may be carried chilled. A 10% carbon dioxide level has been found beneficial in cargo spaces for chilled meat. Fruit and vegetables are regarded as live cargoes until consumed, because they continue to ripen albeit slowly under refrigerated conditions. Fruit and vegetables continue a separate existence during which oxygen is absorbed and CO_2 is given off, with the generation of heat. The rate of respiration varies with the type of fruit and also directly with the temperature. Apples can produce CO_2 at the rate of 0.06 m³/tonne/day at carrying temperature and evolve heat at a rate of some 12 W/tonne/h in the process. Each commodity has its own specific storage condition for the best result. Table 11.1 gives some carrying temperatures.

Vapour compression cycle

The basic components of any refrigeration system (Figure 11.1) working on the vapour compression cycle, are the compressor, condenser, expansion valve,

Table 11.1   *Carrying temperatures, °C*

Apples	1–2	Beef	
Bacon		frozen	−9
Cured	1–5	chilled	−2
uncured	−9	Lamb	−9
Bananas	11–14	Pork	−9
Butter	−9	Lemons	12–13
Cheese	1–10	Oranges	3–7
Eggs		Pears	<0
frozen	−9	Plums	
chilled	0–1	4 days at	<0
Fish	−10	thereafter	7–10
Grapes	0–1		

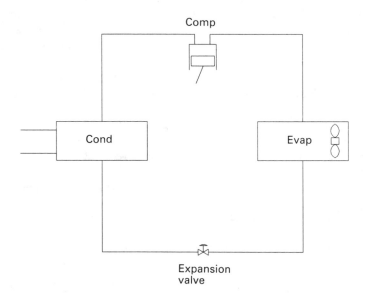

Figure 11.1   *Simple refrigeration circuit*

evaporator and the refrigerant fluid which is alternately vaporized and liquefied during the refrigeration cycle. The temperature at which a fluid boils or condenses, is known as the saturation temperature and varies with pressure. The compressor in a refrigeration system in raising the pressure of the vaporized refrigerant, causes its saturation temperature to rise, so that it is higher than that of the sea water or air, cooling the condenser. The compressor also promotes circulation of the refrigerant by pumping it around the system.

In the condenser, the refrigerant is liquefied by being subcooled to below the saturation temperature relating to the compressor delivery pressure, by the circulating sea water (or air for domestic refrigerators). Latent heat, originally from the evaporator, is thus transferred to the cooling medium. The liquid refrigerant, still at the pressure produced by the compressor, passes to the receiver and then to the expansion valve.

The expansion valve is the regulator through which the refrigerant flows from the high pressure side of the system to the low pressure side. Its throttling effect dictates the compressor delivery pressure which must be sufficient to give the refrigerant a saturation temperature which is higher than the temperature of the cooling medium.

The pressure drop through the regulator causes the saturation temperature of the refrigerant to fall, so that it will boil at the low temperature of the evaporator. In fact, as the liquid passes through the expansion valve, the pressure drop makes its saturation temperature fall below its actual temperature. Some of the liquid boils off at the expansion valve, taking latent heat from the remainder and causing its temperature to drop.

The expansion valve throttles the liquid refrigerant and maintains the pressure difference between the condenser and evaporator, while supplying refrigerant to the evaporator at the correct rate. It is thermostatically controlled in modern systems.

The refrigerant entering the evaporator coil, at a temperature lower than that of the surrounding secondary coolant (air or brine) receives latent heat and evaporates. Later the heat is given up in the condenser, when the refrigerant is again compressed and liquefied.

For a small refrigerator the evaporator cools without forced circulation of a secondary coolant. In larger installations, the evaporator cools air or brine which are circulated as secondary refrigerants.

Choice of refrigerant

Mechanical refrigeration makes possible the control of the pressure and therefore, the temperature at which a refrigerant boils (within the limits of critical pressure and temperature). The closed circuit ensures repeated use of the same refrigerant with little or no loss to atmosphere. Theoretically, almost any liquid can be used as a refrigerant if its pressure/temperature relationship is suitable for the conditions. Although no perfect refrigerant is known, there are certain factors which determine a refrigerant's desirability for a particular duty and the one selected should possess as many as possible of the following characteristics:

1   Moderate condensing pressure, obviating the need for heavily constructed compressors, condensers and high pressure piping.
2   High critical temperature, as it is impossible to condense at a temperature above the critical, no matter how much the pressure is increased.
3   Low specific heat of the liquid. This is desirable as throttling at the expansion valve causes liquid refrigerant to be cooled at the expense of partial evaporation.
4   High latent heat of vaporization, so that less refrigerant may be circulated to perform a given duty.

5   The refrigerant should be non-corrosive to all materials used in the construction of the refrigerating machinery and systems.
6   It should be stable chemically.
7   It should be non-flammable and non-explosive.
8   World wide availability, low cost and ease of handling are desirable.
9   The problem of oil return to the compressor crankcase is simplified when an oil-miscible refrigerant is used, by the admixture of the oil and the refrigerant in the system. With non-miscible refrigerants, once oil has passed to the system, its return to the crankcase can only be effected with difficulty.
10 The current concern with depletion of the ozone layer has resulted in a new requirement that refrigerants should be environmentally friendly. Strong pressure exerted to phase out CFCs and in particular R12, resulted in the Montreal Protocol adopted in 1987 after ratification by 70 countries and additional conventions seeking to phase out these gases.

Finally the refrigerant should preferably be non-toxic, have satisfactory heat transfer characteristics, and leakages should be easy to detect either by odour or by the use of suitable indicators.

It is not proposed to list or deal with all known refrigerants, only those likely to be encountered on board. These refrigerants are referred to by their trade name, chemical name or their internationally recognized numbers.

Commonly used refrigerants and possible replacements

R11 is a CFC which is included in the Montreal protocol having an ozone depletion potential (ODP) of 1 and a greenhouse potential (GP) of 3300. The formula for R11 is CCl_3F and it has the chemical name trichloromonofluoromethane. It was found suitable for air conditioning installations.

R12 is a CFC which, with an ODP of 1 and a GP of 10 000 is also to be phased out by the Montreal protocol. The formula for R12 is CCl_2F_2 and the chemical name dichlorodifluoromethane.

R22 is an HCFC with a much lower ODP of 0.05, is much less of a threat to the ozone layer and has been used in place of R12 in some recent installations. The formula for R22 is $CHClF_2$ and its chemical name is monochlorodifluoromethane. It has a GP of 1100.

R502 is an azeotropic mixture composed of 48.8% R22 and 51.2% of R115. Details of R22 are given above and R115 is a CFC with the formula $CClF_2CF_3$ which is used only in R502. As a CFC, with an ODP of 0.6 and a GP of 25 000 the R115 is included in the Montreal convention and this implies that R502 is also included. R502 is particularly suited for use with hermetic compressors.

Replacement refrigerants

A number of new refrigerants have been developed by the large chemical companies as potential replacements for refrigerants commonly in service.

Three refrigerants considered as candidates to replace R12 are referred to as HFC-134a, Blend MP33 and Blend MP39. The two blends are both made up from the same three other compounds but in different proportions by weight. Thus Blend MP33 consists of 40% HCFC-22 plus 17% HFC-152a plus 43% HCFC-124 and Blend MP39 is made up from 52% HCFC-22 plus 15% HFC-152a plus 33% HCFC-124. These gases are stated to have very low ozone depletion potential when compared with R12 and the two blends are less toxic with TLVs of 750 ppm for Blend MP33 and 800 for MP39. The HFC-134a has a TLV of 1000 ppm which is the same as that of R12.

Any replacement refrigerant must compare in performance with existing gases, for direct economy of operation as well as in respect of the amount of fossil fuel consumed.

There have always been minor problems with refrigerant and oil miscibility. Some problems with lubricants are being experienced with the new gases.

R134a is seriously considered as the best replacement for R12. It is an HFC with the chemical formula CF_3CH_2F an ODP of 0 and a GP of 900.

R123 is being considered as a replacement for R11. It is an HCFC with the chemical formula $CHCl_2CF_3$ an ODP of 0.02 and a GP of only 50.

R125 is being tested as a potential replacement for R22 and the gas with similar properties R502. The gas R125 is an HFC with the chemical formula CHF_2CF_3 an ODP of 0 and a GP of 1900.

R717 ammonia

The ammonia used for refrigeration systems based on the use of a compressor, condenser, expansion valve and an evaporator (Figure 11.2) is dry (anhydrous) in that there is no water in solution with it. It has the chemical formula NH3 but as a refrigerant, it is coded with the number R717. The good qualities of ammonia as a refrigerant have been offset by its toxicity, flammability and pungent odour, so that carbon dioxide and then CFCs (which replaced CO_2) were used at sea in preference. Now that R12 is to be phased out in the short term and R22 at a later date, ammonia is being considered as a replacement, because despite its local harmful effects and disadvantages, it is ozone friendly.

The upper and lower explosive limits for pure ammonia in air, are 27% and 16% by volume, respectively. With oil contamination, the latter may reduce to 4%. Long-term exposure to ammonia should be restricted to the current threshold limit value (TLV). Exposure to higher concentrations of 1500 ppm will result in damage to body tissue and death may result at 2500 ppm. However, ammonia leaks are instantly detected at less than 10 ppm concentration by the pungent odour and this is a safety feature. Very few people can endure ammonia when its concentration exceeds the TLV. In the liquid form, ammonia causes chemical and frost burns.

Corrosion of brass, bronze and similar alloys, occurs in ammonia systems if there is any water present. These materials are avoided with steels being used instead.

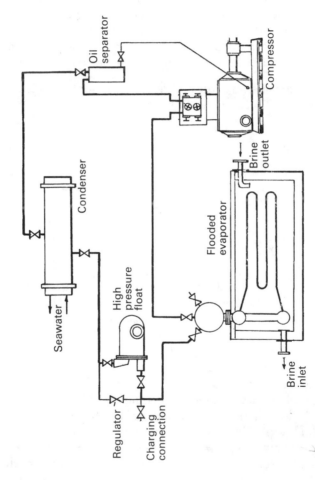

Figure 11.2 *Ammonia circuit*

Ammonia is highly soluble in water with which it forms ammonium hydroxide, a weak base. About 1300 volumes of ammonia can be dissolved in 1 volume of water at low temperature. However it is easily expelled by boiling. This action makes the vapour absorption refrigerator possible. Refrigerators, of this type do not require a compressor for operation, only a heat source.

R744 carbon dioxide

When carbon dioxide (CO_2) is used as a refrigerant the working pressures are high, being about 70 bar at the compressor discharge and 20 bar at the compressor suction. The machinery and system must therefore be of substantial construction. The critical temperature is low (31°C) and this causes problems in areas with high sea-water temperature. It also has a low coefficient of performance. Poor miscibility with oil required that a drain system be provided to remove oil continuously from the evaporator.

The gas is not explosive or flammable but a leak is potentially dangerous because it can displace air and asphyxiate. It is also toxic. The liquid is stored in steel bottles at high pressure, ideally in a cool space. A temperature rise will cause a pressure rise in the bottles which is relieved by the rupturing of a safety disc and release of gas.

Automatic freon system

The circuit shown in Figure 11.3 contains the basic compressor, condenser, expansion valve, evaporator and also typical controls for automatic operation.

The compressor is started and stopped by the LP (low pressure) controller in response to changes of pressure in the compressor suction. There is also an HP (high pressure) cut-out with a hand re-set which operates to shut down the compressor in the event of excessively high discharge pressure. The compressor can supply a number of cold compartments through thermostatically controlled solenoids. Thus as each room temperature is brought down, its solenoid will close off the liquid refrigerant to that space. When all compartment solenoids are shut, the pressure drop in the compressor suction will cause the compressor to be stopped through the LP controller. A subsequent rise of compartment temperature will cause the solenoids to be re-opened by the room thermostats. A pressure rise in the compressor suction acts through the LP controller to restart the compressor.

Each cold compartment has a thermostatic expansion valve, which acts as the regulator through which the correct amount of refrigerant is passed.

On large systems a master solenoid may be fitted. If the compressor stops due to a fault, the master solenoid will close to prevent flooding by liquid refrigerant and possible compressor damage.

The sketch is for a three compartment system but shows only the detail of one. Each room has a solenoid, regulator and evaporator. Air blown through the evaporator coils acts as the secondary refrigerant. Regular defrosting by

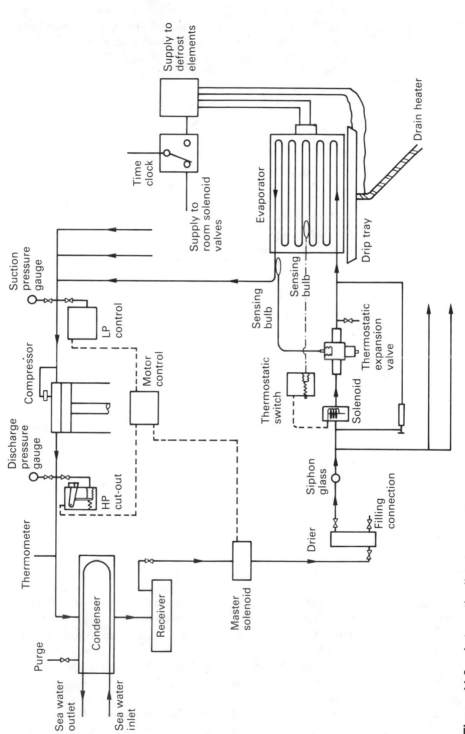

Figure 11.3 *Automatic direct expansion refrigeration*

means of electric heating elements keeps the evaporator free from ice. The automatic defrost time switch de-energizes the solenoids to shut down the system and diverts electrical current to the heaters, for a set period.

System components

The components described below may be found in refrigeration circuits. Some of the component descriptions apply to items in the automatic direct expansion system described above.

Pressure gauges

The pressure gauge on the compressor discharge shows the gas pressure and also has marked on it the relative condensing temperature. This is not the actual temperature of the gas which is higher and shown by the thermometer. The pressure gauge should show a pressure with an equivalent temperature about 5° or 6°C above the condenser sea-water inlet. The pressure gauge on the compressor suction has its scale similarly marked and should also indicate an operating pressure with a related temperature less than the evaporator temperature by 5° or 6°C.

Compressors

Refrigeration compressors are usually either reciprocating, or of the rotary screw displacement type. Centrifugal and rotary vane compressors have also been used.

Reciprocating compressors

Reciprocating compressors for systems cooling domestic store rooms are usually of the vertical in-line type. The larger reciprocating compressors (Figure 11.4) have their cylinders arranged in either V or W formation with 4, 6, 8, 12 or even 16 cylinders. Compressor speeds have been increased considerably over the years from 500 rev/min to the high speed of 1500 to 2000 rev/min. The stroke/bore ratio has diminished to the point of becoming fractional because of improvements in valve design and manufacture.

Provision is made for unloading cylinders during starting and for subsequent load control, by holding the suction valves off their seats (Figure 11.5) by suitable oil-pressure operated mechanisms. With this control the compressors can be run at constant speed which is an advantage with a.c. motors.

A bellows device, actuated by suction pressure can serve to cut out one or more cylinders. Thus a falling suction pressure, indicating a reduced load on the

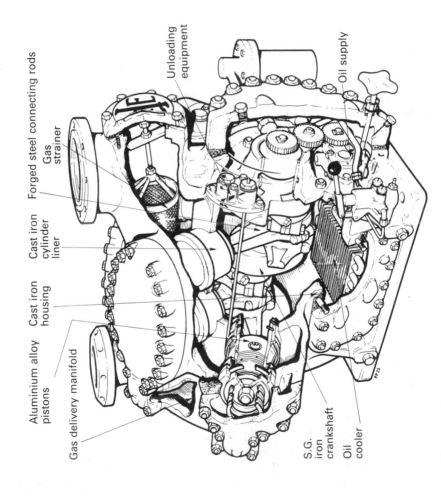

Aluminium alloy
pistons

Cast iron
housing

Cast iron
cylinder
liner

Forged steel connecting rods
Gas
strainer

Unloading
equipment

Oil supply

Gas delivery manifold

S.G.
iron
crankshaft

Oil
cooler

Figure 11.4 *Reciprocating compressor*

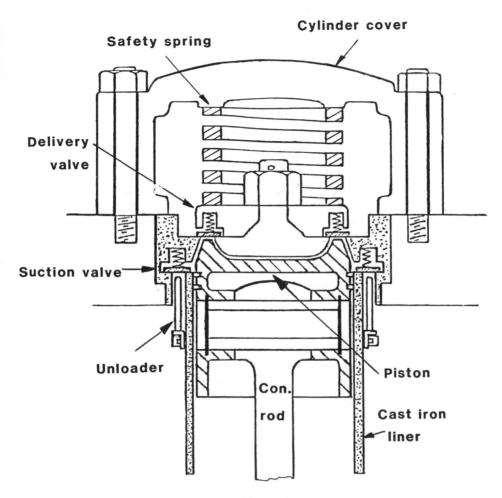

Figure 11.5 *Cylinder unloading mechanism*

system, can be used to reduce automatically the number of working cylinders to that required to deal with the existing load. Nearly all compressors of this type are fitted with plate type suction and delivery valves, whose large diameter and very small lifts offer the least resistance to the flow of refrigerant gas.

Each crank of the spheroidal graphite cast iron crankshaft for the W configuration compressor shown carries four bottom ends. The aluminium alloy pistons operate in cast iron liners, which are honed internally. Piston rings may be of plain cast iron but special rings having phosphor-bronze inserts are sometimes fitted. These assist when running in. Connecting rods are H section steel forgings with white metal lined steel top end bushes. Liners are of high tensile cast iron and the crankcase and cylinders comprise a one-piece iron casting. The two throw crankshaft is of spheroidal graphite cast iron. Each throw carries four bottom ends as mentioned above but in other machines the number of banks of cylinders may be less. Main bearings are white metal lined steel shells.

Gas from the evaporator passes through a strainer housed in the suction connection of the machine. This is lined with felt to trap scale and impurities scoured from the system by the refrigerant particularly during the running-in period. Freons are searching liquids, being similar to carbon tetrachloride. they tend to clean the circuit but the impurities will cause problems unless removed by strainers. Any oil returning with the refrigerant drains to the crankcase through flaps at the side of the cylinder space.

The valve assembly is shown in Figure. 11.5 in more detail. The delivery valve is held in place by a safety spring which is fitted to allow the complete valve to lift in the event of liquid carry over to the compressor. The delivery valve is an annular plate with its inside edge seated on the mushroom section and its outside edge on the suction valve housing. The suction valve passes gas from the suction space around the cylinder.

The control system includes a high pressure cut-out but a safety bursting disc is also fitted between the compressor discharge and the suction. This may be of nickel with a thickness of 0.05 mm. A ruptured disc is indicated by suction and discharge pressures equalizing.

Shaft seal

Where motor and compressor casings are separate, a mechanical seal is fitted around the crankshaft at the drive end of the crankcase. This prevents leakage of oil and refrigerant from the crankcase. The type shown (Figure. 11.6) consists of a rubbing ring with an oil hardened face against which the seal operates. The seal is pressed on to the face by the tensioning spring and being attached to bellows, it is self-adjusting. The rubbing ring incorporates a neoprene or duprene ring which seals it to the shaft.

The mechanical seal is lubricated from the compressor system and can give trouble if there is insufficient or contaminated oil in the machine. Undercharge may be caused by seal leakage (sometimes due to oil loss). When testing the seal for gas leakage, the shaft should be turned to different positions if the leak is not apparent.

Lubrication

Oil is supplied to the bearings and crankshaft seal by means of a gear pump driven from the crankshaft. The oil is filtered through an Auto-Klean strainer and/or an externally mounted filter with isolating valves. A pressure gauge and sight glass are fitted and protection against oil failure is provided by a differential oil pressure switch. Oil loss from the compressor is sometimes the result of it being carried into the system by the refrigerant.

Oil pressure is about 2 bar above crankcase pressure and the differential oil pressure switch is necessary to compare oil pressure with that of the gas in the crankcase. There is a relief valve in the oil system set to about 2.5 bar above crankcase pressure.

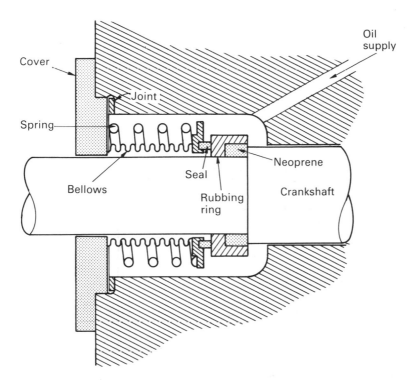

Figure 11.6 *Shaft seal*

Cylinder walls are splash lubricated and some of the oil is carried around with the refrigerant. If oil and refrigerant are miscible (i.e. if they form a mutual solution) the oil tends to return via the circuit to the crankcase. When the refrigerant is not miscible with the oil, the latter may be precipitated in the system, usually in the evaporator. A float controlled oil trap (Figure.11.7) may be fitted or an automatic drain for the evaporator.

Oil must have a low pour point (additives are used) and with freons a wax-free oil is necessary to avoid precipitation of wax at low temperatures.

Oil separators

Oil separators (Figure 11.7) of the impingement type, may be fitted in hot gas discharge lines from the compressor. The type shown, is a closed vessel fitted with a series of baffles or a knitted wire mesh through which the oil-laden vapour passes. The reduction in velocity of the vapour as it enters the larger area of the separator allows the oil particles, which have greater momentum, to impinge on the baffles. The oil then drains by gravity to the bottom of the vessel where a float valve controls flow to the compressor crankcase.

The chief disadvantage of separators placed in the hot gas discharge line is the possibility of liquid refrigerant passing from the separator to the compressor crankcase when the compressor is stopped. In order to minimize

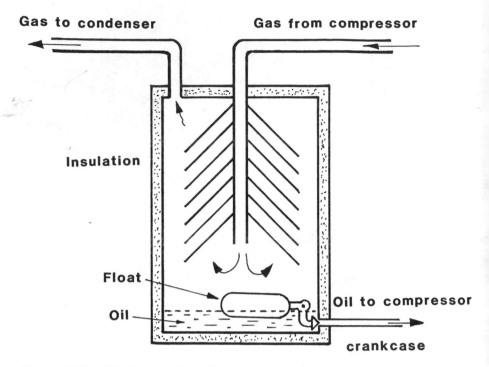

Gas to condenser

Gas from compressor

Insulation

Float

Oil

Oil to compressor

crankcase

Figure 11.7   *Float controlled oil trap*

this risk, the separators should be placed in the warmest position available. It is good practice to drain the oil from the separator into a receiver containing heating elements, where the liquid refrigerant boils off to the compressor suction line and allows the oil to drain to the compressor crankcase through a float valve.

Oil rectifier

Some refrigerants are miscible with the compressor lubricating oil which means that the two substances form a mutual solution. Because oil carry over does occur, the miscibility is actually beneficial because the oil tends to be taken around the refrigeration circuit and back to the compressor, by the refrigerant. If the oil and refrigerant are not miscible, as is the case with oil and CO_2, then there may be a loss of lubricant from the compressor sump and accumulation in the system. With CO_2 the oil collected in the evaporator and was drained off regularly by hand, then returned to the compressor after being strained. The oil rectifier (Figure 11.8) can be fitted to drain oil from the evaporator, automatically, and return it to the compressor.

The oil is automatically bled from the evaporator to a heat exchanger in which liquid refrigerant mixed with the oil is vaporized. The heat for vaporizing the refrigerant is obtained by passing warm liquid refrigerant from

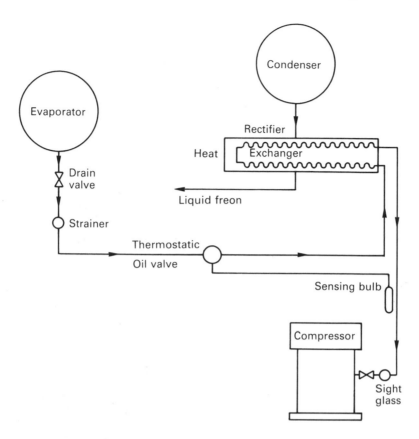

Figure 11.8 *Oil rectifier*

the condenser, through the heat exchanger. Vapour and oil are passed to the compressor where the oil returns to the sump while the freon passes to the compressor suction. The regulator is a thermostatically controlled valve which operates in the same way as the expansion valve in the main system. It automatically bleeds the oil from the evaporator so that the gas leaves the rectifier heat exchanger in a superheated condition.

Screw compressors

The capacity, range and use of screw compressors (Figures 11.9 and 11.10) has increased over the years. Their economy in space and weight and their capacity for long periods of uninterrupted running are inherent advantages. They can be oil-free, when used within their limited pressure range when oil contamination of the gas cannot be tolerated. For higher pressure range work they are oil-injected. In both types, two steel rotors are mounted in a gas-tight casing, usually of a high quality cast iron such as meehanite. Lobes and mating flutes on

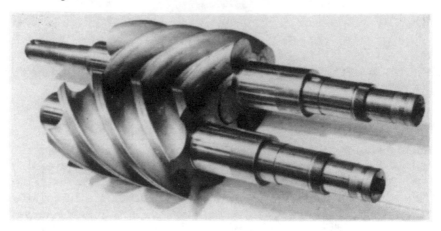

Figure 11.9   *A pair of compressor rotors (Howden Group Ltd)*

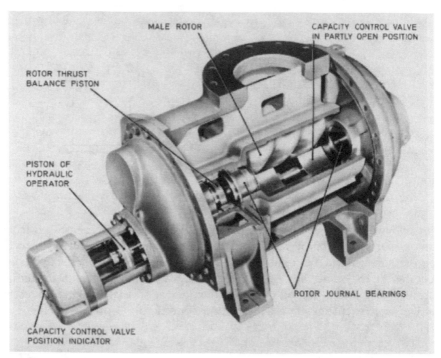

Figure 11.10   *Cut-away section of an oil-injected compressor (Howden/Godfrey Ltd)*

the male and female rotors are machined helically and so dimensioned that they mesh like helical gears.

Oil-free or dry machines are fitted with timing gears to ensure synchronization of the rotors, without contact. In oil-injected or wet machines, the female rotor is driven by the male counterpart, with surfaces being separated by the oil, which serves also for sealing and cooling. Capacity is varied by incorporation of a slide valve which, moving axially within the casing, varies the effective length of the rotors. Figure 11.9 shows a pair of rotors and Figure 11.10 a cut-away section of an oil-injected compressor, with the female rotor removed to show the slide valve and the actuating piston in its cylinder.

As will be apparent, large, effective oil separators are necessary when rotary oil-injected compressors are used.

Hermetic compressors

Hermetic and semi-hermetic machines have conventional compressors but the driving motor is enclosed within the same chamber or casing and is cooled by the passage of the refrigerant on the suction side. The motor heat is, of course, added to the refrigeration load. Total enclosure removes the risk of gas leakage through the vulnerable shaft seal.

These compressors are not intended to be overhauled on board but to be removed and replaced after a predetermined period of operation. This reduces the risk of moisture ingress into the system during overhaul in far from ideal conditions.

Condensers

Marine condensers are generally of the shell and tube type, designed for high pressures. There may a few coil-in-casing or other types still in use. The coolant passes through the tubes with refrigerant condensing on the outside.

Condenser tubes provide heat transfer surfaces in which heat from the hot refrigerant on the outsides of the tubes passes through the walls of the tubes to the cooling water inside. Sea water is the usual cooling medium for shell and tube condensers, but fresh water from central cooling systems is increasingly being used. The refrigerant vapour is cooled first to saturation point, then to the liquid state. The design of condensers is largely dictated by the quantity and cost of the circulating water, and, where water is plentiful as at sea, a large number of short circuits may be used, to keep pressure drop to a minimum. Water velocity is restricted to prevent erosion of the tubes, usually being kept below 2.5 m/s.

Marine condensers are very susceptible to corrosion and much research has been done to lengthen their useful life. With halogenated hydrocarbon refrigerants, the use of aluminium brass or cupro-nickel tubes and brass tube plates is acceptable and has reduced the rate of corrosion on the sea-water side.

Refrigerant 717 (ammonia) tends to corrode these materials however, and it has been found necessary to use stainless steel or even bimetallic tubes and clad tube plates, with ferrous metal being in contact with the refrigerant, and non-ferrous with the sea water. The use of sacrificial anodes in the water ends of the condensers is common and sometimes a short length of ungalvanized pipe is fitted between the condenser and the galvanized steel pipe system.

The use of fresh water from a central cooling system closed circuit, avoids the problem of sea-water corrosion.

Evaporators

In the direct expansion system, shown in Figure 11.3, evaporation takes place in air coolers consisting of pipe grids, plain or finned, enclosed in a closely fitting casing, through which air from the holds or chambers is circulated by forced or induced draught fans. This type of evaporator can be operated either partly flooded, fully flooded, with incorporated accumulators, or dry. In the latter, the refrigerant flow is controlled at the expansion valve in such a way that, as it passes through the grids, it is completely vaporized and slightly superheated.

Where brine is used as the secondary refrigerant, evaporators may be of the shell and tube type. In a shell and tube evaporator, the area of tube surface in contact with the liquid refrigerant determines its performance. The brine to be cooled may be circulated through the tubes with the refrigerant being on the outside of them. This involves either a high liquid level in the shell, or the placing of the tubes in the lower part of the shell only, the upper part then forming a vapour chamber. Modern flooded evaporators incorporate finned tubes.

Shell and tube evaporators (Figure 11.11) may also be of the dry expansion type, in which the refrigerant passes through the tubes, and the brine is

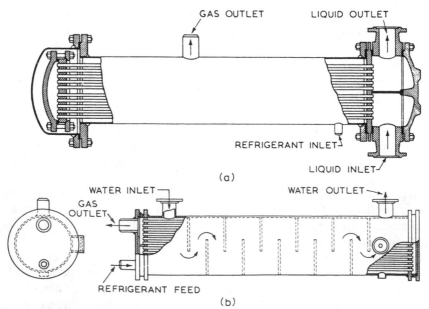

Figure 11.11

(a) Flooded-type evaporator   (b) Dry expansion-type evaporator

circulated through the shell. The advantages of this type are a smaller refrigerant charge and a more positive return of lubricating oil to the compressor.

Expansion valves

The expansion valve is the regulator through which the refrigerant passes from the high pressure side of the system to the low pressure side. The pressure drop causes the evaporating temperature of the refrigerant to fall below that of the evaporator. Thus, for example, the refrigerant can be boiled off by an evaporator temperature of $-18°C$ because the pressure drop brings the evaporating temperature of the refrigerant down to say $-24°C$.

The liquid refrigerant leaves the condenser with a temperature just above that of the sea-water inlet, say $15°C$. As it passes through the expansion valve the evaporating temperature decreases to $-24°C$ and some of the liquid boils off taking its latent heat from the remainder of the liquid and reducing its temperature to below that of the evaporator. There are six basic types of refrigerant controls or expansion devices, which can be summarized as follows.

Manually operated expansion valves

These were used for CO_2 refrigeration installations where the compressor was started and stopped by a watchkeeper. The compressor was started with the expansion valve open. The valve was then closed in to bring up pressure on the condenser side until the saturation or condensing temperature for the pressure (shown on the gauge) was five or six degrees above that of the cooling sea water. After the manual expansion valve had been set in this way, the gauge on the compressor suction (or evaporator side) was checked. Equivalent saturation or boiling temperature shown for the suction or evaporator pressure had to be about five or six degrees lower than the brine temperature. Any discrepancy indicated undercharge due to CO_2 leakage or overcharge due to over enthusiastic topping up with gas.

Many modern refrigeration systems have an emergency expansion valve which can be set manually in a similar way.

Manually operated expansion valves have the disadvantage of being unresponsive to changes in load or sea-water temperature and must be adjusted frequently. The valve itself is a screw down needle valve dimensioned to give fine adjustment.

Automatic expansion valves

These consist of a needle with seat and a pressure bellows or diaphragm with a torsion spring capable of adjustment. Operated by evaporator pressure their chief disadvantage is their relatively poor efficiency compared with other types. Constant pressure in the evaporator also requires a constant rate of vaporization, which in turn calls for severe throttling of the liquid. There is also the danger of liquid being allowed to return to the compressor when the load falls below a certain level.

This type of valve is used principally in small equipment with fairly constant loads, such as domestic storage cabinets and freezers.

Thermostatic expansion valves

These valves are similar in general design to automatic valves, but having the space above the bellows or diaphragm filled with the liquid refrigerant used in the main system and connected by capillary tube to a remote bulb. This remote bulb is fixed in close contact with the suction gas line at the outlet from the evaporator and is responsive to changes in refrigerant vapour temperature at this point. These valves are the most commonly employed (as in the automatic freon system Figure 11.3) and are suitable for the control of systems where changes in the loading are frequent.

Unlike the automatic valve, based on constant evaporator pressure, the thermostatic valve is based on a constant degree of superheat in the vapour at the evaporator outlet, so enabling the evaporator at any load to be kept correctly supplied with liquid refrigerant without any danger of liquid carry over to the suction line and thence to the compressor.

The aperture in the expansion valve is controlled by pressure variation on the top of a bellows. This is effective through the push pins (Figure 11.12) and tends to open the valve against the spring. Spring pressure is set during manufacture of the valve and should not be adjusted. The pressure on the bellows is from a closed system of heat sensitive fluid in a bulb and capillary connected to the top of the bellows casing. The bulb (Figure 11.13) is fastened to the outside of the evaporator outlet so that temperature changes in the gas leaving the evaporator are sensed by expansion or contraction of the fluid. Ideally the gas should leave with 6° or 7°C of superheat. This ensures that the refrigerant is being used efficiently and that no liquid reaches the compressor. A starved condition in the evaporator will result in a greater superheat which through expansion of the liquid in the bulb and capillary, will cause the valve to open further and increase the flow of refrigerant. A flooded evaporator will result in lower superheat and the valve will decrease the flow of refrigerant by closing in as pressure on the top of the bellows reduces.

Saturation temperature is related to pressure but the addition of superheat to a gas or vapour occurs after the latent heat transaction has ended. The actual pressure at the end of an evaporator coil is produced inside the bellows by the equalizing line and this is in effect more than balanced by the pressure in the bulb and capillary acting on the outside of the bellows. The greater pressure on the outside of the bellows is the result of saturation temperature plus superheat. The additional pressure on the outside of the bellows resulting from superheat overcomes the spring loading which tends to close the valve.

A hand regulator is fitted for emergency use. It would be adjusted to give a compressor discharge pressure such that the equivalent condensing temperature shown by the gauge at the compressor outlet was about 7°C above that of the

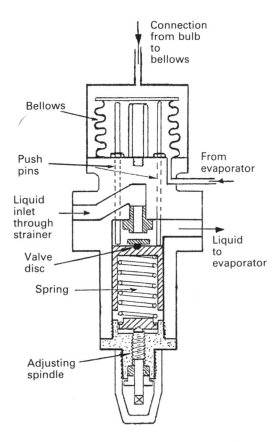

Figure 11.12 *Thermostatic expansion valve*

sea-water temperature and the suction gauge showed an equivalent evaporating temperature about the same amount below that of the evaporator.

Low pressure float controls

The mechanisms are similar to most other float controls, and act to maintain a constant level of liquid refrigerant in the evaporator by relating the flow of refrigerant to the rate of evaporation. It is responsive only to the liquid level which it will keep constant, irrespective of evaporator temperature or pressure.

This type of valve is usually provided with a manually operated bypass valve, so that the system can be kept in operation in the event of a float valve failure or float valve servicing.

High pressure float valves

These valves are similar to low pressure valves in that they relate flow of liquid into the evaporator to the rate of vaporization. The low pressure valve controls

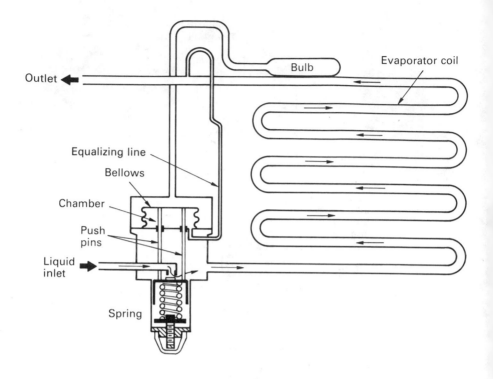

Figure 11.13   *Thermostatic expansion valve connection*

the evaporator liquid level directly. The high pressure valve located on the high pressure side of the system, controls the evaporator liquid level indirectly, by maintaining a constant liquid level in the high pressure float chamber.

As vapour is always condensed in the condenser at the same rate as liquid is vaporized in the evaporator, the high pressure float valve will automatically allow liquid to flow to the evaporator at the same rate as it is being evaporated, irrespective of the load on the system.

Capillary tube control

This is the simplest of all refrigerant controls and consists of a length of small diameter tubing inserted in the liquid line between the condenser and the evaporator. For a given tube bore and length the resistance will be constant, so that the liquid flow through the tube will always be proportional to the pressure difference between the condensing and evaporating pressures of the system.

Although self-compensating to some extent, this type of control will only work at maximum efficiency under one set of operating conditions, and for this reason is principally employed on close coupled package systems using hermetic or semi-hermetic compressors.

High pressure cut-out

In the event of overpressure on the condenser side of the compressor (Figure 11.3) the high pressure cut-out will cause the compressor to shut down. The device is re-set by hand. There are a number of faults which cause high discharge pressure, including loss of condenser cooling, air in the system and overcharge.

The bellows in the cut-out (Figure. 11.14) is connected by a small bore pipe between the compressor discharge and the condenser. The bellows tends to be expanded by the pressure and this movement is opposed by the spring. The adjustment screw is used to set the spring pressure.

During normal system operation, the switch arm is held up by the switch arm catch and holds the electrical contact in place. Excessive pressure expands the bellows and moves the switch arm catch around its pivot. The upper end slips to the right of the step and releases the switch arm so breaking the electrical contact and causing the compressor to cut-out. The machine cannot be restarted until the trouble has been remedied and the switch re-set by hand.

Room temperature control

The temperature of the refrigerated spaces with a direct expansion system (Figure 11.3) is controlled between limits through a thermostatic switch and a solenoid valve which is either fully open to permit flow of refrigerant to the room evaporator, or closed to shut off flow.

The solenoid valve (Figure 11.15) is opened when the sleeve moving upwards due to the magnetic coil hits the valve spindle tee piece and taps the

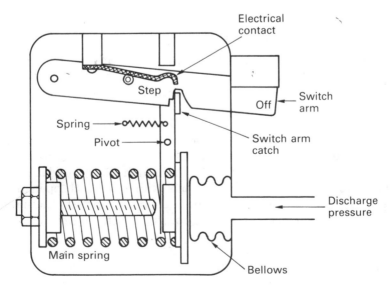

Figure 11.14 *High pressure cut out*

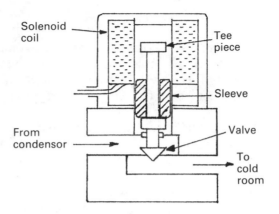

Figure 11.15   *Solenoid valve*

valve open. It closes when the coil is de-energized and the sleeve drops and taps the valve shut. Loss of power therefore will cause the valve to shut and a thermostatic switch is used to operate it through simple on/off switching.

The thermostatic switch contains a bellows which expands and contracts under the influence of fluid in a capillary and sensing bulb attached to it. The bulb is filled with freon or other fluid which expands and contracts with the temperature change in the space in which it is situated. As the temperature is brought down to the required level, contraction of the fluid deflates the bellows. The switch opens and the solenoid is de-energized and closes. A temperature rise operates the switch to energize the solenoid which opens to allow refrigerant through to the evaporator again. The switch is similar in principle to the high pressure cutout and low pressure controller.

Low pressure controller

The low pressure control (Figure 11.3) stops the compressor when low suction pressure indicates closure of all cold compartment solenoids. When the pressure in the compressor suction rises again due to one or more solenoids opening, the low pressure control restarts the compressor.

The controller shown (Figure 11.16) is of the Danfoss type operated through a bellows which monitors pressure in the compressor suction. A pressure differential between cut out and cut in settings is necessary to avoid hunting. The push pin operates the switch through a contact which is flipped open or closed through a coiled spring plate. With the contacts open the spring is coiled as shown. Outward movement of the pin compresses the spring and this then flips the contact to close the compressor starting circuit.

Pipelines and auxiliary equipment

Refrigerant piping may be of iron, steel, copper or their alloys but copper and brass should not be used in contact with Refrigerant 717 (ammonia).

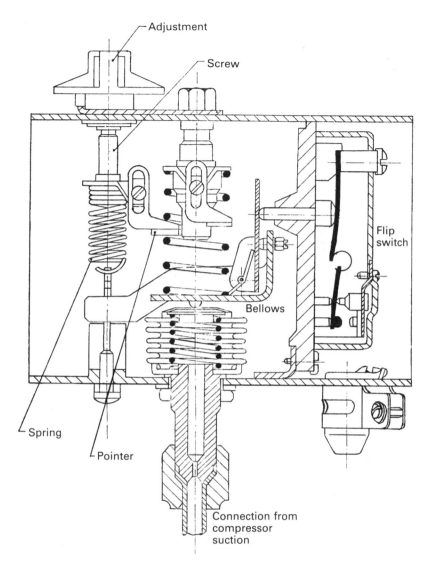

Figure 11.16 *Danfoss type LP controller*

The design of piping for refrigerating purposes differs a little from other shipboard systems in that the diameter of the piping is determined principally by the permissible pressure drop and the cost of reducing this. However, any pressure drop in refrigerant suction lines demands increased power input per unit of refrigeration and decreases the capacity of the plant. Pressure drop in these lines should be kept to a minimum.

To ensure continuous oil return, horizontal lines are usually dimensioned to give a minimum gas velocity of 230 m/min and vertical risers to give 460 m/min. The pressure drop normally considered allowable is that equal to about 1°C change in saturated refrigerant temperature. This means a very small loss in low temperature systems as the pressure change at 244 K for a one degree saturation temperature change, is only one half of that consequent upon the same temperature change at 278 K.

Horizontal pipelines should be pitched downstream to induce free draining and where the compressor is 10 m above the evaporator level, U-traps should be provided in vertical risers.

Welding, or in the case of non-ferrous piping, soldering and brazing, are practically universal in pipe assembly, and except where piping is connected to removable components of the system, flanges are rarely used.

Liquid indicators

These can be either cylindrical or circular glasses installed in the liquid line, providing a means of ascertaining whether or not the system is fully charged with refrigerant. If undercharged, vapour bubbles will appear in the sight glass.

To be most effective indicators should be installed in the liquid line as close to the liquid receiver as possible. Some types incorporate a moisture indicator which, by changing colour indicates the relative moisture content of the liquid passing through.

Driers

Where halogenated hydrocarbon refrigerants are used it is absolutely essential that driers are fitted in the refrigerant piping and most Classification Societies make this mandatory. Water can freeze on the expansion valve so causing excess pressure on the condenser side and starvation of refrigerant to the evaporator. When this occurs, the compressor will cut out due to operation of the high pressure cut-out or low pressure controller. The presence of a small amount of water can have an effect on plant performance and driers are essential. These are usually simple cylindrical vessels, the refrigerant entering at one end and leaving at the other. For modern installations the strainer/drier pack is replaced complete after opening the bypass and isolating the one to be replaced. Older systems are likely to have a strainer/drier partly filled with renewable drying agent. The drier, usually silica gel or activated alumina, is supported on a stiff gauze disc, overlaid with cotton wool with a similar layer above. In most installations the driers have bypasses so that they can be isolated without interfering with the running of the plant and the drying agent renewed or re-activated (by the application of heat).

If the drier is located in the liquid line it should be arranged so that the liquid enters at the bottom and leaves at the top. This is to ensure that there is uniform contact between the liquid refrigerant and the drying agent and that any entrained oil globules will be floated out without fouling the particles of the drying agent. If located in the suction line, the gas should enter at the top and leave at the bottom so that any oil can pass straight through and out.

Chamber cooling arrangements

The refrigerant which boils off from the evaporator removes latent heat and provides the cooling action of the refrigerator circuit. The cooling effect provided by the evaporator can be used directly as described below in direct expansion grids but for better efficiency, the cooling effect is applied by circulation of air through the evaporator or direct expansion batteries. To avoid having an extended refrigeration circuit for cargo cooling, a brine system can be used. The brine is cooled by the evaporator and in turn cools grids or batteries. Grids provide cooling which relies on convection and conduction but air circulated through brine batteries provides a positive through cooling effect.

Direct expansion grids

Direct expansion grids (Figure 11.17) provide a simple means of cooling a small refrigerated chamber. Such a system could be costly in terms of the quantity of refrigerant required and the cooling would rely on convection currents. Leakage of refrigerant into the cargo space could be a problem. A further objection would be that multiple circuits of liquid refrigerant could give control problems.

Cold brine grids

The pipe grids for this type of system (Figure 11.18) were arranged so that they cover as much as possible of the roof and walls of the chamber. The greatest coverage was needed on those surfaces which formed external boundaries and the least on divisional bulkheads and decks. As the actual cooling of the cargo also depended on movement of air by natural convection, this type of chamber cooling required good, careful and ample dunnaging of the cargo stowage. This appreciably diminished the amount of cargo that could be carried so that the system is no longer favoured.

Brine as a cooling medium (or secondary refrigerant) is cheap and easily regulated.

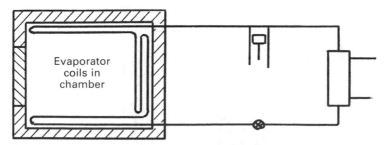

Figure 11.17 *Direct expansion grids (R. C. Dean)*

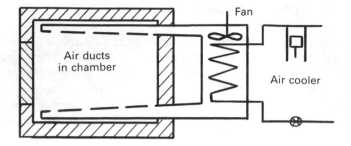

Figure 11.18   *Cold brine grids*

Direct expansion batteries and air

This is a commonly used system (Figure11.19) where the refrigerant circulates through batteries enclosed in trunkings or casings. Air from the refrigerated chambers, is circulated through the batteries by fans. Its great advantages are economy in space, weight and cost, and also the use of circulated air as the cooling medium or secondary refrigerant.

Brine battery and air

This system, in which brine instead of primary refrigerant is circulated through the batteries, continued to be employed for reefer ships carrying such cargoes as chilled meat or bananas where extremely close control of temperature was required when direct systems were gaining favour elsewhere. Brine is relatively easy to regulate. The system shown (Figure 11.20) is arranged with two separate refrigeration and brine circuits with connections from both brine systems to the air cooler batteries (or grids). (A more detailed diagram of a brine distribution system is shown in Figure 11.24.)

Brine is inexpensive, being made with calcium chloride and fresh water to a gravity of about 1.25. Sodium dichromate or lime may be added to maintain the brine in an alkaline condition.

Systems have been designed in which brine is replaced by one of the Glycols, for example ethylene glycol. The glycols have the advantage of being

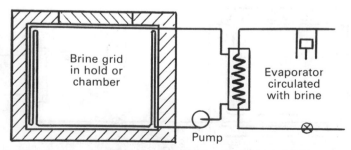

Figure 11.19   *Direct expansion battery with air circulation (R. C. Dean)*

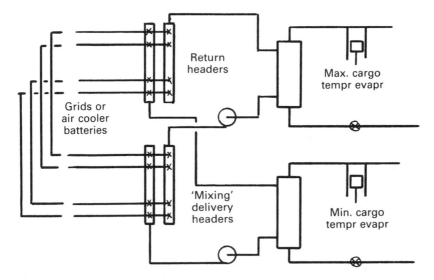

Figure 11.20 *Dual temperature brine system*

non-corrosive, and may be used at much lower working temperatures than brine. Trichlorethylene has also been used as a secondary refrigerant, but has the disadvantages of being toxic and a solvent of many of the synthetic rubbers and other materials normally used as jointing.

Air circulation systems

The design of an air circulating system is dictated principally by the allowable temperature spread in the cargo spaces and is not influenced by the type of air cooler in use. Brine and direct expansion systems have similar air circuits.

The air cooler and fan unit are mounted behind a deck-to-deckhead screen or trunking at one side of the chamber with air being delivered and returned via trunking, false decks or deckheads provided with suitable openings. The delivery openings are arranged with the largest furthest away from the fan where the air pressure is at its lowest and the smallest nearest to the fan. Correct cargo stowage is important as voids in the stow could allow the air to short circuit to the suction side of the cooler. Cargo stowed adjacent to air inlets can become desiccated (dried out). For this reason and also to remove hot spots in the cargo, air circulation should be reversible. With ripening (live) cargoes, CO_2 tends to accumulate in cargo spaces and it is necessary to limit the level by freshening the circulating air. Vents must be provided for this purpose.

Container cooling

Systems designed for the cooling of refrigerated containers employ trunkings (Figure 11.21) arranged so that containers stowed in stacks between built-in

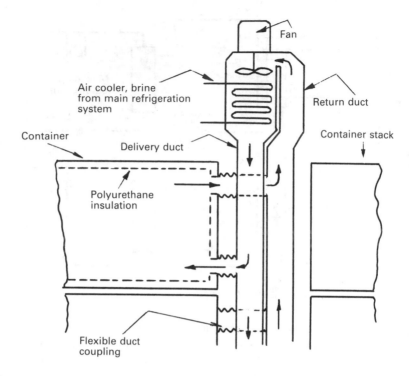

Figure 11.21 *Container cooling system (R. C. Dean)*

guide rails, can be connected to the suction and delivery air ducts of the ship's refrigeration plant by bellows pieces operated pneumatically. The air is cooled either by brine or direct expansion batteries and the containers are arranged so that one cooler can maintain a stack of containers at a given temperature. The temperature of the return air duct for each container is monitored. Provision of a cooler and trunking system for maintaining container temperatures must also be provided at container terminals.

Individual containers with their own refrigeration plant (Figure 11.22) are connected to the 440 or 220 a.c. sockets provided on deck. These containers may be arranged for ships' systems with either 440 or 220 V by provision of a direct connection for a 220 V supply to the self-contained refrigerator and a 440 V connection through a step down transformer.

Air cooler fans

Fans may be either centrifugal or of the propeller type; the air circulation systems being based on a pressure requirement of about 50 mm W.G. (water gauge). All of the electrical energy of the fan motors is dissipated in the form of heat and has to be removed by the refrigerating plant. Fan output should be variable so that it can be reduced as heat load diminishes. There was no problem with d.c. motors but with a.c. either the motors are two speed, or each

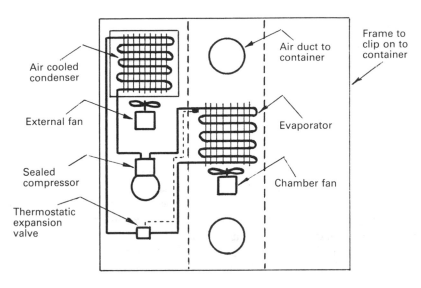

Figure 11.22 *Container with refrigeration unit*

cooler has a number of fixed speed fans which can be switched off individually to suit the load. In the latter case, provision must be made to blank off the stopped fans to prevent air loss.

The capacity of the fans is determined by the number of air changes per hour required in the cargo chambers, and this is influenced by the maximum calculated heat load. In a system using air coolers and fans, all the heat load must be carried away by the circulating air and the difference between delivery and suction air temperatures is directly proportional to the weight of air being circulated. Since the temperature difference is limited by the allowable temperature spread in the cargo chambers and maximum temperature spread in the cargo chambers and maximum load can be estimated, the selection of suitable fans is straightforward.

In most installations the number of air changes required per hour, based on an empty chamber, varies between 40 for dead cargoes such as frozen meat or fish and 80 for fruit cargoes, such as bananas which evolve heat freely.

Instruments

It is essential to measure and log the temperature of refrigerated cargo to ensure that the correct conditions are maintained and also to provide a record should there be complaints from a shipper. Mercury or spirit thermometers suspended from screwed plugs in vertical steel tubes with perforations hung in the cold chamber have been replaced by remote reading devices.

Electrical resistance and electronic self-balancing thermometers use the principle of the Wheatstone bridge. The former rely on a galvanometer to indicate a balance. In the latter the unbalanced current causes an electric motor to adjust the resistance. All necessary cargo temperature readings are obtained on modern reefers and container ships on a data logger which makes an automatic record.

The temperatures and pressures relating to refrigerant gas and liquid, cooling water, brine and the ambient are also required. Most of these are obtained from direct reading instruments.

Carbon dioxide measurement

Carbon dioxide concentration in the cargo chamber is important when fruit or chilled beef is carried. The electrical CO_2 indicator (Figure 11.23) operates on the principle that CO_2 is a better heat conductor than air. A sample of air with CO_2 content, is passed over platinum resistance wires carrying a constant heating current. Between the sample chambers CO_2 is absorbed to give a differential reading. The wire temperature is less when CO_2 content is higher. The temperature difference is detected on a Wheatstone bridge circuit through a suitably calibrated milliammeter which gives a direct CO_2 reading.

Defrosting

This very necessary operation presents no difficulty when the cooling medium is brine. All that is required is a brine heater (Figure 11.24) with brine pump and circuits to circulate hot brine through the coolers.

In direct expansion systems, defrosting can be effected by separate electric heaters installed in the evaporator grids (see Figure 11.3) or by providing a means of bypassing the condenser so that hot gas from the compressor circulates the evaporator directly.

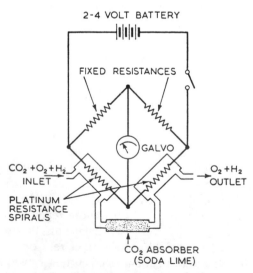

Figure 11.23   *Wheatstone bridge circuit for CO₂ indicator*

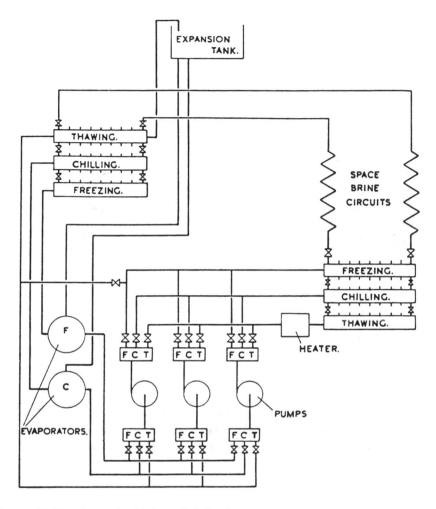

Figure 11.24 *Layout of brine distribution system*

Heat leakage and insulation

The total load on a cargo refrigerating plant is the sum of:

1 surface heat leakage from the sea and surrounding air;
2 deck and bulkhead edge leakage from the same sources;
3 heat leakage from surroundings into system pipes;
4 heat equivalent of fan and some brine pump power;
5 cooling of cargo not precooled at loading;
6 respiratory heat of live cargoes;
7 heat introduced by air refreshment of live cargoes.

The load arising from 1, 2 and 3 can be much reduced by the efficient use of insulation. A number of materials are used for this including slab cork, glass and mineral wools, expanded plastics, aluminium foil and polyurethane. The latter, although generally most costly, is the best insulator, having the lowest coefficient of conductivity, with the further advantages of being impervious to air leaks and almost impervious to the passage of vapour, when the material is foamed *in situ*.

Materials which contain CFCs should not be specified. Some rigid urethane foams (polyisocyanates and polyurethanes) and expanded polystyrene or phenolics may contain CFCs. These materials are used for their low thermal conductivity, high resistance to the passage of vapour, good mechanical properties and ease of construction. They can be produced so as to be free of CFCs but with higher thermal conductivity.

All of the materials mentioned have to be enclosed by linings for protection and the prevention of air leakage. The design and construction of the linings makes a greater contribution to the efficiency of an installation.

Insulation test

The heat balance test which replaced an earlier unsatisfactory version, was introduced by the major Classification Societies in 1947. In this trial, temperatures in the refrigerated spaces are reduced to a specified figure and then after a lapse of time sufficient to remove all residual heat from the insulation and structure, the spaces are maintained at constant temperature for at least six hours by varying the compressor output. During this period all temperatures and pressures, speeds and electrical consumption of compressors, fans and pumps are carefully logged and the compressors' output is noted from appropriate tables.

From this information it is possible to compare the efficiency of the insulation with the theoretical estimate made during the design stage and also to decide whether or not the installation can maintain these temperatures in maximum tropical sea and ambient conditions. Obtaining the theoretical estimate entails taking each external surface of the individual chambers separately and considering all factors affecting the heat leakage. These factors include the pitch, depth and width of face of all beams, frames and stiffeners buried in the insulation, the type of grounds securing the linings, the presence of which have their effect in reducing the effective depth of the insulation.

Hatches, access doors, bilge limbers, air and sounding pipes also have their effect on heat leakage and must come into consideration. It should be noted that in these calculations the laboratory value of the insulation is generally increased by about 25% to allow for deficiencies in fitting.

It has been found that the overall co-efficient of heat leakage in well insulated installations can vary between 0.454 W/m²/°C for 'tweendecks in small lightly framed ships and 0.920 W/m²/°C for fully refrigerated moderate sized ships having deep frames with reverse angles. Where there are also buried air ducts, the effective depth of the insulation may reduce to little more than zero.

Further reading

Lawson, C. C. (1991) Performance of alternative refrigerants for CFC-12 in stationary refrigeration equipment, International Congress of Refrigeration, Montreal 1991.

12

Heating, ventilation and air conditioning

Good ventilation is vital to the health and well-being of those on board ship and the general requirements for ventilation, formulated before the universal installation of air conditioning systems, still apply. Heating, always necessary for the colder areas of the world, has in the past been provided by local radiators or by heating coils incorporated with ventilation units. With extremes of low temperature, these primitive methods of heating increased the capacity of the air to absorb moisture and caused excessive evaporation with discomfort to crew and passengers due to drying of the nasal passages, throat and skin. Air conditioning is based on the ventilation requirement for accommodation and incorporates heating with any necessary humidification and importantly, cooling with de-humidification as necessary. Comfortable conditions depend on the temperature and humidity but are also sensitive to air movement, air freshness and purity.

Legionella bacteria

A type of pneumonia which may be fatal to older people, has been blamed on the presence of a bacteria associated with the air conditioning plant of large buildings. Because the outbreak which heralded the disease, occurred at a convention for American ex-servicemen (The American Legion), the identified cause of the problem, was labelled legionella bacteria and the sickness is referred to as legionnaires disease.

There is a risk that the bacteria could flourish in the air conditioning systems of ships and consequently a Department of Transport M Notice (1) has been issued to give warning and to recommend preventative measures.

The M notice explains that the organisms breed in stagnant water or in wet deposits of slime or sludge. Possible locations for bacteria colonies, are mentioned as being at the air inlet area and below the cooler (stagnant water), in the filter, in humidifiers of the water spray type and in damaged insulation. Provision of adequate drainage is recommended to remove stagnant water.

Guidance is given for regular inspections and cleaning as necessary of filters and other parts, using a 50 ppm super-chlorinated solution as the sterilizing agent. The solution is to be used also on the cooler drain area at not more than three month intervals. Regular sterilization is necessary for water spray type humidifiers (steam humidifiers being preferred).

Air conditioning

A very significant factor affecting an air conditioning system is the rapidly changing climatic conditions. The equipment has to perform within these variations and has to meet the differing requirements of the occupied spaces of the ship.

The early air conditioning systems were rather bulky because designs were based on low air velocities in the distribution ducts, with velocities in the order of 10 m/s or less. In later years there was extensive standardization with very substantial increases in air velocities, reaching a maximum of abut 22.5 m/s in the ducts and producing a large reduction in the space occupied by the equipment. Increased operating costs as a result of higher velocities have to be set against reduced installation costs and the value of space saving, but the owner is usually disposed favourably towards the high velocity system with its lower initial cost.

Basic standards

The designer and user of air conditioning plant must study the physiological factors involved. The terms used to define the atmospheric conditions are fairly well known, but are reviewed here since it is essential to know exactly what they mean before proceeding further.

Dry bulb (d.b.) temperature is the temperature as measured by an ordinary thermometer which is not affected by radiated heat.

Wet bulb (w.b.) temperature is the temperature registered by a thermometer with wetted fabric around the bulb. (When moisture evaporates from a surface, i.e. the skin, the latent heat required, is drawn from the surface causing it to be cooled. If a thermometer bulb is covered by a wetted fabric and exposed to the air, the rate of evaporation will depend upon the humidity of the surrounding air. As the heat required must come from the bulb, this results in a lower temperature reading than if the bulb was dry.)

Psychometric chart or table is used to find relative humidity from dry bulb and wet bulb readings taken at the same location in a space. (The thermometers may be in a fixed position or in a football rattle type device.)

Relative humidity (r.h.) of the air indicates the amount of moisture carried by the air at a particular temperature as a percentage of the maximum amount that could be carried at the particular temperature. (The capacity of the atmosphere to hold water vapour is dependent upon its temperature. At higher temperatures this is much greater than at the lower temperatures. When the maximum is reached at a given temperature, the air is said to be saturated. Saturated air has 100% relative humidity.)

Dewpoint (d.p.) is the temperature to which unsaturated air must be cooled to bring it to saturation point and to cause moisture to precipitate. (If an unsaturated mixture of air and water vapour is cooled at constant pressure, the temperature at which condensation of water vapour begins is known as the dewpoint. [Moisture from air starts to condense on a cold window or glass when the air near the cold surface reaches its dewpoint. As the air is further cooled, more moisture is deposited.]

Ideal conditions

The condition of the air in a space depends on its temperature, humidity and movement. The effect of the air on people in a space, is dictated by their metabolism, state of health, acclimatization, degree of exertion and the amount of clothing being worn.

The ideal conditions for comfort vary considerably between one person and another, so it is only possible to stipulate a fairly wide zone. In this connection it would obviously be of great value if a single index could be used to define the physiological reaction to the various combinations of factors involved. Among other suggestions the most satisfactory has been the effective temperature index. It is the temperature of still, saturated air which would produce the same feeling of warmth.

The American Society of Heating and Air Conditioning Engineers carried out a comprehensive series of tests on a large number of people, from which they were able to draw up an effective temperature chart. The comfort chart in

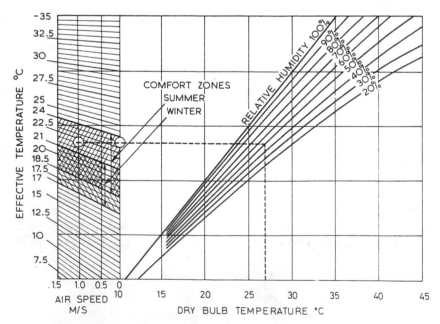

Figure 12.1 *Comfort chart compiled from tests carried out by the American Society of Heating and Air Conditioning Engineers*

Figure 12.1 is based on the results. Taking as an example 27°C d.b. and 50% r.h., for zero air speed, the effective temperature is found to be 23.4°C, which is within the summer comfort zone. With an air speed of 1.0 m/s, the effective temperature is reduced to 22.2°C.

The higher limit of comfort is more critical than the lower as it is closely associated with the essential process of getting rid of body heat. Also it is of considerable significance to the air conditioning engineer, who has to fix the capacity of the refrigerating plant so that it can provide conditions within this limit when the outside conditions are at their most onerous. Protracted investigations have been made by Hall-Thermotank Ltd., to determine the effective temperature which could be said to represent the upper threshold of the comfort zone. The reactions of a large number of persons, mostly mariners, some of whom had just completed a voyage, were analysed. For a person dressed in tropical clothing and at rest, the threshold value was found to be about 25.6°C effective temperature.

Tests showed that most persons tended to sweat when the temperature rose a degree or so above this value, and to cease sweating at the same value as the temperature fell again. The conclusion was therefore reached that there could be a relationship between the threshold of comfort and the onset of sweating. This does not take into account differences in the thermal sensations of men and women due to metabolism, general health and other factors mentioned previously. Conclusions from the study indicated that in the main, the comfortable level of warmth for acclimatized persons of all races is very similar in spite of certain differences in the reaction to heat stress.

At room temperatures above 21°C an air velocity of 0.15–0.2 m/s is desirable, to avoid any feeling of stuffiness, and to provide proof that the space is being ventilated. On the other hand, velocities higher than 0.35 m/s are usually classed as draughts – to be avoided particularly for the person at rest. When a space is heated or cooled, it is impossible to ensure an absolutely uniform distribution of the effect throughout the space. Because warm air rises, the air can be appreciably warmer at the higher levels than at the deck level, giving rise to discomfort unless the air terminals are designed to counteract this effect.

Radiation, planned or otherwise, can contribute to the effective temperature of a space, but in ships' accommodation it is not significant, even in cabins adjacent to engine room bulkheads given high standards of insulation.

Air purification

Outside air must be introduced to all living spaces, although the amount of fresh air necessary to sustain life is very small indeed. Space conditions can vary greatly in a short time. They are governed by factors such as body odours and smoking, which may require a fresh air supply of 12 litre/s per person or more. The actual quantity of fresh air supplied is partially governed by the provision that it must not be less than the total capacity of the accommodation exhaust fans (excluding galley). On this basis the minimum Department of Transport requirement of 7 litre/s is usually exceeded.

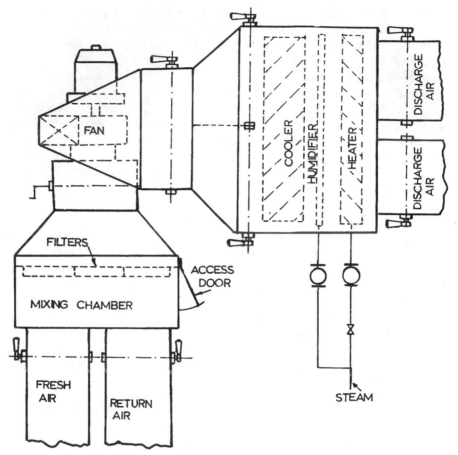

Figure 12.2   *Marine air conditioning unit*

It is desirable that outside air should be cleaned before being introduced to the space, but this is less essential than on land, except when the vessel is in port. Of greater influence are the impurities such as lint carried in the circulating air, which must be filtered out to prevent eventual choking of the heating and cooling elements in the conditioning plant. Filters (Figure 12.2), unless regularly cleaned can provide a breeding ground for the legionella bacteria.

Cooling load

As the temperature of air is reduced, so too is its capacity for carrying water vapour. With the aid of a psychrometric chart (Figure 12.3) it can be shown that air, with an initial dry bulb temperature of 36°C and relative humidity of about 60%, will, when cooled to 27°C dry bulb temperature, have a relative humidity of 100%. The temperature drop reduces the capacity of the air to carry moisture in suspension. Further cooling will cause moisture to be precipitated. Air cooled to a comfortable temperature level of 21°C but having a relative

PSYCHROMETRIC CHART

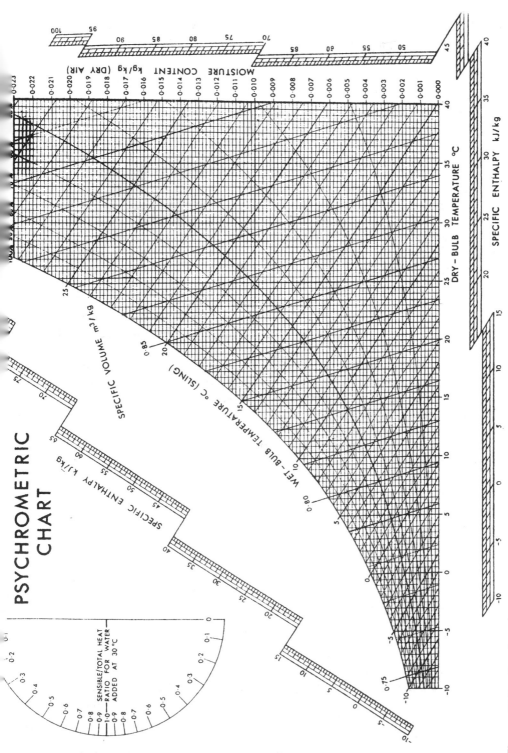

Figure 12.3 Carrier-type psychrometric chart

humidity of 100% would not be able to take up further moisture and perspiration would not be evaporated. People in an atmosphere at 21°C with 100% relative humidity would be uncomfortable.

The remedy of dehumidifying the air is achieved by overcooling to precipitate excess moisture, (removed via the drain – Figure 12.2) so that when air is brought to the correct temperature, its humidity will be at an acceptable level. Thus the air could be overcooled to about 10°C dry bulb temperature so that warming to about 21°C, would bring humidity to about 50%. The air is warmed in the trunking or by contact with warmer air in the space. A zone heater could be used.

An adequate drain is required to remove what can be a considerable flow of water from dehumidification of the air.

It can be seen that a large proportion of the total cooling load is required to reduce the moisture content of incoming fresh air and the relative humidity as well as temperature of outside air can be regarded as of the greatest significance in the specification of outside conditions for design. Surveys have been carried out to determine the conditions over the main trade routes of the world.

The figure commonly taken for outside dry bulb temperature by some designers is 32.2°C, but much higher temperatures may be experienced in tropical zones with offshore winds. These winds are, however, usually dry and do not impose as high a load on the cooling plant. Obviously local conditions will vary.

It is usual to express the conditions inside the accommodation in terms of dry bulb temperature and relative humidity obtained from the psychometric chart. Although people are not sensitive to the degree of humidity over quite a wide range, it is usual to design for between 40% and 60% relative humidity.

The inside design conditions have a very important bearing on the cooling plant power. Another very significant factor is the degree of insulation of the surfaces bounding the air conditioned spaces. At one extreme, the only insulation may be provided by decorative lining of the bulkheads and by the treatment of the engine room bulkhead, while some insulating value would lie in the deck coverings.

At the other extreme, some shipowners specify generous insulation enclosing all sections of the accommodation. A reasonable mean standard would be obtained by assuming the equivalent of 25 mm of high class insulating material, having an insulating value of 1.5 W/m²/°C, over all surfaces normally treated.

Heating load

When the temperature of air is increased, so too is its capacity for carrying water vapour. Using the psychometric chart, it can be shown that air, with a very low initial dry bulb temperature of −5°C and relative humidity of about 50%, will, when heated to 21°C dry bulb temperature, have a relative humidity of about 10%. The temperature rise increases the capacity of the air to carry moisture in suspension. Air heated to a comfortable temperature level of 21°C

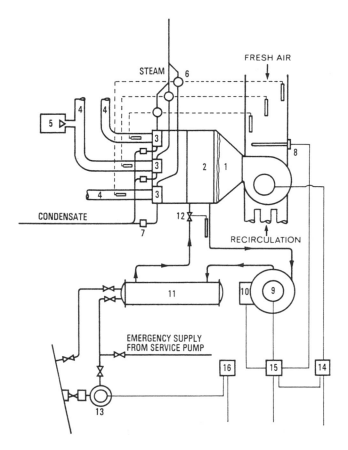

Figure 12.4 *Zone Control System*

1. Filter
2. Cooler
3. One, two or three-zone heaters as required
4. Pre-insulated pipes delivering air to zones
5. Sound attenuating air terminal, with volume control

6. Automatic steam valves. One per zone heater
7. Steam trap. One per zone heater
8. Multi-step cooling thermostat
9. Compressor
10. Automatic capacity control valves

11. Condenser
12. Thermostatic expansion valve
13. Sea water pump
14. Fan starter
15. Compressor starter
16. Sea water pump starter

but having a relative humidity of 10% will readily take up moisture whether from perspiration or from the nasal passages and throat. People in an atmosphere at 21°C but 10% relative humidity, would experience discomfort from dryness in their nose and throat and on the skin.

The remedy is to humidify the air (Figure 12.3) with a hot water or steam spray. This action increases humidity towards 100% relative humidity and also increases the temperature from −5°C and 50% relative humidity to say, +7°C. Straight heating by the zone heater bringing the air to about 21°C, will drop relative humidity to 40%. The humidity will be at an acceptable level but is kept low to minimize condensation on any very cold external bulkheads.

Heating moderately cold outside air will not cause a dryness problem because it mixes with recirculated air and air in the space served. Moisture is continually added to air in accommodation areas from breathing, perspiration and other activities. The humidifier is likely to be necessary only in extremely cold conditions hence the one shown in the sketch has a simple valve for setting by hand.

The heating load includes heat leakage losses through the structure, calculated with the aid of the requisite coefficients for the various materials involved, together with the heat required to raise the outside air temperature to the space temperature. The latter is evaluated from the formula:

$$H = 1.21 \ Q(t_i - t_o)$$

where:

H = heat required, kW,
Q = airflow, m³/s,
t_i = inside temperature, °C,
t_o = outside temperature, °C,

and the density of the air is taken as 1.2 kg/m³ at 20°C.

The outside temperature chosen may not be the extreme minimum for the trading routes of the vessel, but a value chosen within a range from $-20°C$ to $0°C$. The inside condition would be from 18°C to 24°C depending upon the type of accommodation.

Evaluation of heating and cooling loads and air quantities

The cooling load has a great influence on the design of the equipment since it influences the quantity of air to be circulated and determines the size of the refrigerating plant. The following sensible heat gains must be balanced to maintain the required inside temperature, when cooling is in operation:

1   Heat transmission through the structure. This is dependent on the physical properties of the materials surrounding the air conditioned spaces and the relative humidity to be maintained inside. Allowance has also to be made for the effect of sun heat on exposed surfaces. This is very difficult to define with any accuracy, and is usually computed with the aid of tables and charts based on experience.
2   Body heat. Account must be taken of the heat gain in the space due to the occupants.
3   Lighting heat. This can be a significant factor on board ship, where lighting is in use almost continuously.
4   Fan heat. The energy applied to the air is converted to heat in the passage of the air through the system.

The air delivered conveys the cooling effect to the spaces. This air must be delivered at a temperature below that desired in the accommodation fixed by the moisture content of the air. The air passing through the cooling coils becomes saturated on cooling and gives up moisture as its temperature falls. When the air leaves the cooler its moisture content remains unchanged until it enters the accommodation. Once inside the accommodation, the temperature of the air rises and the relative humidity (but not the moisture content) drops but it then also mixes with the resident atmosphere.

The quantity of air must be so arranged that the temperature rises to the specified inside conditions. The quantity is given by the formula:

$$Q = \frac{H}{1.21} (t_i - t_e)$$

where:

Q = total volume of air circulated, m³/s,
H = total heat gain in the spaces, kW,
t_i = inside temperature, °C,
t_e = temperature of entering air, °C.

It invariably happens that this quantity is considerably greater than the fresh air requirements discussed previously, so that the balance is recirculated in order to economize in the cooling load. In practice, usually about two-thirds of the air delivered to the space is recirculated.

Types of air conditioning systems

Air conditioning systems may be divided into two main classes – the central unit type in which the air is distributed to a group of spaces through ducting, and the self-contained type, installed in the space it is to serve.

The central unit type is the most widely used, in one or other of a number of alternative systems, characterized by the means provided to meet the varying requirements of each of the spaces being conditioned. The systems in general use are as follows:

1 Zone control system;
2 Double duct system;
3 Reheat system.

Zone control system

This is the most popular because of its basic simplicity. The accommodation is divided into zones, having different heating requirements. Separate air heaters for each zone are provided at the central unit as shown in Figure 12.4.

The main problem is to obtain a typical sample of air for thermostatic control of the heaters, for it may not be possible to choose a location which is uninfluenced by local factors. This has led to the general adoption of a compromise solution, which is to vary the temperature of the air leaving the heater in accordance with the outside temperature prevailing. This can be effectively performed by a self-actuating regulator controlled by two thermostat sensors, one in the air leaving the heater, the other outside. Air quantity control in each room served gives individual refinement. In summer, air temperature is controlled by a multi-step thermostat in the recirculating air stream, which governs the automatic capacity control of the refrigerating plant.

The regulation of temperature by individual air quantity control in this system can give rise to difficulties unless special arrangements are made. For instance, a concerted move to reduce the air volume in a number of cabins would cause increased air pressure in the ducts, with a consequent increase in air flow and possibly in noise level at other outlets. This can be avoided but economic factors usually place a limit on this. Some degree of control is possible through maintaining a constant pressure at the central unit, but since most of the variation in pressure drop takes place in the ducts, the effect is very limited. A pressure-sensing device some way along each branch duct, controlling a valve at the entry to the branch, strikes a reasonable mean, and is fairly widely applied.

Double duct system

In this system, two separate ducts are run from the central unit to each of the air terminals, as shown in Figure 12.5. In winter two warm air streams, of differing temperatures, are carried to the air terminals, for individual mixing. The temperatures of both air streams are automatically controlled. In summer the air temperature leaving the cooler is controlled by a multi-step thermostat in the recirculating air stream, which governs the automatic capacity control of the refrigerating plant, as with zone control. Steam is supplied to one of the heaters, so that two air streams are available at the air terminals for individual mixing.

Reheat system

In winter, the air is preheated at the central unit, its temperature being automatically controlled. The air terminals are equipped with electric or hot water heating elements, as shown in Figure 12.6. These raise the temperature of the air to meet the demands of the room thermostats which are individually set.

In the case of electric reheat, fire protection is provided by overheat thermostats which shut down the heaters in the event of air starvation, while a fan failure automatically cuts off the power supply. In summer, the air temperature is controlled by a multi-step thermostat in the recirculating air stream, which governs the automatic capacity control of the refrigerating plant, as in the other system.

Self contained air conditioner

In the early days of air conditioning, there was a demand for self-contained units to serve hospitals and some public rooms where the advantages of air conditioning were very obvious. At first these units were rather cumbersome,

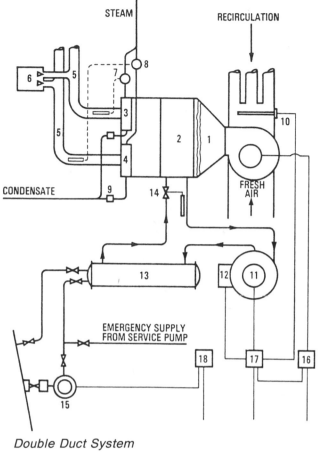

Figure 12.5 *Double Duct System*

1. Filter
2. Cooler
3. Low-duty heater
4. High-duty heater
5. Pre-insulated air pipes
6. Sound attenuating air terminal with volume and temperature control
7. Automatic steam valve for tempered air stream
8. Automatic steam valve for warm air system
9. Steam traps
10. Multi-step cooling thermostat
11. Compressor
12. Auto-capacity control valves
13. Condenser
14. Thermostatic expansion valve
15. Sea water pump
16. Fan starter
17. Compressor starter
18. Sea water pump starter

but with the advent of hermetically sealed components and other developments associated with land applications, the modern cooling unit can usually be accommodated within the space it is to serve. By taking full advantage of the available height, the deck space required is relatively small.

Self-contained units may be used instead of a central unit system in new or existing ships where space is not available for the latter. The term 'self-contained' is only relative, since fresh air and cooling water are required and provision must be made for removal of condensate.

The S-type Thermo-Unit is widely used on board ship. It combines a

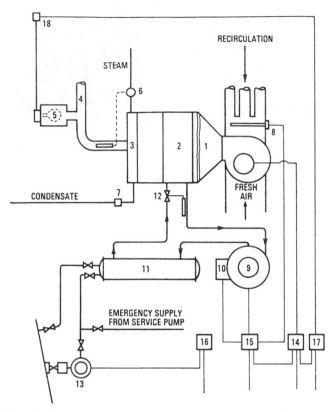

Figure 12.6   *Reheat system*

1. Filter
2. Cooler
3. Pre-heater
4. Pre-insulated air pipe
5. Sound attenuating air terminal containing electric re-heater and overheat thermostat

6. Automatic steam valve
7. Steam trap
8. Multi-step cooling thermostat
9. Compressor
10. Auto capacity control valves
11. Condenser

12. Thermostatic expansion valve
13. Sea water pump
14. Fan starter
15. Compressor starter
16. Sea water pump starter
17. Heater contractor
18. Room type thermostat

compact arrangement of the elements with the accessibility which is essential for marine use. To facilitate installation, the unit is divided into an upper and a lower section which can be taken apart readily.

The self-contained unit is ideally suited to the engine control rooms of automated ships. With the additional heat load coming from the equipment housed within the room, cooling may be required at the same time as the accommodation requires heat from the central unit system.

Conversion units

For a ship fitted originally with mechanical ventilation only, a good case may be made out for the provision of full air conditioning if the ship has a reasonable span of life ahead of it. This can be done by mounting a conversion unit on the deck, embracing the essential features of a central cooling plant. The unit is so designed that it can be coupled to the existing fan, heater and air distribution system.

The central unit

The elements of a central unit are fan, filter, cooler, heaters and plenum chamber. Normally these are all housed within a single casing, with the possible exception of the fan. It is possible to carry this further by including the refrigerating plant in a single assembly thus providing a complete package. Apart from the obvious saving in space and economy in pipework, the possibility of refrigerant leakage is minimized by having the circuit sealed in the factory. Figure 12.7 shows a central unit of this type.

The filter, which is essential to keep the heat transfer elements clean, is usually formed of a terylene fibre mat, easily removed for periodic cleaning. The cooler is of the fin tube type, as are the heaters, usually steam. The air passes from the heaters into a plenum chamber, and from there into the pipes or ducts leading to the various spaces. The plenum chamber, acoustically lined, acts as a very effective silencer for the fan noise which otherwise would be transmitted along the ducts.

Air distribution

Friction and eddy losses in the ducts make up the greater part of the pressure required at the fan, hence the design of the duct system affects the fan power very considerably. The fan power is a function of the air quantity and the pressure, and is expressed as follows:

$$\text{Fan power (kW)} = \frac{\text{volume } (\text{m}^3/\text{s}) \times \text{pressure (mbar)}}{10 \times \text{efficiency}}$$

The efficiency is static or total, depending on whether the pressure is static or total. The total pressure is the algebraic sum of the static and velocity pressures.

The system is sized for the longest duct branch, so that artificial resistances must be inserted in other branches to balance the air distribution. In designing the system, account is taken of static pressure regain to reduce the rate of fall along the ducts. This regain results from a reduction in velocity when the volume of air in the duct is reduced after an outlet is passed, and can amount to about 75% of the fall in velocity pressure.

High velocity distribution

Over the years, the most significant development has been the introduction of high velocity air distribution, made possible by the reduced air quantity

Figure 12.7 *Central unit (Hall-Thermotank International Ltd)*

required. In other words, the inevitable increase in fan power associated with higher velocities (and hence higher pressures) has been kept within reasonable limits by a reduction in air quantity.

High velocity distribution has a number of clear advantages among these being:

1 Ducting costs much reduced.
2 Standardizing on a few diameters of round ducting up to about 175 mm instead of a great variety of widths and depths of rectangular ducting.
3 Standardized bends and fittings, having improved aerodynamic efficiencies.
4 Use of automatic machines for fabrication of ducts, with a spiral joint giving great stiffness.
5 Greatly reduced erection costs resulting from light weight and small bulk of ductwork.

6   Considerable space saving in the ship.
7   Possible reduced fire risk with smaller duct sections.

Against all these advantages must be set the increase in fan power already referred to. Thus an older system with duct velocities of the order of 8 m/s might require a fan pressure of no more than 50 mm water gauge, whereas with high velocities around 22.5 m/s the fan pressure could exceed 230 mm water gauge. The ratio of fan power increase is not so great as this, however, since high efficiency centrifugal fans of the backward-bent blade type are suited to high pressure operation, but for low pressures the relatively inefficient forward-bent blade type of fan must be used, since the high efficiency type fans would too bulky if designed for low pressures.

With the increase in friction loss due to high velocity, the reheating of the air can result in an appreciable increase in the cooling load, when compared with a low velocity system, and this could be a limiting factor in the choice of the duct velocity.

The design of the air terminals is very important with high velocity distribution, in order to minimize noise and prevent draughts.

Duct insulation

Duct insulation is standard practice, being particularly necessary in installations where the policy has been to reduce the volume of air handled to a minimum, resulting in greater temperature differentials. The ideal is to integrate the insulation with the duct manufacture, or at least to apply it before the ductwork is despatched to the ship.

There are a number of high-class fire resistant insulating materials on the market, such as mineral wool and fibreglass. These, of course, must have a suitable covering to resist the entry of moisture and to protect the material from damage. Jointing of the duct sections is usually by sleeves, with external adhesive binding.

Air terminals

The best designed air conditioning system is only as good as the means of delivering the air to the spaces. The main function of the air terminal is to distribute the air uniformly throughout the spaces without draughts. It is not possible to provide ideal conditions for both heating and cooling from the same outlet. Too low a discharge velocity in the heating season can result in stratification, the air at ceiling level remaining warmer than the air at the floor. Even when cooling, a low velocity stream could fall through to lower levels in localized streams, without upsetting the stratification.

Careful selection of the discharge velocity and direction of flow in the design stages can provide an acceptable compromise between good distribution and draught free conditions. Generally it is found that the ceiling is the most convenient location for their terminal, although in large public spaces extended slot type outlets on the bulkhead, with near horizontal discharge, are satisfactory and blend well with decorative features. The usual recirculation outlet at the bottom of the door normally ensures a good distribution of the air in the space.

With high air velocities, some control of the noise level in the system becomes essential, and it is true to say that equipment design, particularly as applied to the terminals, has been influenced more by this than by any other factor.

Figure 12.8 shows a typical layout of ducts and terminals for the accommodation in a tanker or bulk-carrier.

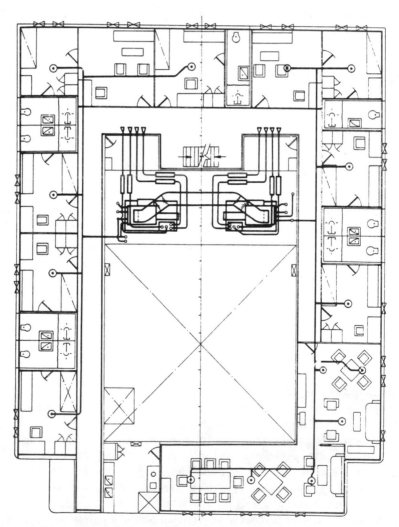

Figure 12.8 *Layout of air conditioning for tanker or bulk carrier accommodation*

Ventilation of boiler and engine rooms

Due to the large amount of heat picked up by the air in these spaces it would be impracticable to maintain ambient conditions within the comfort zone by air conditioning or any other means. The practice is to provide copious mechanical ventilation; in boiler rooms, the quantity is equated to the combustion requirements while in a motorship engine room the supply may be 25–50% in excess of the requirements of the engines.

The axial flow fan is particularly suited to handle these large air volumes at the moderate pressures required, while of course the 'straight-through' flow feature places it at an advantage over the centrifugal fan.

The increasing adoption of automation, with the provision of a separate control room makes less significant the fact that comfort conditions cannot be maintained in the engine room all the time.

Typical specification for air conditioning installation in a tanker or bulk carrier

Installation serving the accommodation aft

The deck officers', engineer officers' and crew accommodation is served by two air conditioning units, with a direct expansion refrigerating plant of the Freon 22 type, capable of maintaining an inside condition of 26.7°C (d.b.) and 20.0°C (w.b.) (55% relative humidity) when the outside condition is 32.2°C (d.b.) and 28.9°C (w.b.) (78% relative humidity). Finned tube type steam heating coils are fitted to maintain 21°C in the space when the outside temperature is − 20.5°C. The schedule in Table 12.3 indicates the rates of air changes to be provided.

A proportion of the air would be recirculated except from the hospital, galley, pantries, laundry, shower rooms and toilets. The recirculated air is to be withdrawn through wire mesh grids mounted on mild steel ducting at deckhead level, located in the alley ways adjoining the treated space.

With heating in operation, automatic temperature regulating valves in the steam supply lines give independent temperature control in the following spaces:

Officers:
   navigating bridge deck;
   bridge deck;
   boat deck;
   part poop deck.

Crew:
   part poop deck;
   upper deck.

The refrigerating plant comprises a compressor driven by a marine type motor with automatic starter, shell and tube type condenser, evaporator/air cooling coils within the air conditioning unit casing, piping and fittings, safeguards, automatic cylinder unloading gear for cooling capacity control, and the initial charge of refrigerant. Mechanical and electrical spare gear is supplied.

Air conditioning units are installed on boat deck aft, with refrigerating machinery remotely situated in the engine room at middle flat level port side, approximately 7 m above ship's keel.

A sea-water circulating pump to serve the condenser is provided. The shipbuilder would supply and fit sea-water piping, and valves between the ship's side and pump, from pump to condenser and from condenser overboard.

Air delivery to the spaces is by means of distributors, with sound attenuating chambers and air volume regulators, mounted on mild steel ducting at high level.

The cargo control room is served by a branch duct delivering 0.18 m³/sec through the distributors. Cooling at selected spots in the galley is provided by the conditioning units.

Non-return valves are fitted in ducts serving hospital and laundries. This prevents odours reaching the accommodation should the fan unit be stopped for any reason.

Mechanical supply ventilation

Gyro and electronics room and motor generator room on bridge deck, switchboard room and telephone exchange on boat deck, motor room, storerooms and bedding store on upper deck, galley on poop deck are ventilated at atmospheric temperature by two axial flow supply fans.

Air is delivered through diffusing type punkah louvres and domed diffusers fitted on mild steel ducts at deckhead level.

Mechanical exhaust ventilation

Galley, pantry, laundries, drying rooms, oilskin lockers, overall lockers, gyro and electronics room, motor gear room, switchboard room, telephone exchange and motor room and all private and communal toilets, washplaces, bathrooms and toilets, are ventilated by five axial flow exhaust fans.

Hospital, medical locker, hospital toilet and bathroom must be independently ventilated by an axial flow exhaust fan.

Vitiated air is withdrawn through domed extractors and regulating type grid openings mounted on mild steel ducting at deckhead level. Canopies over main galley range, should be supplied by the shipbuilder.

Air filtering equipment

Filter screens of washable nylon fibrous material are supplied with air conditioning and supply fan units. Spare screens would be provided.

Technical data

No. of air conditioning units	2
No. of axial flow supply fans	2
No. of axial flow exhaust fans	6
Total fan motor power of above	27.5 kW
Approx. steam consumption for heating and humidification purposes (with 50% recirculation)	0.13 kg/s @ 4.15 bar gauge
No. of direct expansion refrigerating plants	1
Capacity of plant	208 kW
Compressor power	63.3 kW
No. of sea water circulating pumps	1
Capacity of pumps	16.4 litre/s at 1.72 bar gauge
Power of above pump	7.25 kW
Approx. weight of installation	12 000 kg

General

All natural ventilation, together with shut-off valves as required, operating in conjunction with the mechanical system and serving spaces not mechanically ventilated, should be supplied and fitted by the shipbuilder.

The shipbuilder provides the necessary air inlet and outlet jalousies to the unit compartment.

To prevent excessive leakage of conditioned air, all doors leading from the conditioned spaces to the outside atmosphere, machinery casing, etc., should be of the self-closing type and reasonably airtight.

Cooling load calculations have been based on the understanding that there would be no awnings and that all accommodation would have airtight panelling on deckheads and house sides, and ship surfaces would be insulated by the shipbuilder as follows.

Approximate amount of insulation to be supplied and around ducting by shipbuilder:

In accommodation	300 m² of 25 mm thick glass fibre slabs with vapour seal (or equivalent).
Exposed deckheads	68.5 mm thick wood deck or equivalent thickness of insulation.
Exposed sides	25 mm thick insulation suitably finished fitted on back of panelling or on steelwork around beams and stiffeners.
Galley surfaces	Suitable insulation applied adjacent to conditioned spaces
Surfaces of machinery spaces adjacent to conditioned spaces	50 mm thick insulation suitably finished.

All unit rooms in proximity to the accommodation should be suitably sound-insulated. The above proposal is exclusive of any ventilation whatsoever

Table 12.1 *Fan particulars*

Unit No.	Volume m³/s	Static pressure mbar	Dia. mm	Power kW	Speed rev/sec	Encl. with Class 'B' insulation	Spaces served
Air conditioning unit No. 1	3.40	26.2	–	14.9	30	TEFC	Officers' accommodation
Air conditioning unit No. 2	1.87	22.4	–	7.5	30	TEFC	Crew's accommodation
Axial flow supply fan No. 1	0.57	4.0	240	0.6	60	TE	Motor generator room, electronic and switchboard rooms, etc.
Axial flow supply fan No. 2	1.08	4.2	380	0.8	60	TE	Galley and stores
Axial flow exhaust fan Nos. 1, 2, 3 and 6	0.57	4.0	240	0.6	60	TE	Toilet spaces
Axial flow exhaust fan No. 4	1.65	3.0	610	0.8	30	TE	Galley and handling space
Axial flow exhaust fan No. 5	0.16	1.9	152	0.14	60	TE	Hospital, W.C., bathroom and medical locker
Axial flow exhaust fan Nos. 1–2 P.	1.41	3.1	445	1.1	20	TEFC	Cargo pump room

All the above electric motors would be suitable for operation on a ship's voltage of 440 V, 3 ph, 60 Hz.

Table 12.2 *List of motor and starter spares*

Unit	Motors	Starters
Air conditioning unit No. 1	1 14.9 kW motor	1 set contacts, coils and springs
	1 set bearings	
Air conditioning unit No. 2	1 7.5 kW motor	1 set contacts, coils and springs
	1 set bearings	
Supply fan No. 1	1 fan unit, complete	1 Auto-West switch
Supply fan No. 2	1 0.8 kW motor	1 set contacts, coils and springs
	1 set bearings	
Exhaust fan No. 1	1 0.6 kW motor	1 set contacts, coils and springs
Exhaust fan		
No. 2		
No. 3	1 fan unit, complete	
No. 4		
No. 5		
No. 6		
Exhaust fan No. 1 P	1 set bearings	1 set contacts, coils and springs
Exhaust fan No. 2 P		

Note All starters 0.75 kW and above would have running lights and in addition those 3.75 kW and above would have ammeters fitted. The above spares would be suitably packed for stowage on board.

to the forward pump, paint and lamp rooms and stores and also to the emergency generator room and battery room, book stores and CO_2 bottle room.

From Table 12.3 it will be observed that in certain of the air conditioned spaces, the air changes are less than those required by the Department of Transport First Schedule Regulation No. 1036, 1953, for ventilated spaces. Since the internal atmosphere achieved by conditioning the air would be superior to any under ventilation without artificial air cooling, it is understood from communication with the DTp that relaxation from the above requirements would be made by them under the terms of regulation 38(3) which provides for such relaxation in this event.

The fitting of an installation as described above would, therefore, be subject to final approval of the Department of Transport.

Installation serving the engine control room

The engine control room would be served by three self-contained room

Table 12.3   *Schedule of air changes*

		Air changes/hour	
Spaces served	Deck	Supply	Exhaust
Wheelhouse	Nav. Bridge	6	–
W.C.		–	15
Captains, chief officer's and deck officers' accommodation		9	–
Radio office		12	–
Motor gen. room, gyro and electronics room	Bridge	15/30*	15/30
Officers' toilets and drying room		–	15
Baggage room		–	10
Officers' smoke room and bar, games room		12	–
Chief engineer's and engineers' accommodation, pilot and owners representatives		9	–
Laundry	Boat	6	15
Switchboard room, telephone exchange		30*	30
Engineers' toilets and change room, pantry, drying room, equipment locker		–	15
Lockers		–	10
Cadets, catering officer, PO's and junior engineers		9	–
Cargo control room, dining salons, crew's mess room, crew's smoke room	Poop	12	–
Hospital		12	0.14 m³/s †
Medical locker		–	10
Galley		20*	40
Cadets', engineers', catering officer's and PO's toilets		–	15
Games room		12	–
Crew accommodation		9	–
Laundry		6	15
Motor room		30*	30
Bedding, beer, dry provision and bonded and equipment stores	Upper	10*	–
Crew toilets, change room, laundry, drying rooms		–	15
Handling space		–	12.5

*Air at atmospheric temperature   † Through WC and bathroom.

conditioners, to operate on 440 v, 3-phase, 60 Hz supply. Power consumption 1.2 kW per unit.

At least 7 litres/s of fresh air per occupant of room should be taken from machinery space system. The shipbuilder should pipe condenser cooling water from ship's services. Erosion of condenser piping may be prevented by fitting a 'Constaflo' control valve in each unit to limit flow to 0.15 litre/s per unit.

This proposal is exclusive of heating. The shipbuilder should supply and fit suitable bases for units, provide switches, wire up to motors and switches, and supply and fit cooling water drain piping and fittings.

Installation serving the cargo pump room

The cargo pump room is ventilated by two axial flow exhaust fans each to extract 1.4 m³/s against 3.1 mbar static, requiring a total of 2.25 kW. The fans should be of split case bevel gear driven type enabling the fan motor to be mounted in the adjacent engine room. Motor and controller spare gear must be provided. Air is extracted from two low level points through wire mesh grids fitted on galvanized steel ducting.

The shipbuilder must supply and fit air outlet jalousies with hinged water tight covers, and natural ventilation to operate in conjunction with the mechanical system.

Further reading

Merchant Shipping Notice No. M1215 Contamination of Ships' Air Conditioning Systems by Legionella Bacteria.

13

Deck machinery and cargo equipment

The operation of mooring a vessel has traditionally required the attendance of a large number of deck crew fore and aft. Supervision of the moorings was also necessary to maintain correct tension through changes due to the tides and the loading or unloading of cargo. The installation of constant tension mooring winches, which maintain tension in ropes through any rise and fall, has removed the need for constant attendance and equipment is available for tying up which is designed for operation by as few as two men. Large container ships may have four mooring winches on the after deck; each of the self-tensioning type with its own rope drum. Controls are duplicated and are situated at each side of the vessel, giving a clear view of the operation. Mooring ropes are paid out directly from the drums as they are hauled by the heaving lines from the quay. With the loop in place on the bollard, the capstan is set on auto-tension after slack is taken up and the ship is correctly moored. A common arrangement forward is for two similar winches plus rope drums for auto-tensioning on each windlass.

The introduction of steel hatchcovers not only speeded up the operation of opening and closing the covers but also reduced the number of personnel required for the task. Rolling and folding covers may be operated by a pull wire or hydraulically. Covers for large container ships may be lifted bodily by crane and there are now hatchcoverless container ships in service.

Cargo handling may be by winches and derricks or cranes. Some geared bulk carriers have overhead cranes arranged to travel on rails.

Most deck machinery is idle during much of its life while the ship is at sea. In port, cargo equipment will be in use for one or more days but the machinery for anchoring and mooring is used for a very limited time. Deck machinery with a restricted and intermittent duty may be designed with drives with a rating limited from 30 minutes to one hour. Despite long periods of idleness, often in severe weather conditions, machinery must operate immediately, when required. Cooling vents, open when machinery is working, must be closed for the sea passage.

It is essential that deck machinery should require minimum maintenance. Totally enclosed equipment with oil bath lubrication for gears and bearings is now standard but maintenance cannot be completely eliminated and routine checking and greasing should be carried out on a planned basis.

There are many instances where remote or centralized control is of great advantage, for example, the facility for letting go anchors from the bridge under emergency conditions; the use of shipside controllers with mooring winches; or the central control positions required for the multi-winch slewing derrick system.

The machinery on the deck of an oil tanker is limited to that used for anchor handling and mooring plus pumproom fans and equipment for handling the gangway and stores. Power was universally provided in the past by steam. Hydraulic equipment is now common, sometimes with air motors for gangway duties. The availability of safe electrical equipment means that electric motor drives can be used where appropriate.

Liquefied gas carriers and product or chemical tankers have similar deck machinery installations but the drive motor for deepwell pumps may be an induction motor of the increased or enhanced (Ex e) safety type.

Either electric or hydraulic drives are installed for the deck machinery of dry cargo vessels.

Electric drives

Electric motors on vulnerable deck areas may be protected against ingress of water by being totally enclosed in a watertight casing. Vents are provided on some winches, which must be opened when the motor is operating in port.

The direct current (d.c.) motor, although it is relatively costly and requires regular brush gear maintenance, is still used for deck machinery because it has a full speed range with good torque at any speed. The control of d.c. motors by contactor-switched armature resistances, common in the days when ships' electrical supplies were d.c., has long been replaced by a variety of Ward-Leonard type systems which give a better, more positive regulation particularly for controlled lowering of loads. The Ward-Leonard generator is normally driven by an a.c. motor.

An important feature of the d.c. drive is its efficiency, particularly in comparison with a.c. drives, when operating at speeds in the lower portion of the working range. The d.c. motor is the only electric drive at present in production which can be designed to operate in a stalled condition continuously against its full rate torque and this feature is used for automatic mooring winches of the 'live motor' type. The majority of d.c. winch motors develop full output at speeds of the order of 500 rev/min and where necessary are arranged to run up to two to four times this speed for light line duties. Windlass motors on the other hand do not normally operate with a run up in excess of 2 : 1 and usually have a full load working speed of the order of 1000 rev/min.

Direct current motors may also be controlled by static thyristor converters which convert the a.c. supply into a variable d.c. voltage of the required magnitude for any required armature speed. These converters must be of a type capable of controlled rectification and inversion with bi-directional current flow if full control is to be obtained (Figure 13.1).

Alternating current induction motors, of either the wound rotor or of the cage type are also in common use. With these the speed may be changed by

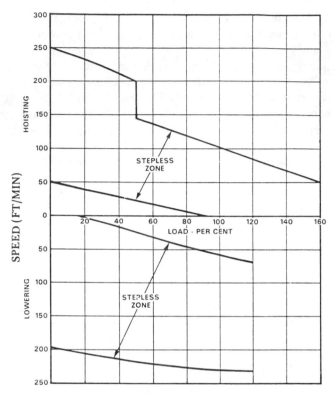

Figure 13.1   *Load/speed characteristic of Ward-Leonard thyristor controlled winch. See Figure 13.2 for conversions to m/sec (Clarke, Chapman Ltd)*

means of pole changing connections and in the case of the wound rotor induction motor, also by changing the value of the outside resistance connected in the rotor circuit. The pole change method involves the switching of high currents at medium voltage in several lines simultaneously, requiring the use of multi-pole contactors. The pole change speed control method offers a choice of perhaps three discrete speeds such as 0.65, 0.325 and 0.1025 m/s corresponding to 4, 8 and 24 pole operation. The wound rotor motor is flexible when hoisting a load, because the starting resistances can be reintroduced into the rotor circuit and the load will cause the motor to slip. The slip gives a range between the speeds dictated by the pole arrangement. As with resistance controlled d.c. motors, difficulty is experienced when providing speed control of an overhauling load, i.e. lowering a suspended load. The disadvantages must be balanced against lower cost, particularly of the cage type induction motor, in comparison with the more flexible d.c. motor. Typical performance curves are shown in Figure 13.2.

Another form of induction motor control system is based on the relationship between output torque and applied voltage, the torque being proportional to the voltage squared. The controller takes the form of a three-phase series regulator with an arm in each supply line to the motor. A stable drive system

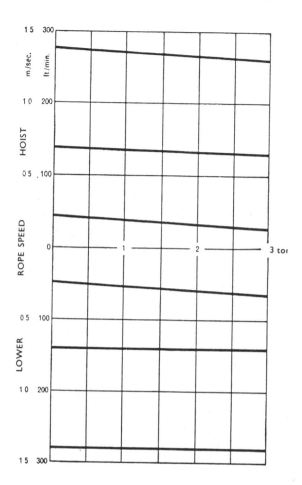

Figure 13.2 *Performance curves of a 3 tonne winch. AC pole-changing 'cage' motor. (Clarke, Chapman Ltd)*

can only be achieved by this means if a closed loop servo control system is used in conjunction with a very fast acting regulator which automatically adjusts the output torque to suit the load demand at the set speed. Control of an overhauling load is made possible by using injection braking techniques. A combined system employing both these control principles can provide full control requirements for all deck machinery.

The a.c. drives described operate at the supply frequency and consequently rapid heating of the motor will occur if the drive is stalled when energized.

The majority of a.c. motors on deck machinery run at a maximum speed corresponding to the 4 pole synchronous speed of 1800 rev/min on a 60 Hz supply. These speeds are similar to the maximum speeds used with d.c. drives and the bearings and shaft details tend to be much the same. The motor bearings are normally grease lubricated. However, where the motor is flange mounted on an oil bath gearcase, the driving end bearing is open to the gearcase oil and grease lubrication is not required.

Hydraulic systems

Hydraulic systems provide a means of distributing power and of obtaining it from a constant speed and constant direction drive such as an a.c. electric motor. The oil pressure can be used to provide variable speed drives through hydraulic motors and power for actuating devices. Hydraulic power is used extensively for deck machinery and remote control of valves.

Hydraulic systems

The three essential components for a hydraulic circuit, are the hydraulic fluid held in a reservoir tank, a pump to force the liquid through the system and a motor or cylinder actuator to convert the energy of the moving liquid into a working rotary or linear mechanical force. Valves to control liquid flow and pressure are required by some systems.

Hydraulic fluid

Water was the original hydraulic fluid and is still used for heavy duty such as operation of lock gates or moving bridges. The disadvantages with water are that it promotes rusting and other forms of corrosion, it is not a good lubricant and it has a limited temperature range.

Hydraulic oils may be straight mineral or special additive oils. Properties of these, enhanced by additives, include oxidation stability, film strength, rust prevention, foam resistance, demulsibility and anti-wear characteristics to enable the fluid to stand up to the higher operating temperatures and pressures of modern systems. Pour point depressants are used to prevent freezing in low temperature conditions. Other fluids used in hydraulic systems may be synthetics or emulsions. Emulsions have been used in systems such as the telemotor, where force is applied and received by pistons. Oils are preferred for systems using rotating pumps and motors, where good lubrication is essential.

In an emergency where short term expediency is the criterion, any thin oil could be used in a system.

Deterioration of hydraulic oils

Hydraulic fluids which are basically mineral oils, will degenerate very slowly over time due to oxidation. The factors which encourage oxidation are the heating and agitation of the oil in the presence of air and metal, particularly copper. The process of oxidation is accelerated by overheating and also by contamination with products of corrosion or the presence of metal wear particles. Oxidation products, both soluble and insoluble, increase the oil's viscosity and cause sludge to be deposited. Oxidation tends to encourage the formation of emulsions with any water from leakage or condensation. Acidic products of oxidation will cause corrosion in the system.

Contamination of oils

Water promotes rusting of steel and must be excluded from hydraulic systems. Rust can be detached and when carried around a circuit can cause the jamming of those valves with fine operating clearance, as well as hastening deterioration of the oil. Sea water can enter through the shaft seals of deck machinery and via system coolers. Condensation on the cold surfaces of reservoir tanks which are open to the atmosphere, is a common source of contamination by water. Tanks should not be constructed such that cold hull plating forms one wall.

Metal wear is inevitable and fine filters are installed to remove these and corrosion particles together with any other grit or dirt that finds its way into the system. Care is necessary with hoses, funnels and oil containers used for filling and topping up reservoir tanks, to ensure that they are clean.

Fine metal wear particles can act as abrasives causing further wear. All particles could cause blocking of small passages or the jamming of valves.

Systems and components

Pump and motor systems are used for powering deck machinery such as winches and windlasses. Pump and actuating cylinders are normally employed for hatch covers. One or more pumps will be used to supply the volume of fluid at the pressure required to operate one or more motors. Pumps may be classified into two groups:

1   those with a fixed delivery when running at a given speed;
2   those with a variable delivery at a given speed.

Fixed delivery pumps can have their constant output bypassed via control valves until required or output can be matched to requirements by incorporating a relief/accumulator, then stopping and starting, varying speed, or connecting a variable delivery pump in parallel.

Variable delivery pump output can be controlled to give full flow in either direction, and volume output can be varied from maximum down to zero.

Fixed delivery pumps

Constant output pumps of the gear or lobe type (see Chapter 5) are precision made to provide high pressure with minimum back leakage. The former operate on the principle that as gears revolve, fluid is carried around the outside between the gear teeth and the housing from the suction to the discharge side of the pump. Fluid from the discharge side is prevented from returning to the intake side by the close meshing of the two gears and the small clearances between the gears and housing. At the discharge side the fluid is discharged partly by centrifugal effect and partly by being forced from between the teeth as they mesh.

Gear pumps may be of the conventional kind or of the type with meshing internal and external gears. Lobe pumps (Figure 5.28, p. 173) are a variation of the latter.

Axial cylinder pumps can be made to deliver a fixed output by setting the swash plate for continuous full stroke operation.

Variable delivery pumps

Variable delivery pumps are used in hydraulic installations as the means of regulating pump output to suit demand. Steering gears are controlled directly by varying the pump output and swash plate pumps are used to supply a range of hydraulic deck machinery. Automatic stroke control can be used to adjust the output.

Constant delivery pump systems

Hydraulic steering gears which are fitted with constant volume or fixed output pumps may have a simple control valve arrangement which either delivers full pump output to the steering gear or bypasses pump output completely. System pressure rises sharply when oil is channelled to the gear. The fixed output pumps of Woodward type hydraulic engine governors, supply to accumulators, which maintain system pressure and hold a reserve of operational oil against demand which may temporarily exceed pump capacity.

For general hydraulic systems where the pump delivers a constant volume of oil, speed control of the hydraulic motor can be obtained by delivering the required amount of oil to the motor through a control valve and diverting the remainder through a bypass to the pump suction. The pump discharge pressure is determined by the load. Speed and direction of rotation are controlled by a lever operated balanced spool valve.

Unit type of circuit

The basic components of a hydraulic system of the Norwinch design are shown in Figure 13.3. The pump in this case is of the vane type which consists of a slightly elliptical case with a cylindrical rotor. The latter has radial slots containing closely fitting rectangular vanes which are forced out against the casing by centrifugal effect and oil pressure. As the rotor turns, the expanding and contracting clearance between it and the casing produces a pumping action. Both mechanical and magnetic filters and a relief valve are provided. The expansion tank contains a reserve of oil. The hydraulic motor is also of the vane type, with vanes mounted in a cylindrical rotor working in a housing

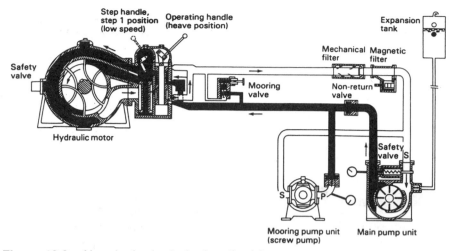

Figure 13.3 *Norwinch single hydraulic drive*

which incorporates two pressure chambers. When the motor is required to exert maximum torque, oil flow from the pump is directed into both chambers. For lighter loads an operating lever is actuated to direct the full flow to only one of the pressure chambers. This system provides two variable speed ranges. The system shown is for mooring winches which are self-tensioning.

Pumps for hydraulic installations, such as the one described, run at constant speed and are driven by an electric motor or directly by a prime mover. With the pump running there is a continuous flow of oil through the system whether the motor is in operation or not. When the winch is not in use the oil merely passes through the operating valve, bypassing the hydromotor and returns to the pump.

Oil pressure is negligible when the hydromotor is idle, reducing power required to a minimum. Oil in the pipelines to and from the motor always flows in the same direction. At the motor controls the flow direction can be reversed to change the rotation of the winch.

Many of the hydraulic systems, fitted to deck machinery are of the 'unit' type, with one pump driving one motor, but there are great advantages to be gained by the use of a ring main system. With the latter type of system, one centrally located hydraulic pump is able to cater for the needs of a number of auxiliaries which can work simultaneously or alternately at varying loads. As the equipment powered from this central pumping installation need not be restricted to deck machinery or to one type of equipment, the system offers considerable savings on capital cost.

Variable displacement pump systems

The hydraulically operated steering gear with an axial piston (vsg type) or radial piston (Hele-Shaw type) variable delivery pump, is an example of variable displacement pump system. The pump itself controls the liquid flow to move a ram or vane steering gear, so that operational control valves are not required. The variable delivery pump is driven at constant speed by an a.c. induction motor; the pump and motor being referred to in the regulations as a power unit. The rate of oil flow from the power unit controls the speed of movement of the steering gear and rudder. A small movement of the telemotor linkage puts the pump on part stroke and the gear moves through a small distance, slowly. When a large movement of the rudder is required, the telemotor linkage puts the pump on full stroke and initially the gear moves rapidly to take the rudder to the desired angle. As the rudder moves, the hunting gear gradually brings the pump control towards neutral, lessening the pump stroke, so that the rate of movement reduces.

A variable displacement system can be used for deck machinery such as windlasses, winches and capstans and also for cargo pumps. The power unit for such circuit may be an axial piston (vsg type) pump with swash plate control to maintain constant pressure in the system. To match the demand of the hydraulic motors being supplied the swash plate control servo-motor monitors system pressure and automatically adjusts pump output to keep pressure constant. Oil cooling is provided by conventional sea-water circulated, tube type heat exchangers.

System design

Careful system design and contamination control are required during manufacture and installation of equipment. The number of joints and pipes are kept to a minimum to reduce the possibility of leakage. Materials are selected that will produce the least quantity of contaminating particles in the system. Filters capable of taking out particles down to a specified size are necessary. Shaft glands or seals must prevent leakage of oil from the machinery and they must also keep contamination out whether the plant is running or shut down.

It is important with all hydraulic systems to ensure that interlocking arrangements provided for pump or motor control levers are in the neutral position before the pump driving motor can be started, in order to avoid inadvertent running of unmanned machinery. Overload protection on hydraulic systems is provided by use of the pressure relief valves set between 30–50% in excess of rated full load pressures.

Anchor handling

The efficient working of the anchor windlass is essential to the safety of the ship. An anchor windlass can expect to fulfil the following:

1   The windlass cablelifter brakes must be able to control the running anchor and cable when the cablelifter is disconnected from the gearing when 'letting go'. Average cable speeds vary between 5 and 7 m/s during this operation.

2   The windlass must be able to heave a certain weight of cable at a specified speed. This full load duty of the windlass varies and may be as high as 70 tonne; figures between 20 and 40 tonne are not unusual. Commonly the load is between 4 and 6 times the weight of one anchor. The speed of haul is at least 9 m/min and up to 15 m/min.
3   The braking effort obtained at the cable lifter must be at least equal to 40% of the breaking strength of the cable.

Most anchor handling equipment incorporates warpends for mooring purposes and light line speeds of up to 0.75 to 1.0 m/s are required. The conventional types of equipment in use are as follows.

Mooring windlasses

This equipment is self contained and normally one electric or hydraulic motor drives two cablelifters and two warpends. The latter may not be declutchable and so will rotate when the cablelifters are engaged. There is some variation in the detailed design of cablelifters and in their drives. Figure 13.4 shows a typical arrangement. Due to the low speed of rotation required of the cablelifter whilst heaving anchor (3–5 rev/min) a high gear reduction is needed when the windlass is driven by a high-speed electric or hydraulic motor. This is generally obtained by using a high ratio worm gear followed by a single step of spur gears between the warpend shaft and cablelifters, typically as shown in Figure 13.5. Alternatively, multi-steps of spur gear are used.

Anchor capstans

With this type of equipment the driving machinery is situated below the deck and the cablelifters are mounted horizontally, being driven by vertical shafts as shown in Figure 13.6. In this example a capstan barrel is shown mounted above the cablelifter (not shown) although with larger equipment (above 76 mm dia. cable) it is usual to have only the cablelifter, the capstan barrel being mounted on a separate shaft.

Winch windlasses

This arrangement utilizes a forward mooring winch to drive a windlass unit thus reducing the number of prime movers required. The port and starboard units are normally interconnected, both mechanically and for power, in order to provide a stand-by drive and to utilize the power of both winches on the windlass should this be required.

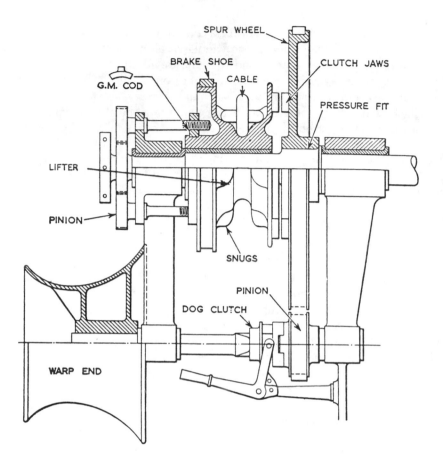

Figure 13.4 *Part plan of windlass dog-clutch-type lifter*

Control of windlasses

As windlasses are required for intermittent duty only, gearing is designed with an adequate margin on strength rather than on wear.

Slipping clutches (Figure 13.7) may be fitted between the drive motors and the gearing to avoid the transmission of inertia in the event of shock loading on the cable when, for example, the anchor is being housed.

Windlasses are normally controlled from a local position, the operator manually applying the cablelifter brake as required to control the speed of the running cable, and whilst heaving anchor the operator is positioned at the windlass or at the shipside so that he can see the anchor for housing purposes. It is quite feasible, however, to control all the functions of the windlass from a remote position. The spring applied cablelifter brakes are hydraulically released, and to aid the operator the running cable speed and the length paid out are indicated at the remote position during letting go. The cablelifter can

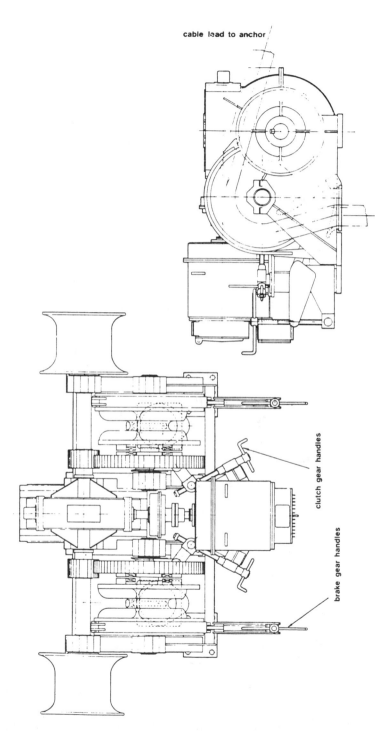

cable lead to anchor

clutch gear handles

brake gear handles

Figure 13.5 *Typical electrically driven mooring windlass*

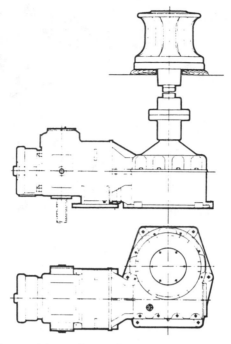

Figure 13.6　*Anchor cable and warping capstan*

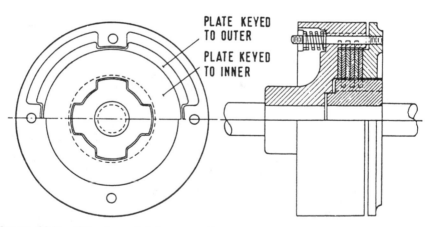

Figure 13.7　*Slipping clutch*

also be engaged from the remote position so that the anchor can be veered out to the waterline before letting go or heaved in as required.

The windlass is in the most vulnerable position so far as exposure to the elements is concerned and maintenance demands should be an absolute minimum. Normally primary gearing is enclosed and splash lubricated, maintenance being limited to pressure grease points for gunmetal sleeve bearings. However, due to the large size of the final of the bevel or spur reduction gears, and the clutching arrangements required, these gears are often of the open type and are lubricated with open gear compounds.

Mooring equipment

Full load duties of warping capstans and mooring winches vary between 3–30 tonnes at 0.3 to 0.6 m/s and twice full load speed is normally provided for recovering slack lines.

The size of wire rope used on mooring winch barrels is governed by the weight of wire manageable by the crew; this is currently accepted as 140 mm circumference maximum. The basic problems associated with the use of wire ropes is that they are difficult to handle, do not float and when used in multi-layers, due to inadequate spooling, the top, tensioned layer cuts down into the underlying layers causing damage. To counteract this problem a divided barrel can be used such that the wire may be stored on one portion and a single layer of wire transferred to the second portion when tensioned. Low density, high breaking strength synthetic ropes (polypropylene, nylon or terylene) offer certain advantages over wire, its main disadvantage being a tendency to fuse if scrubbed against itself or the barrel.

Winches

Mooring winches provide the facility for tensioning the wire up to the stalling capacity of the winch, usually 1.5 times full load thereafter the load is held by the motor brake, or by the barrel brake when the power is shut off. The winch cannot pay out wire unless the brake is overhauled or recover wire unless manually operated, thus wires may become slack.

Automatic mooring winches provide the manual control previously described but in addition incorporate control features such that, in the automatic setting, the winch may be overhauled and wire is paid off the barrel at a pre-determined maximum tension; also wire is recovered at a lower tension should it tend to become slack. Thus there is a certain range of tension, associated with each step of automatic control, when the wire is stationary. It is not practical to reduce this range to the minimum possible as this results in hunting of the controls.

It should be noted that the principal reason for incorporating automatic controls with the features described is to limit the render value of the winch and avoid broken wires; also to prevent mooring wires becoming slack. Load sensing devices are used with automatic mooring winches, e.g. spring-loaded gearwheels and torsion bars are widely used with steam and electric winches; fluid pressure sensing, either steam or hydraulic oil pressure, is also used where appropriate.

Mooring winches are usually controlled at the local position, i.e. the winch. For vessels of unusually large beam or where docking operations are a frequent occurrence e.g. in ships regularly traversing the St. Lawrence Seaway, remote and shipside controllers are of great advantage. As mooring techniques vary widely, the position and type of control must be engineered to suit the application. It is considered, especially on vessels where mooring lines may be long and ship position critical, that the greatest asset to the operator is knowledge of the wire tensions existing during the mooring operation coupled with an indication of the amount of wire paid off the barrel. It is quite feasible to record these at a central position and mooring lines would then only have to be adjusted periodically as indicated by the recording instruments.

The majority of automatic mooring winches are spur geared to improve the backward efficiency of the gear train for rendering, the gearing and bearings being totally enclosed and lubricated from the oil sump. On larger mooring winches were a barrel brake is fitted, it is now common practice to design the brake to withstand the breaking strength of the mooring wire. Worm geared automatic mooring winches are uncommon as the multi-start feature required to improve gear efficiency reduces the main advantage of the worm gear i.e. the high gear ratio.

Cargo handling

The duty of a deck winch is to lift and lower a load by means of a fixed rope on a barrel, or by means of whipping the load on the warp ends, to top or luff the derricks, and to warp the ship. In fulfilling these duties it is essential that the winch should be capable of carrying out the following requirements:

(a)   lift the load at suitable speeds;
(b)   hold the load from running back;
(c)   lower the load under control;
(d)   take up the slack on the slings without undue stress;
(e)   drop the load smartly on the skids by answering the operators application without delay;
(f)   allow the winch to be stalled when overloaded, and to start up again automatically when the stress is reduced;
(g)   have good acceleration and retardation.

In addition when the winch is electrically driven the requirements are:

(a)   prevent the load being lowered at a speed which will damage the motor armature;
(b)   stop the load running back should the power supply fail;
(c)   prevent the winch starting up again when the power is restored until the controller has been turned to the correct position.

Hydraulic winch systems are quite common but electric drives for cargo winches and cranes are most widely used. For the conventional Union Purchase cargo handling arrangements or for slewing derrick systems handling loads up to 20 tonne, standard cargo winches are normally used for hoist, topping and slewing motions, the full load duties varying from 3–10 tonne at 0.65–0.3 m/s. For the handling of heavy loads, although this may be accomplished with conventional derrick systems using multi-part tackle, specially designed heavy lift equipment is available. The winches used with these heavy lift systems may have to be specially designed to fit in with the mast arrangements and the winch duty pull may be as high as 30 tonne.

Cargo winches

It is usual to select the number and capacity of winches and to group them in such a way that within practical limits, all hatches can be worked simultaneously and having regard to their size (and the hold capacity beneath them) work at each can be carried out in the same period.

Reduction of the cycle time during cargo handling is best accomplished by the use of equipment offering high speeds say from 0.45 m/s at full load to 1.75 m/s light, the power required varying from 40 kW at 7 tonnes to 20 kW at 3 tonnes; this feature is available with electro-hydraulic and d.c. electric drives as they offer an automatic load discrimination feature. However, the rationalization of electrical power supply on board ship has resulted in the increased use of a.c. power and the majority of winch machinery now produced for cargo handling utilizes the pole-change induction motor. This offers two or more discrete speeds of operation in fixed gear and a mechanical change speed gear is normally provided for half load conditions. Normally all modern cargo handling machinery of the electric or electro-hydraulic type is designed to fail safe. A typical application of this is the automatic application of the disc brake on an electric driving motor should the supply fail or when the controller is returned to the OFF position.

Derricks

Most older ships have winches in conjunction with derricks for working cargo. The derricks may be arranged for fixed outreach working or slewing derricks may be fitted. A fixed outreach system (Figure 13.8) uses two derricks, one 'topped' to a position over the ship's side and the other to a position over the hold. The usual arrangement adopted, is known as the Union Purchase rig. The disadvantages of the fixed outreach system are that firstly if the outreach requires adjustment cargo work must be interrupted, and secondly the load that can be lifted is less than the safe working load of the derricks since an indirect lift is used. Moreover considerable time and manpower is required to prepare a ship for cargo working.

The main advantages of the system are that only two winches are required for each pair of derricks and it has a faster cycle time than the slewing derrick system.

The slewing derrick system, one type of which is shown in Figure 13.9 has the advantages that there is no interruption in cargo work for adjustments and that cargo can be more accurately placed in the hold. However in such a system three winches are required for each derrick to hoist, luff and slew.

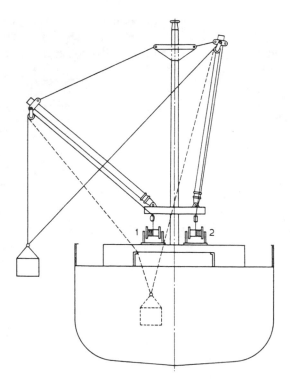

Figure 13.8   *Union Purchasing rig (Clarke, Chapman & Co Ltd)*

Deck cranes

A large number of ships are fitted with deck cranes. These require less time to prepare for working cargo than derricks and have the advantage of being able to accurately place (or spot) cargo in the hold. On container ships using ports without special container handling facilities, cranes with special container handling gear are essential.

Deck-mounted cranes for both conventional cargo handling and grabbing duties are available with lifting capacities of up to 50 tonnes. Ships specializing in carrying very heavy loads, however, are invariably equipped with special derrick systems such as the Stulken (Figure 13.10). These derrick systems are capable of lifting loads of up to 500 tonnes.

Although crane motors may rely on pole-changing for speed variation, Ward-Leonard and electro-hydraulic controls are the most widely used. One of the reasons for this is that pole-change motors can only give a range of discrete speeds, but additional factors favouring the two alternative methods include less fierce power surges since the Ward-Leonard motor, or the electric drive motor in the hydraulic system, run continuously and secondly the contactors required are far simpler and need less maintenance since they are not continuously being exposed to the high starting currents of pole-changing systems.

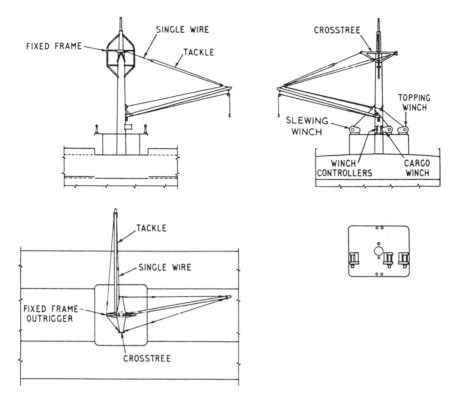

Figure 13.9 *Slewing derrick (Clarke, Chapman Ltd)*

Deck cranes are required to hoist, luff and slew, and separate electric or hydraulic motors will be required for each motion. Most makes of crane incorporate a rope system to effect luffing and this is commonly rove to give a level luff – in other words the cable geometry is such that the load is not lifted or lowered by the action of luffing the jib and the luffing motor need therefore only be rated to lift the jib and not the load as well.

Generally, deck cranes of this type use the 'Toplis' three-part reeving system for the hoist rope and the luffing ropes are rove between the jib head and the superstructure apex which gives them an approximately constant load, irrespective of the jib radius. This load depends only on the weight of the jib, the resultant of loads in the hoisting rope due to the load on the hook passes through the jib to the jib foot pin (Figure 13.11a). If the crane is inclined 5° in the forward direction due to heel of the ship the level luffing geometry is disturbed and the hook load produces a considerable moment on the jib which increases the pull on the luffing rope (Figure 13.11b). In the case of a 55 tonne crane the pull under these conditions is approximately doubled and the luffing ropes need to be over-proportioned to meet the required factor of safety. If the inclination is in the inward direction and the jib is near minimum radius, there is a danger that its weight moment will not be sufficient to prevent it from luffing up under the action of the hoisting rope resultant. Swinging of the hook will produce similar effects to inclination of the crane.

Figure 13.10 *Stulken derrick (Blohm and Voss)*

In the Stothert & Pitt 'Stevedore' electro-hydraulic crane the jib is luffed by one or two hydraulic rams. Pilot operated leak valves in the rams ensure that the jib is supported in the event of hydraulic pressure being lost and an automatic limiting device is incorporated which ensures that maximum radius cannot be exceeded. When the jib is to be stowed the operator can override the limiting device. In the horizontal stowed position the cylinder rods are fully retracted into the rams where they are protected from the weather.

Some cranes are mounted in pairs on a common platform which can be rotated through 360°. The cranes can be operated independently or locked together and operated as a twin-jib crane of double capacity, usually to give capacities of up to 50 tonnes.

Most cranes can, if required, be fitted with a two-gear selection to give a

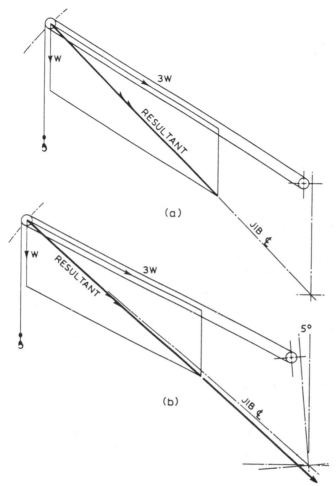

Figure 13.11 *Rope lift cranes – resultant loads when hoisting. (Stothert & Pitt Ltd)*

choice of a faster maximum hoisting speed on less than half load. For a 5 tonne crane full load maximum hoisting speeds in the range 50–75 m/min are available with slewing speeds in the range 1–2 rev/min. For a 25 tonne capacity crane, maximum full load hoisting speeds in the range 20–25 m/min are common with slewing speeds again in the range 1–2 rev/min. On half loads hoisting speeds increase by two to three times.

Drive mechanism and safety features

In both electric and electro-hydraulic cranes it is usual to find that the crane revolves on roller bearings. A toothed rack is formed on the periphery of the supporting seat and a motor-driven pinion meshes with the rack to provide drive. Spring-loaded disc or band brakes are fitted on all the drive motors.

These are arranged to fail safe in the event of a power or hydraulic failure. The brakes are also arranged to operate in conjunction with motor cut-outs when the crane has reached its hoisting and luffing limits, or if slack turns occur on the hoist barrel.

In the case of the electro-hydraulic cranes it is normal for one electric motor to drive all three hydraulic pumps and in Ward-Leonard electric crane systems the Ward-Leonard generator usually supplies all three drive motors.

Cargo access

Although not strictly a machinery item, the mechanical complexity of present-day cargo hatch covers – whose periodic maintenance may fall in the domain of the ship's Engineering Department – warrants some mention in this chapter. Many types of mechanically operated hatch covers can now be found at sea. The principal ones are listed in Table 13.1.

The single pull hatch cover (Figure 13.12) consists of a number of transverse panels which span the hatchway and are linked together by chains. In the closed position, the panel sides sit firmly on a horizontal steel bar attached to the top of the hatch coaming. Just inside the side plates is a rubber gasket housed in a channel on the underside of the hatch cover and which rests on a steel compression bar to form a weathertight seal (Figure 13.13). When closed the covers are held on to the seals by a series of peripheral cleats. Rollers are arranged on the sides of the covers to facilitate opening and closing.

To open a single pull cover the securing cleats are first freed and each panel is raised off its compression bars by hydraulic jacks. The cover wheels, which are arranged on eccentrics, are rotated through 180° and locked into position. The jacks are then removed and the cover can be pulled backwards or forwards as required.

The rail arrangement on both sides at the end of the hatch coaming, is designed to turn the panels so that they are left stacked upright in the space provided.

Table 13.1   *Types of mechanically operated hatch covers*

Operating mode	Type
Rolling and tipping	single pull
Rolling	end rolling
	side rolling
	lift and roll ('piggy-back')
Roll stowing	rolltite
Folding	hydraulic folding
	wire-operated folding
	direct pull
Sliding/nesting	tween-deck sliding

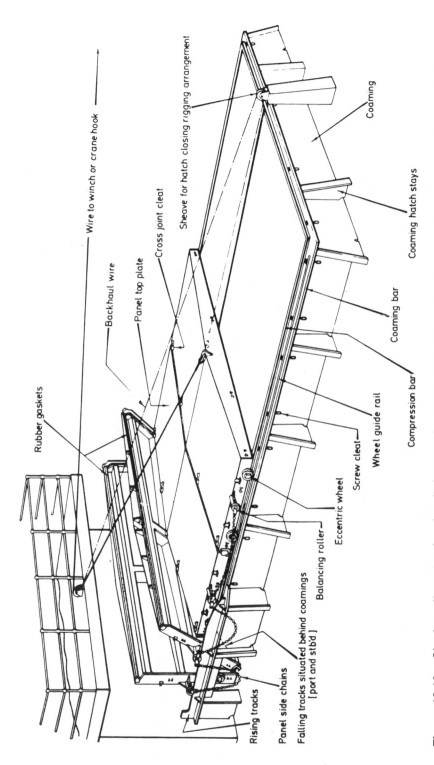

Figure 13.12 Single pull cover showing fittings and opening arrangements (The Henri Kummerman Foundation)

Wire to winch or crane hook

Sheave for hatch closing rigging arrangement

Cross joint cleat

Panel top plate

Backhaul wire

Rubber gaskets

Rising tracks

Panel side chains

Falling tracks situated behind coamings [port and stbd]

Balancing roller

Eccentric wheel

Screw cleat

Wheel guide rail

Compression bar

Coaming bar

Coaming hatch stays

Coaming

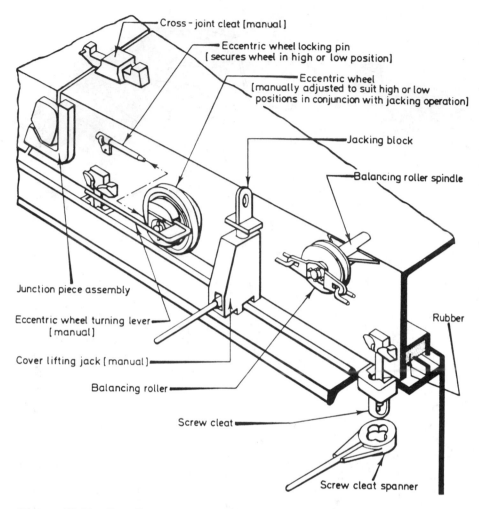

Cross-joint cleat [manual]

Eccentric wheel locking pin
[secures wheel in high or low position]

Eccentric wheel
[manually adjusted to suit high or low
positions in conjuncion with jacking operation]

Jacking block

Balancing roller spindle

Junction piece assembly

Eccentric wheel turning lever
[manual]

Cover lifting jack [manual]

Balancing roller

Screw cleat

Rubber

Screw cleat spanner

Figure 13.13 *Detail of single-pull cover showing scaling arrangement and jacking system (The Henri Kummerman Foundation)*

An alternative arrangement uses a fixed chain drive on the periphery of the hatch, complete with its own electric or hydraulic motor.

Folding covers

Covers of the folding type, are hinged together and arranged to be opened by a wire pull or operated hydraulically. The multi-panel, end-folding, hydraulically operated cover shown in Figure 13.14 has a part panel at each end with hinged arms. As the system is acted upon by the hydraulic cylinder, they fold and travel on the wheels to stow on the end extensions of the rails.

Other folding covers may have fewer panels but work in a similar way, with hydraulic cylinders supplying the effort. The hydraulic system for the hatch

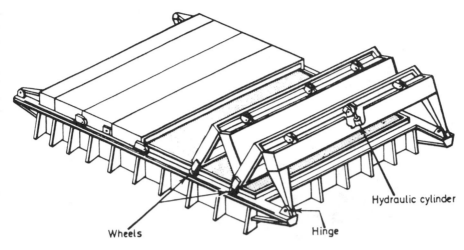

Figure 13.14 *A multi-panel end folding hydraulic cover for weather deck use*

covers may be based on a fixed displacement pump the output from which is bypassed until required for cylinder operation. When pump output is channelled to the cylinders, system pressure and pump motor load rise. Protection is provided by system relief valve and motor protective devices.

Figure 13.15 shows an interesting hydraulic arrangement termed the Navire Hydratorque hinge. It incorporates a pair of helixes attached to two pistons. When hydraulic pressure is applied between the two pistons it forces them and the helixes apart thus causing rotation through the mating helixes and operation of the hinge. Pressure applied to the outside of the pistons creates a torque in the opposite direction.

Maintenance

Hatch cover equipment like the other deck machinery, has to exist in a very hostile environment and the importance of regular maintenance cannot be over-emphasized. Drive boxes and electrical enclosures should be checked regularly for water-tightness. Drive chains, trolleys and adjusting devices such as peripheral and cross-joint cleats should be cleaned and greased regularly. Seals, compression bars and coamings should be inspected and cleaned at each port. Drain channels should be cleared regularly.

On the subject of seals and cleats it is important not to overtighten cleats. The seal should be compressed but not beyond the elastic limit of the gasket material. Standard rubber gaskets can be expected to last from four to five years of normal service. In freezing conditions special grease or commercial glycerine should be spread over the surface of all gaskets to prevent them from sticking to their compression bars. Quick-acting cleats are fitted with thick neoprene washers arranged to exert the correct degree of compression. After a time these lose their elasticity and the cleat must be adjusted or replaced.

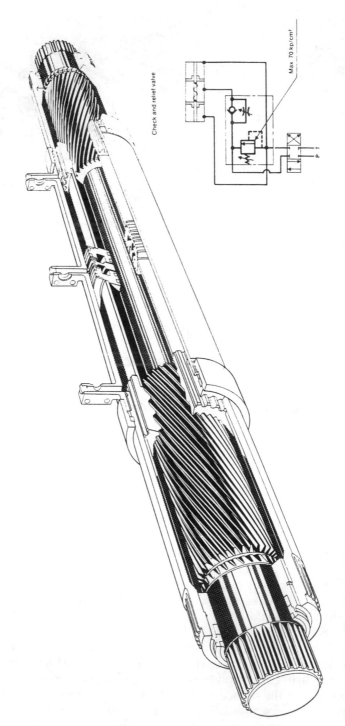

Check and relief valve

Max 70 kp/cm²

Figure 13.15 *The Navire Hydratorque rotary actuator and hinge (Navire Cargo Gear International AB, Sweden)*

Hydraulic systems

The most important thing about any hydraulic system is to ensure that the hydraulic oil remains clean (regular inspection of filters). Any protective boots fitted over rams should be periodically examined as also should flexible hoses. Hydraulic hoses should have their date of manufacture printed on them and can be expected to have a life of about five years.

14

Fire protection

Fire protection on ships is provided by detection and fire-fighting equipment together with structural features which are intended to contain an outbreak of fire and the employment when required of non-combustible materials to prevent its spread.

Air supply

An outbreak of fire requires a source of ignition, the presence of combustible material and ample oxygen. Of the three factors, oxygen is provided in large quantities in machinery spaces, accommodation, dry cargo holds and tanker pumprooms by ventilation fans. Air supply trunkings are not only a source for a supply of oxygen to feed the fire but also have potential for carrying smoke from one area to another. Emergency stops must be fitted so that ventilation fans can be stopped from a position external to the space served. Trunkings are provided with flaps which can be used to isolate various areas as necessary. Provision should be made so that all openings which could admit air can be closed off from a safe position external to the space or from the deck.

Fire main

Whilst the various types of portable extinguishers form the front line of attack against a fire detected in its early stages, the fire main or one of the other fixed fire-fighting installations is used if a fire becomes established. The fire main extends to the full length of the ship and from the machinery spaces to the highest levels. Hydrants served by the main, are situated so that with suitable hoses any area on the ship can be reached.

Water is the chief fire fighting medium on a ship and the fire main is the basic installation for fighting fires. The system shown for the cargo ship (Figure 14.1) has two independently powered pumps which are also used for general service and ballast. These pumps supply engine room hydrants and the deck main through the screw down isolating valve which must be accessible from outside of the machinery space. The latter is required to prevent loss of water through damaged pipework in the engine room if, to maintain the deck supply, the emergency fire pump has to be used. The emergency fire pump is shown as being situated in a tunnel, with a supply to the deck main through the tunnel

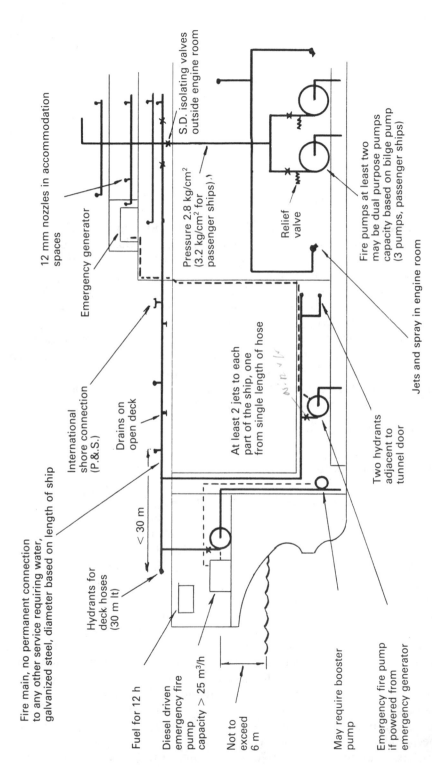

Figure 14.1 *Cargo ship fire main (R. C. Dean)*

escape and also to the twin hydrants in the shaft tunnel by the engine room watertight door. The deck main has a drain at the lowest position so that the pipe can be emptied (particularly of fresh water) in cold weather. If this is not done, the pipe can be damaged by the water freezing but more importantly, it will be blocked by the ice and not usable.

The fire main for a passenger ferry (Figure 14.2) provides cover for the vehicle deck in addition to that furnished by the fixed drencher system. Foam may be provided for foam branch pipes which can be operated through hoses from the fire main hydrants.

Fire pumps

Two independently powered pumps must be provided in all cargo ships of 1000 tons gross and over and in passenger ships of less than 4000 tons gross. Larger passenger vessels and passenger ferries must have three such pumps. The pumps are fitted with non-return valves if they are of the centrifugal type, to prevent loss of water back through open valves when not running. A relief valve is necessary in the system if the pumps are capable of raising the pipeline pressure so that it is greater than the design figure. With centrifugal pumps the relief valve is unlikely to be needed but it is essential for a positive displacement fire pump. Minimum line pressures and capacities are governed by the regulations. Whilst fire pumps may be used for other duties such as ballast or general service they should not normally be used for pumping oil. Changeover arrangements may be fitted allowing a pump to be used for oily bilges.

Emergency fire pump

Normally, cargo vessels are provided with emergency fire pumps because a fire in the engine room could put all of the other pumps out of action. Such a pump is indicated in Figure 14.1 and is located away from the engine room in the shaft tunnel, steering gear or in the forward part of the ship. The suction lift of any pump is limited and for this reason emergency fire pumps are restricted to being at a maximum of 6 m from the water level at light draught conditions. Ideally they are installed below the waterline to guarantee avoidance of suction problems. There have been difficulties in the past with some steering gear located emergency fire pumps when the ship was in the ballast condition.

If the location of a centrifugal type emergency fire pump is the steering flat then, because of the high suction lift involved, a priming pump is fitted. This may be friction driven from the fire pump flywheel and once the fire pump is running the priming pump drive wheel, normally held away from the flywheel by a spring, must be held against it until the fire pump is primed. The fire pump (centrifugal type) discharge valve is shut while the pump is being primed and opened gradually as the suction is taken up.

On large vessels a special two stage pump arrangement may be used

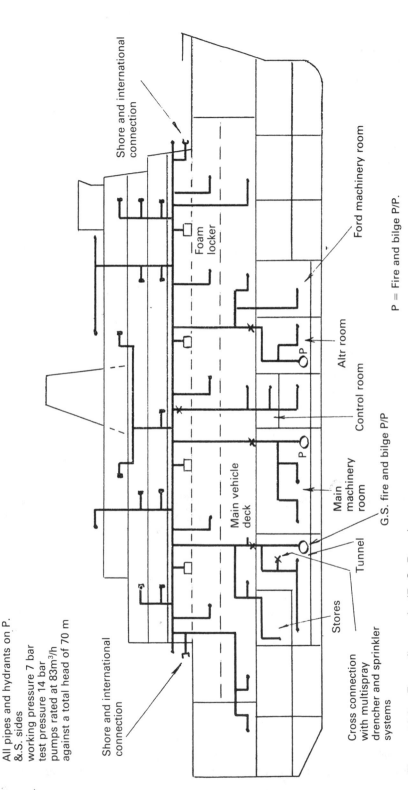

All pipes and hydrants on P. &.S. sides
working pressure 7 bar
test pressure 14 bar
pumps rated at 83m³/h
against a total head of 70 m

Shore and international connection

Shore and international connection

Foam locker

Main vehicle deck

Cross connection with multispray drencher and sprinkler systems

Stores

Tunnel

G.S. fire and bilge P/P

Main machinery room

Control room

Altr room

Ford machinery room

P = Fire and bilge P/P.

Figure 14.2 *Ferry fire main (R. C. Dean)*

(Figure 14.3) The first stage below the waterline is driven by a hydraulic motor. The second stage and the hydraulic power unit are driven by a prime mover (or electric motor from the emergency supply) which can be positioned at more than the normal distance from the waterline.

If the fire pump is driven by a water-cooled diesel engine supplied with cooling water from the fire pump a header tank will be provided to ensure that the engine is cooled while the fire pump is being primed. The engine could have a closed circuit fresh-water system, with the water being cooled in a radiator. It is usual however, to fit an air-cooled diesel engine.

Where a closed-circuit fresh-water cooled engine is installed anti-freeze may be needed for the radiator in cold weather.

An emergency pump has an independent diesel drive or some alternative such as an electric motor powered from the emergency generator.

Pipelines

Where steel pipes are used, they are galvanized after bending and welding. Their diameter is between 50 mm and 178 mm depending on the size and type of ship. Engine room hydrants must have hoses and nozzles for jet and fog or dual purpose nozzles.

International shore connection

The international shore connection (Figure 14.4) is a standard sized flange with nuts, bolts and washers and a coupling for the ship's fittings. The dimensions are shown. The fitting and joint must be suitable for a working pressure of at least 10.5 bar.

Four bolts are required of 16 mm diameter and 50 mm length, also eight washers.

Hoses and nozzles

Fire hoses must be of approved materials. They are positioned adjacent to hydrants together with suitable nozzles. Dual purpose nozzles can be adjusted by rotation of the sleeve to produce a jet or spray. These are an alternative to having available separate jet and spray or fog nozzles.

Foam adaptor

Foam branch pipes (Figure 14.5) which operate in a similar manner to those used in deck installations for tankers, are fitted for use with the hydrants in some machinery spaces and in particular for passenger ferry car decks. These are available in various sizes for operation at a range of pressures and outputs.

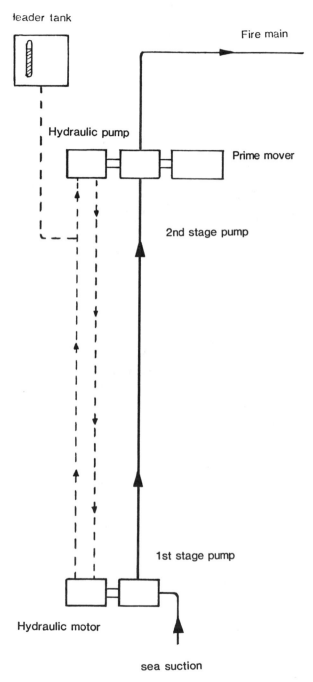

Figure 14.3 *Two stage fire pump*

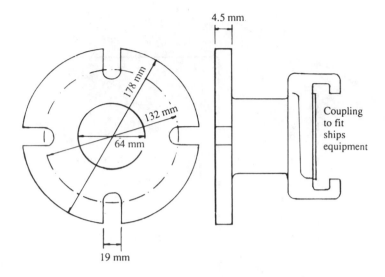

Figure 14.4   *International shore connection*

The foam branch pipe is connected through a hose to the hydrant and the water flow produces a venturi effect which draws up the foam producing liquid through the pick-up tube, from the container. The action also draws in air. Mixing of the three components in the nozzle, causes formation of a jet of foam. Initially, only water issues from the branch pipe and the nozzle is directed away from the fire until foam appears. When the foam compound is exhausted, water will again appear at the nozzle. Foam continuity is achieved by dropping the pick-up tube in a bucket and keeping the bucket topped up with foam liquid.

Machinery space systems

Machinery spaces are protected by fixed fire extinguishing installations, the fire main and extinguishers. Any statutory fixed installation must be operable from a position outside of the space. Any emergency stops for machinery and vent fans, means of securing openings and fuel tank valve shut-off devices, must also be located external to the space.

Machinery space fires

Engine room fires have been started by neglected oil leaks with the combustible material, in the form of fuel or lubricating oil dripping on to and being ignited by hot exhaust manifolds. There are a number of examples of combustible materials and potential ignition sources in machinery spaces. A fire when it starts is usually small enough to be dealt with by a vigilant watchkeeper using a portable fire extinguisher. A fire which develops undetected in an unmanned machinery space (or one where the watchkeeper is

Figure 14.5 *Chubb fire FB5X foam making branch pipe (Chubb Fire Security Ltd)*

in a control room) could require complete shutdown, evacuation and employment of the inert gas system. With unmanned spaces, an efficient alarm system for early detection is vital.

Detectors for unmanned machinery spaces

A variety of devices are available for detecting fire in unmanned machinery spaces but each has an ability to detect basically only one aspect. Thus, smoke detectors based on the ionization chamber are able to recognize combustion products but will not register radiation from a flame or heat. A smoke detector, based on the interruption of light reception by a photo-electric cell, will only identify the shadow effect of dark smoke when it passes through its chamber. It will not identify unseen combustion products, heat or flame. Heat sensors and rate of temperature rise sensors based on the differential expansion of bimetal strips will detect only heat. Flame detectors may not detect a flame which is hidden by equipment and are sometimes caused to operate by sources of radiation other than from a flame.

Each device has a limited ability to detect a feature associated with fire so that systems are likely to have more than one type of device. Watchkeepers can detect fire by seeing flames and smoke, smelling combustion products and feeling heat.

Ionization chamber combustion products detectors

Within machinery spaces, ionization chamber combustion products detectors (Figure 14.6) are the type most used. These devices monitor the electrical change which occurs when combustion particles reach the open chamber of the detector.

Air in the open and closed chambers conducts a small electric leakage level current from the low pressure d.c. supply, because it is ionized by alpha particles emitted from a small piece of enclosed radium. Any combustion products reaching the open chamber, reduce the conductivity. The increased resistance to current flow across the open chamber air gap, causes a rise in potential which triggers the electronic device and switches on a full current flow in the alarm circuit. Operation of any or all of the devices will operate the

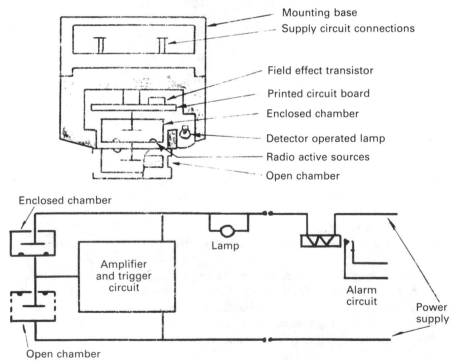

Figure 14.6 *Ionization chamber combustion products detectors (R. C. Dean)*

one alarm. The indicator light shows which device has functioned so that the source of combustion products (sometimes not visible) can be located. The indicator light provided on the sensor body is also used for testing with gas from a test canister or cigarette smoke.

The appropriate number of detector heads are sited in strategic positions above equipment which is a fire risk and at deckhead level in the protected compartment.

Obscuration type smoke detectors

A simple photo-cell and light arrangement (Figure 14.7) can be used as a detector for visible smoke but this type of device would not detect invisible combustion products.

Flame detectors

Photo conductors sensitive to infra-red light (Figure 14.8) or photo-emmissive cells (Figure 14.9) which are sensitive to ultra-violet light, can both be used for flame detection. Light may be directed through a filter, on to the detecting

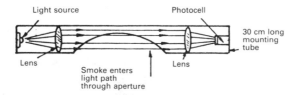

Figure 14.7   *Obscuration type smoke detector (R. C. Dean)*

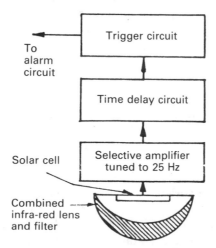

Figure 14.8   *Infra-red flame detector (R. C. Dean)*

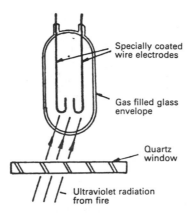

Specially coated
wire electrodes

Gas filled glass
envelope

Quartz
window

Ultraviolet radiation
from fire

Figure 14.9   *Ultra-violet flame detector*

surface by a wide angle lens. A rotating mirror can be incorporated to extend
the area under surveillance. The radiation increases conductivity so that current
flow increases and causes an alarm circuit to be triggered. To reduce the chance
of false alarms from light sources other than flame, a capacitor and a coil are
connected in series, such that only current fluctuations, with a frequency similar
to that of flames, are passed.

Heat sensors

Bimetal strips made up from brass and INVAR will deflect when heated due to
the high expansion coefficient of the brass and the low expansion coefficient of
the INVAR, which is 36% nickel steel. Deflection increases with temperature
and could in itself be used to complete an electrical circuit and operate an alarm
at some specific figure. However, rate of rise can be detected using two bimetal
strips (Figure 14.10) of different thickness. These are set up parallel to each
other and arranged to deflect in the same direction. Rapid temperature rise will
cause the thinner element to deflect more quickly than the other, so causing
contact to be made and the alarm circuit to be completed. A slow rise causes a
similar deflection in both elements, so that a gap between the contacts persists,
until a certain maximum temperature is reached when contact is made because
of other differences in the make up of the strips.

Detection system

The control equipment associated with the installation may be accommodated
in the wheel house or in a fire control centre and comprises a fire indicating
cabinet to which the detector heads are connected, a power unit, to convert the
incoming ship's supply to the voltage appropriate to the equipment and a
standby battery unit. The cabinet will indicate in which space a fire has been

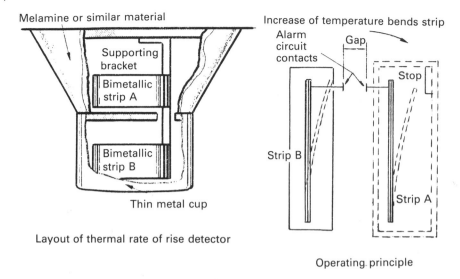

Layout of thermal rate of rise detector

Operating principle

Figure 14.10 *Rate of rise heat detector (R. C. Dean)*

detected and will also monitor the system and indicate whether a fault has developed. It will also instigate an audible alarm.

CO_2 fire extinguishing installations (machinery spaces)

Fire extinguishing installations employing CO_2 stored under pressure at ambient temperature are extensively used to protect ships' cargo compartments, boiler rooms and machinery spaces. When released the CO_2 is distributed throughout the compartment, so diminishing the relative oxygen content and rendering the atmosphere inert.

The quantity of CO_2 required is calculated from the gross volume of the largest cargo space or machinery compartment, whichever is the greater of the two. Additional CO_2 may be required for machinery spaces containing large air receivers. This is because air released from the cylinders through fusible plugs or over pressure release, would increase oxygen content in the space.

The high pressure carbon dioxide (CO_2) system shown (Figure 14.11) is supplied from storage bottles of CO_2 which are opened by a servo-piston operated gang release. A safety feature to protect against accidental release is provided by the master valve on the pipe to the engine room distribution nozzles.

The CO_2 system is used if a fire is severe enough to force evacuation of the engine room or to prevent entry. An alarm is sounded by an alarm button as the CO_2 cabinet is opened and in some ships there is also a stop for the engine room fans incorporated (Figure 14.12). Before releasing the CO_2, personnel must be accounted for and the engine room must be in a shut down condition with all openings and vent flaps closed. It is a requirement that 85% of the

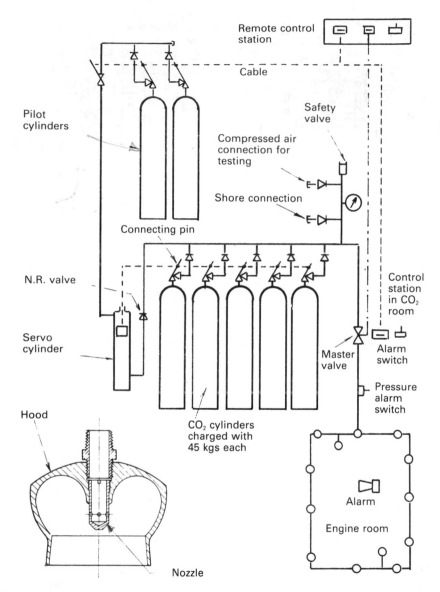

Figure 14.11 *Carbon dioxide system*

required quantity of gas is released into the space within two minutes of actuating the system release.

In the installation shown, the actuating handle opens pilot cylinders of CO_2 and the gas from these pushes the piston in the servo-cylinder down, to operate the gang release for the other bottles. To avoid sticking, all the handles must be in good alignment. The bottle valves may be of the quick-release type (Figure 14.13) where the combined seal and bursting disc is pierced by a cutter. The latter is hollow for passage of liquid CO_2 to the discharge pipe. An

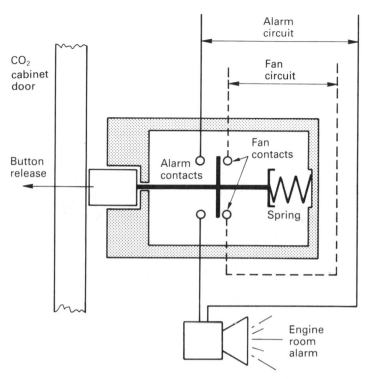

Figure 14.12 *CO_2 cabinet alarm*

alternative type of quick release valve (Figure 14.14) is held in place by a hinged linkage arrangement until released.

Bottle pressure is normally about 52 bar (750 lb/in²) but this varies with temperature. Bottles should not be stored where the temperature is likely to exceed 55°C. The seal/bursting discs are designed to rupture spontaneously at pressures of 177 bar produced by a temperature of about 63°C. The master valve prevents CO_2 released in this way from reaching the engine room. Gas is released by the relief arrangement on the manifold, into the CO_2 space where, in the event that the release was caused by a fire in the compartment, the fire would be extinguished.

Rapid injection of CO_2 is necessary to combat an engine room fire which has attained such magnitude that the space has to be evacuated. Hence the rule that 85% of the gas must be released within two minutes. The quantity of gas carried (a) must be sufficient to give a free gas volume equal to 40% of the volume of the space except where the horizontal casing area is less than 40% of the general area of the space, or (b) must give a free gas volume equal to 35% of the entire space, whichever is greater. The free air volume of air receivers may have to be taken into consideration.

The closing of all engine room openings and vent flaps will prevent entry of air to the space. All fans and pumps for fuel, can be shut down remotely as can valves on fuel pipes from fuel service and storage tanks.

CO_2 bottles are of solid drawn steel, hydraulically tested to 228 bar. The

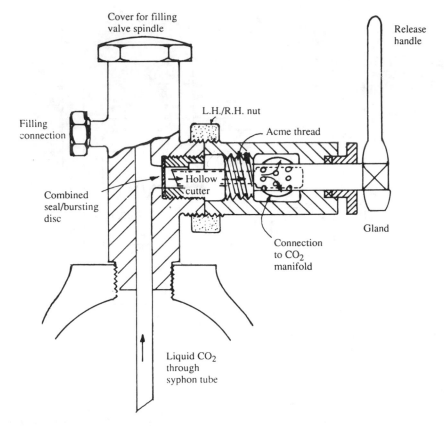

Figure 14.13   *CO₂ cutter type quick release*

contents are checked by weighing or by means of a radioactive level indicator. Recharging is necessary if there is a 10% weight loss.

Pipework is of solid drawn mild steel, galvanized for protection against corrosion. The syphon tube in the bottle ensures that liquid is discharged from the bottles. Without the syphon tube the CO_2 would evaporate from the surface giving a very slow discharge rate and, taking latent heat, would probably cause the remaining CO_2 in the bottle to freeze.

Walter Kidde CO₂ system

Figure 14.15 is a schematic layout of a Walter Kidde CO_2 system in which pilot CO_2 cylinders are used to open the distribution system main stop valve and subsequently the valves on the individual cylinders. The system shown has two banks of cylinders. The pilot CO_2 cylinders are contained in a control box and normally disconnected. To operate the system a flexible pipe fitted with a quick action coupling is plugged into a corresponding socket. When the valve on the pilot cylinder is opened the pilot CO_2 will open the system main stop valve. The stop valve actuator is a piston device and when the piston is fully

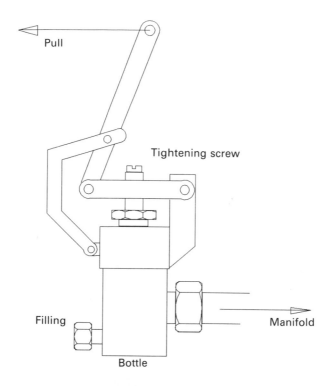

Figure 14.14 *Valve type release*

depressed a second port is exposed which allows the pilot gas to flow to the CO_2 cylinder bank and to operate the cylinder valves. As soon as the control cabinet door is opened to do this, the alarm is initiated. The position of the hoses in the quick-coupling housings prevents the door from being closed.

The pilot CO_2 cylinders and the main CO_2 cylinders for this system, are fitted with Klem valves. An isometric sketch of a Klem valve fitted with a CO_2 actuator is shown in Figure 14.16. The safety pin shown is for transporting the cylinders. When installed the safety pins are removed from the valves, allowing them to be operated manually or by pilot pressure. As soon as mechanical or pilot pressure is removed, the valve will close again. In this system each of the cylinders is fitted with a CO_2 operated actuator.

Low pressure CO_2 storage

In some installations, the CO_2 is stored in low pressure refrigerated tanks. The cylindrical storage vessels (Figure 14.17) are fabricated to the pressure vessel requirements of the authorities. The tanks are of low temperature steel, fully tested and stress relieved. They are mounted on supports designed to withstand shock from collision. The insulation to limit heat ingress into the CO_2, which is stored at a temperature of $-17°C$, is polyurethane with fire resistant additives, foamed *in situ*.

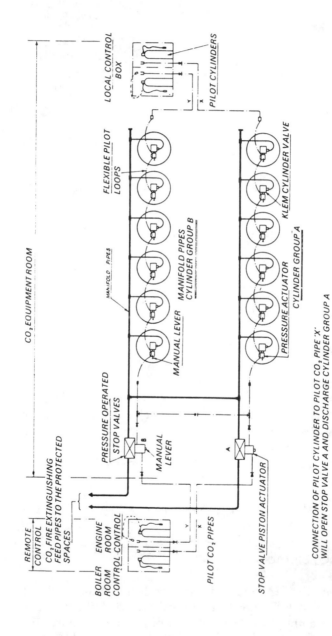

Figure 14.15 A CO_2 fire extinguishing and smoke detector installation (The Walter Kidde Co. Ltd)

CO₂ EQUIPMENT ROOM

REMOTE CONTROL

CO₂ FIRE EXTINGUISHING FEED PIPES TO THE PROTECTED SPACES

BOILER ENGINE ROOM CONTROL CONTROL

PILOT CO₂ PIPES

STOP VALVE PISTON ACTUATOR

PRESSURE OPERATED STOP VALVES

MANUAL LEVER

PILOT CYLINDERS

LOCAL CONTROL BOX

FLEXIBLE PILOT LOOPS

MANIFOLD PIPES

MANUAL LEVER

MANIFOLD PIPES CYLINDER GROUP B

KLEM CYLINDER VALVE

PRESSURE ACTUATOR CYLINDER GROUP A

CYLINDER GROUP A

CONNECTION OF PILOT CYLINDER TO PILOT CO₂ PIPE 'X' WILL OPEN STOP VALVE A AND DISCHARGE CYLINDER GROUP A

CONNECTION OF PILOT CYLINDER TO PILOT CO₂ PIPE "Y" WILL OPEN STOP VALVE B AND DISCHARGE CYLINDER GROUPS B+A

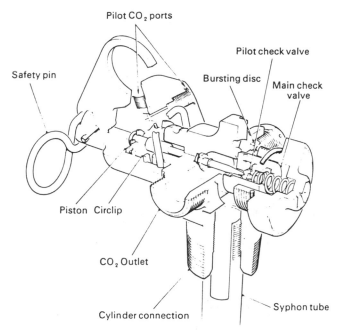

Figure 14.16 *Klem valve with pressure actuator (The Walter Kidde Co. Ltd)*

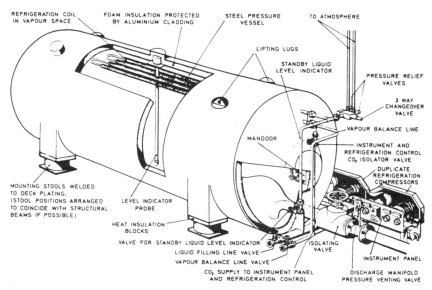

Figure 14.17 *Distillers low-pressure refrigerated storage tank*

Two refrigeration compressors connected to separate evaporator coils mounted inside the pressure vessels, maintain the CO_2 at the required temperature of $-17°C$. One unit is sufficient to deal with heat ingress into the CO_2 the other unit is a stand-by. Either water or air cooling can be arranged.

The duty from main to stand-by unit can be reversed to equalize running hours. The instrument panel contains tank contents gauge, pressure gauge and alarms to indicate CO_2 liquid level and pressure and also contains the refrigeration controls.

To ensure that a dangerous high pressure condition does not exist if a serious refrigeration fault develops, pressure relief valves are fitted discharging directly to atmosphere. The valves are mounted on a changeover valve and are set to discharge CO_2 gas if the pressure in the vessel rises above the design pressure of 23.8 bar. Each valve can in turn be isolated for removal and periodic testing.

The vessels are fitted with a capacitance type continuous indicator together with a stand-by liquid level indicator which ensures that CO_2 liquid level can always be checked approximately by opening the stand-by liquid level indicator valve which will flood the pipe to the same level as the pressure vessel. A frost line will appear due to the low temperature of the liquid CO_2. Closing the valve will cause the CO_2 to vaporize back into the pressure vessel.

The filling and balance lines are normally run to the main deck port and starboard sides for hose connections to be made to a road tanker. The balance line is used to equalize pressure with the tanker during the filling operation.

The liquid CO_2 discharge is through a 150 mm bore pipe fitted with an isolating valve but the quantity of CO_2 discharged into the various spaces is controlled by timed opening of a discharge valve. A relief valve is fitted which will relieve excess pressure in the discharge pipe should the isolating valve be closed with liquid CO_2 trapped in the discharge manifold. Automatic or remote operation can be achieved by utilizing CO_2 gas pressure from the top of the tank as the operating medium.

Because of the considerably reduced amount of steel, the storage tank compared with cylinders gives an approximate 50% weight saving and because low pressure CO_2 has a greater density than CO_2 at ambient temperature, the volume it occupies is considerably less in terms of deck space. Also, low pressure CO_2 usually costs considerably less than CO_2 supplied in cylinders.

A periodical survey of the refrigeration compressors is required by the Classification Societies and this is limited normally to the exchange of pressure relief valves. At intervals of up to 10 years an internal inspection is required via the man-door provided.

Halon systems

Halons are included in the Montreal protocol as gases with ozone depletion potential (ODP) and must not be specified for use in fixed fire fighting installations. Compared with R12 which has been assigned an ODP of 1, Halon 1211 has been given an ODP of 3 and Halon 1301 a figure of 10. Existing fixed halon fire protection systems will need a replacement fire fighting medium for which adaptations will be required.

Halon systems with Halon 1301 have been fitted to a large number of ships.

Halon is the name for halogenated compounds made by the replacement of hydrogen in methane or ethane with one of the halogens. Fluorine, chlorine and bromine are halogens.

Halon 1301 has the chemical formula CF3 Br being known as bromo-trifluoromethane. It is a colourless, odourless gas with a density five times that of air and extinguishes fire by breaking the combustion chain reaction.

Other halogenated hydrocarbons such as methylbromide and carbon tetrachloride have been used in the past as fire extinguishing agents but have been banned by various authorities because of their extremely toxic nature. Halon 1301 however, is classed by Underwriters Laboratories as 'least toxic' (Group 6) and properly applied discharges of the gas allow people to see and breathe permitting them to leave the fire area with some safety.

It must be pointed out however, that when Halon 1301 is exposed to flame or hot surfaces above 480°C halogen acids and free halogens having a higher level of toxicity are produced. A self-contained breathing apparatus or a fresh air mask is therefore essential equipment when entering a space which has been flooded with Halon 1301.

In some small machinery spaces Halon 1301 systems may be found in which the Halon is stored in a sphere within the machinery space. In larger installations the storage battery is similar to that used in CO_2 systems. The gas may be stored in 67 litre cylinders at a pressure of 40 bars. This equates to 75 kg of Halon 1301.

Figure 14.18 shows the release gear used by Fire Fighting Enterprises (UK)

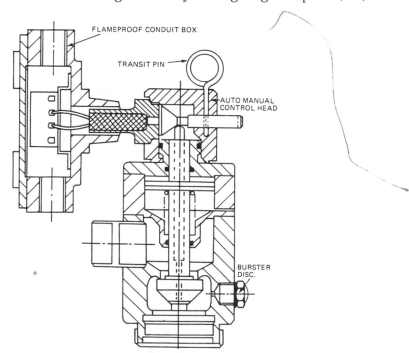

FLAMEPROOF CONDUIT BOX

TRANSIT PIN

AUTO MANUAL CONTROL HEAD

BURSTER DISC.

Figure 14.18 *Halon 1301 cylinder valve with electric actuator (Fire Fighting Enterprises (UK) Ltd)*

Ltd. In this system a cam on the end of the horizontal pin pushes the vertical valve rod downwards, when the pin is actuated. The pin may be moved by a pull wire or by one of a range of electrically or pneumatically operated actuators. As soon as the valve rod opens the cylinder valve, gas from the cylinder is admitted to the top side of the piston and holds the valve open. When the gas pressure falls away a spring below the piston re-seats the valve. Only one cylinder needs to be fitted with the horizontal pin, shown in the illustration. The valves on the other cylinder can be linked to it with small bore copper pipe so that gas from the master cylinder opens all of the other valves simultaneously. Most halon cylinders are fitted with pressure gauges so that leakage can be more readily detected. The cylinders are also fitted with bursting discs.

Halon system operation

The halon release arrangement shown in Figure 14.19 consists of a storage tank, two sets of CO_2 operating cylinders and a manual release cabinet. The halon is stored at a pressure of 14 bar in the holding tank, which has a pressure relief, filling valve and level-indicator.

Release procedure is much the same as for CO_2. When the cabinet is opened, the alarm operates, fans stop and dampers or vent flaps will close. With all entrances closed, the handles (1) and (2) are operated in succession. Handle (2) can only be moved when released by the blocking mechanism.

The contents of the CO_2 bottles opened by handle (1) pressurize the pipeline between the halon tank and the master valve causing the bursting disc to rupture and allowing the halon to flow as far as the master valve. The pressure build up in this line acts on the blocking device to permit operation of handle (2). The latter opens the master valve to the engine room distribution pipe and also opens the CO_2 bottles (2). CO_2 from these ruptures the bursting disc at the top of the storage tank and then assists in expelling the halon.

The discharge must be completed in 40 seconds but the alarm sounds as the cabinet door is opened. Personnel must evacuate the space when warned. The 5% concentration gives a risk because the gas is toxic and must be treated with the caution as CO_2.

Multi-spray system for the machinery spaces

This system is similar to the sprinkler used in accommodation areas but the spray heads are not operated automatically. The section control valves (Figure 14.20) are opened by hand to supply water to the heads in one or more areas. Ready to use hoses can also be supplied. Fresh water is used for the initial charging and the system is brought to working pressure by means of the compressed air connection. The air bottle provides a cushion and prevents cut-in of the pump due to any slight leakage of the water. The pump is automatically operated by pressure drop in the system when the control valve to one section is opened.

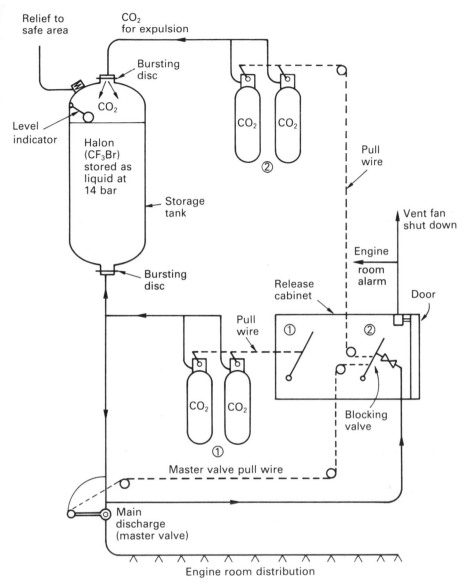

Figure 14.19 *Halon system release*

The pump must have either an independent drive or an electric motor with a supply via the emergency generator switchboard. It must be able to maintain working pressure when supplying all the sections simultaneously in one compartment. It is installed outside the compartment it serves.

Spray nozzles are designed to give the correct droplet size for fires in flammable liquids such as fuels and lubricating oils, when working at the correct pressure. They are located so as to give adequate water distribution over the tank top and all fire risk areas.

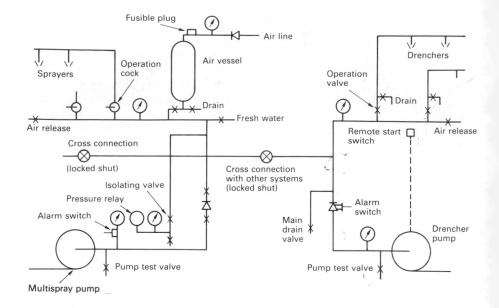

Figure 14.20 *Multi-spray and drencher system (R. C. Dean)*

Water spray is a potentially good fire-fighting medium because

(a) it produces a large quantity of steam which has a smothering action;
(b) in producing the steam, a large amount of heat is required (latent heat) and this gives a cooling effect;
(c) the spray will protect personnel in the compartment;
(d) water is readily available.

Corrosion of the system is minimized by keeping it charged with fresh water. After operation, the pipework is drained of salt water and refilled with fresh water after washing through. Damage to any electrical equipment by the salt water is dealt with by washing with hot fresh water before drying out. The multi-spray, sprinkler or drencher systems, can be connected through a cross connection which is normally locked shut. A car deck drencher system is included in the sketch.

Mechanical foam systems

Foam installations have been fitted in the machinery spaces of all types of ships, to provide fire-fighting capability, usually for areas of specific risk. A system supplying heavy foam will not give the same overall protection as gas smothering because the foam deposits as a layer on a specific surface. High expansion foam can, however, completely fill a space to give a comprehensive extinguishing action. Foam systems are designed specifically to suit the vessel concerned, and must therefore vary in arrangement and capacity.

Pump-operated mechanical foam system

The pump-operated Chubb Fire mechanical foam systems have been installed to extinguish oil fires on the tank top in the machinery space, or above the machinery and over such vulnerable auxiliaries as the oil fuel plant and any other surface in the machinery compartment over which oil is liable to spread. Figure 14.21 shows a twin tank venturi proportionator unit. This installation is typical of the kind of equipment that has been used in the machinery spaces of older ships, or for the decks of older tankers. Two tanks are shown (one being a reserve) together with the foam liquid venturi fed by a dual water supply from the ship's pumps. Sometimes single tank units are fitted. The foam liquid tanks are located at a suitable and approved position outside the machinery space; the capacity of the tanks depends on the surface area to be covered and the depth of foam required.

The venturi to which the dual water supply from the ship's pumps is led is fitted beneath the tanks. Its capacity is also governed by the area of the tank top surface. The water pressure for the venturi, must be at least 7 bar. The duplicated water supply ensures operation of the system should one pump be out of order and is a requirement of the authorities.

Two small bore pipes connect the foam liquid tanks and the venturi, one delivers water under pressure to the top of the tank; the other is connected to the internal syphon tube. Water passing through the venturi takes with it the correct proportion of foam liquid induced from the tank. The resultant solution of water and foam making liquid is led to the foam making nozzles in the machinery space or to the foam monitors on the cargo tank decks of oil tankers.

The capacity of the foam liquid tank is determined by the area of the surface to be covered with foam and is sufficient to deposit foam to a depth of 150 mm

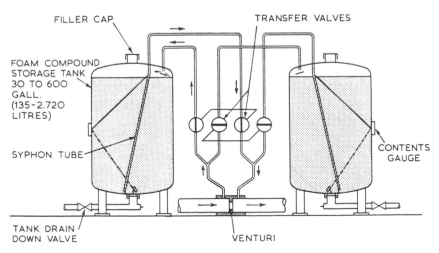

Figure 14.21 *Automatic foam compound proportioning by venturi*

over this area in a machinery space or a specified quantity over the cargo oil tank deck of a tanker to comply with requirements.

A foam liquid level gauge at the side of each tank indicates when the tank is empty of foam liquid and filled with the water which has forced it out. Water will issue from the nozzles unless the water control valve on the inlet side of the venturi is closed at this point.

For vessels having two machinery spaces, one foam-making liquid tank and venturi may be fitted, with distribution valves to discharge the foam to either space, or twin tanks can be installed (one for each space). In these systems the units are usually cross connected so that foam can be discharged into either space from either tank.

Alternative foams

A number of foam concentrates have been developed including Protein, Fluoroprotein and Aqueous Film-Forming Foam solutions (AFFF). Some of these have a specific gravity close to that of water and causing a modification to the tank proportioning system to be necessary. In some modified units (Figure 14.22) the foam liquid is stored in a flexible rubber bag within the steel tank. Water, bypassed via the venturi, feeds into the space between the tank and the bag thus exerting an indirect pressure on the foam liquid and forcing it into the low pressure side of the venturi. The system can have a single or twin tank unit and may be used to feed a variety of low- or high-expansion foam concentrates, although in the latter case the foam generator is somewhat different.

When replenishing any foam concentrate it is essential to use a product compatible with the system.

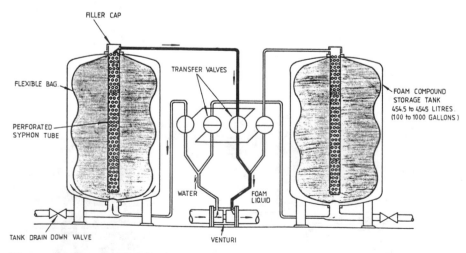

Figure 14.22   *Automatic foam compound proportioning by venturi*

Pressurized pre-mixed solution system

For installation in ships with no pump suitable for operating the above system, a self-contained pressurized pre-mix type may be fitted. This unit contains the water and foam liquid in one tank mixed together in the correct proportion to form the foam making solution. One or more CO_2 cylinders are connected to the tank to expel the solution. A diagrammatic drawing of this type of installation is reproduced in Figure 14.23. When the CO_2 is released into the head-space of the tank, the solution is expelled to the foam distributor network in the machinery or other space to be protected. The capacity of the tank and the number of CO_2 cylinders will depend on the area to be protected.

Automatic foam liquid induction (tankers)

A flexible foam liquid induction unit used in many tankers to supply deck foam-monitor equipment, is shown in Figure 14.24. The automatic inductor on the suction side of the pump maintains the appropriate foam liquid quantity in the water stream. Water pressure and foam liquid suction ports are provided at the side of the automatic foam inductor unit. Water under pressure from the discharge side of the pump enters the pressure port inducing foam liquid from the tank through the foam liquid suction port. A swing paddle fitted in the body of the inductor, in the main water flow, moves backward and forward, according to the rate of flow. The paddle rotates a water metering vane in the water pressure port, bypassing water into the foam liquid port, thus diluting the foam liquid entering the water stream at the correct concentration to meet

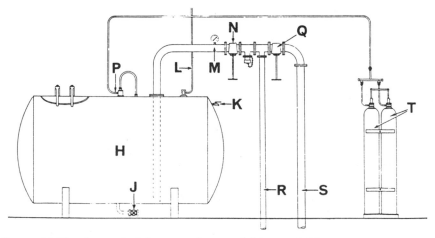

Figure 14.23 *Layout of foam maker sprinkler system*

H Solution storage tank	M Pressure gauge	R Solution supply to branchpipe in
J Drain valve	N Stop value	machinery space
K Level cock	P CO_2 supply from cylinders	S Solution supply to foam-makers
L Safety valve	Q Distribution valve	T CO_2 cylinders

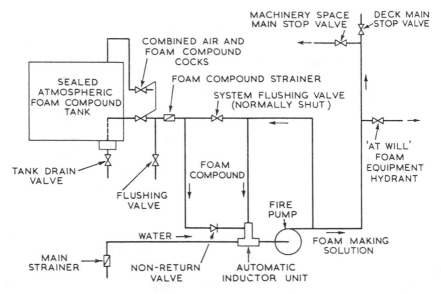

Figure 14.24   *Automatic foam compound induction system*

all flow conditions. The atmospheric type foam liquid tank enables the system to be replenished during operation.

Figure 14.25 shows the Jetmaster foam monitor as installed on the decks of tankers. This monitor requires a minimum water pressure of 8 bar in the deck main, and foam output is approximately 11 800 litre/min. A foam jet radius up to 33–7 metres can be achieved at the working pressure of 8 bar.

Foam branch pipe

As an alternative to introducing the foam-making compound from a bulk storage tank, foam-making nozzles capable of use with the normal fire main but drawing the compound into the nozzle from drums can be used for bulk liquid carrying vessels (also installed in ferries and other vessels). Branch pipes for portable application of foam are made in sizes capable of generating foam at up to 9000 litre/min. Normally an operating pressure of 7 bar is required from the fire hydrant, but lower pressures can be tolerated with a slight drop in foam output. The Chubb Fire FB5X foam branch pipe, (Figure 14.5) will deliver up to 2200 litre/min of foam at a foam jet range of 16/18 m.

Pressure water from the fire main passes through a jet orifice assembly in the water head, inducing foam liquid and air into the water stream in the correct proportion to form foam. Foam liquid is induced from a 20 litre drum placed next to the operator.

High expansion foam

High expansion foam (Figure 14.26) is generated by blowing air through a mesh which has been wetted by a solution of foam concentrate in water. It has

Figure 14.25 *Marine jet-master foam monitor (Chubb Fire Security Ltd)*

been used for hold protection on some container vessels and has been tested for engine room fire fighting.

The mesh is corrugated and its hole size governs the expansion ratio of the foam which is limited to 1000:1 by regulations. The limit is required because the foam is composed largely of air and easily breaks down when in contact with a fire. However, in the 1000:1 foam, the original 1 volume of liquid evaporates and produces enough steam to reduce the percentage oxygen in the steam/air mixture to about 7.5%. This amount of oxygen is below the level normally required for combustion. Heavier high expansion foams can be produced with a different mesh size.

The foam concentrate is metered or mixed with the water to give a 1.5% solution of concentrate in water, and sprayed on to the screen. Air is blown through by an electrically driven fan (water pressure drives have been used). Delivery ducts are necessary to carry the foam to the fire area but normal ventilation trunkings may be acceptable.

Generation of foam must be rapid and sufficient to fill the largest space to be protected at the rate of 1 metre (depth)/minute.

Accommodation

A number of fatal accommodation fires have been started by people falling asleep whilst smoking. The sprinkler system provides protection against this type of incident.

Automatic sprinkler systems

The combination of structural fire protection and an installed sprinkler system which incorporates detection, alarm and fire-fighting capability, has proved very successful in combating the outbreak of fire in passenger ship accommodation. The structural fire protection is based on zones separated by fire proof bulkheads and having fire proof divisions within them. A network of

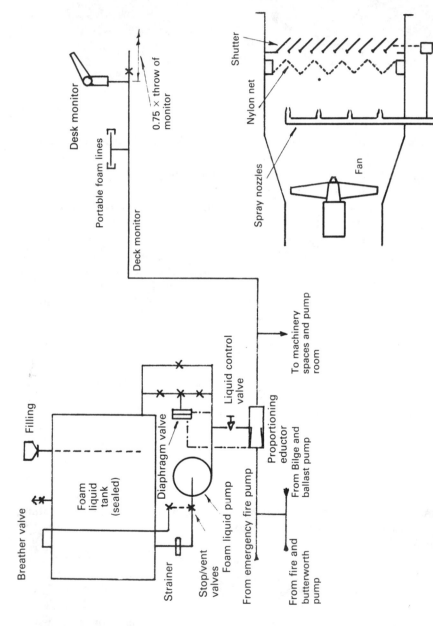

Figure 14.26 *High expansion foam*

sprinkler heads is arranged throughout the spaces to be protected. Each sprinkler head (Figure 14.28a) is normally kept closed by a quartzoid bulb which is almost filled with a liquid having a high expansion ratio. When the liquid is exposed to abnormal heat it expands rapidly to completely fill the bulb. Further expansion is sufficient to shatter the bulb. Water, maintained under pressure by compressed air, is then expelled from the sprinkler head or heads in the form of a heavy spray. Each head adequately showers a deck area of 16 m² and the heads are arranged so that every part of each space requiring protection, can be covered by water spray.

The system shown (Figure 14.27) has a pressure tank which is kept part-filled with fresh water and pressurized to 8 bar by compressed air. When the pressure drops below 5.5 bar, a salt water pump cuts in automatically so that if the sprinklers operate a supply of water is maintained.

Each installation is divided into sections containing up to 200 sprinkler heads and each section has an alarm valve (Figure 14.28b). When a sprinkler comes into operation water flows through the section alarm valve. The water lifts the non-return valve exposing an annular groove which connects to a diaphragm alarm switch. This switch is coupled to an alarm and to an indicator panel on the bridge which gives audible and visual warning that a sprinkler has operated and fire has probably broken out in the section indicated. Fires have frequently been found to have been extinguished by the system alone. When any occurrence has been dealt with the stop valve, which is usually locked open, may be closed to replace the sprinkler head which has operated and to enable the section to be drained of salt water before being filled with fresh from the system. Any maintenance on a section is carried out with the stop valve closed. The test valve can be opened to create flow and cause the non-return section alarm valve to open, to test the alarm.

Regular maintenance of the system consists of greasing the various valves and checking their freedom of movement, logging the pressure gauge reading, before and after each alarm valve (thus checking the tightness of the non-return valves) and checking the alarm system. The latter is done by opening the test valves and checking that the audible and visual alarms work. The pressure tank level is checked and recharged, if necessary, with fresh water and air. The centrifugal salt-water pump should also be tested by closing the isolating valves and draining the pressure switch circuit, when the pump should start automatically. Delivery pressure should be logged. In the event of a fire, when a normal situation is recovered the section and system are drained and flushed out; then recharged with fresh water and air.

Where an automatic system is not fitted in accommodation spaces, it is necessary to install an automatic fire alarm system similar to that used for unmanned machinery spaces. The system would consist of an electric circuit for smoke detectors and possibly bimetallic temperature sensors. Warning is given by an audible alarm with visual indication showing the section in which the fire has occurred.

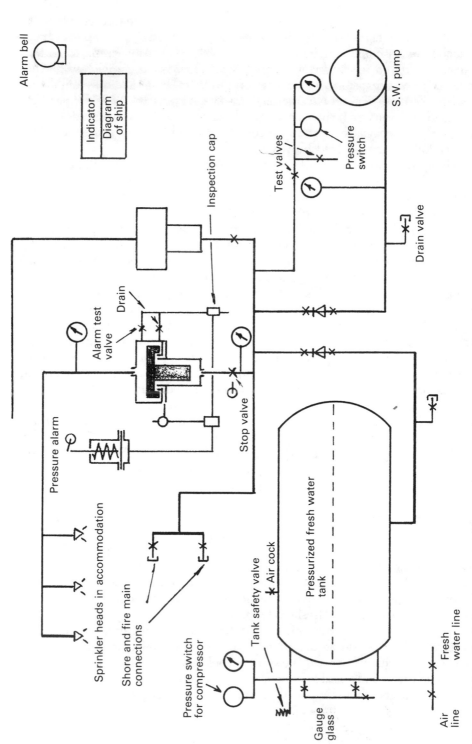

Figure 14.27 *Sprinkler system (R. C. Dean)*

Alarm bell

Indicator

Diagram of ship

Inspection cap

S.W. pump

Test valves

Pressure switch

Drain valve

Drain

Alarm test valve

Pressure alarm

Stop valve

Sprinkler heads in accommodation

Shore and fire main connections

Pressurized fresh water tank

Tank safety valve

Air cock

Pressure switch for compressor

Gauge glass

Fresh water line

Air line

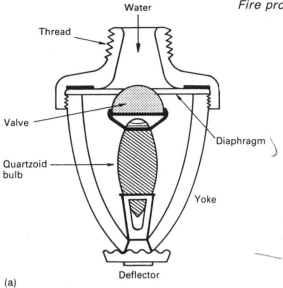

Water

Thread

Valve

Diaphragm

Quartzoid bulb

Yoke

Deflector

(a)

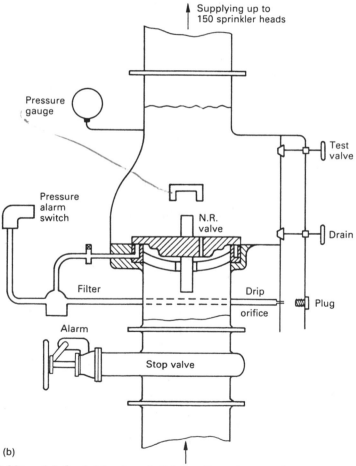

Supplying up to 150 sprinkler heads

Pressure gauge

Test valve

Pressure alarm switch

N.R. valve

Drain

Filter

Drip orifice

Plug

Alarm

Stop valve

(b)

Figure 14.28a *(a) Sprinkler head; (b) Section alarm valve*

Dry cargo holds

Holds for general cargo, have been protected against fire by fixed installations which deliver inert gas from an inert gas generator based on combustion of fuel (similar to the system available for inerting oil tankers) and halon systems. The sea-water fire main is also designed to be used in such spaces if necessary. Fire detection heads like those described for engine rooms have also been used. The type of installation described below, however, is typically used. It provides detection and the means to extinguish a fire.

CO_2 cargo space system with smoke detection

The CO_2 fire extinguishing installation for the cargo holds normally incorporates a smoke sampling system so that fire can be detected in the early stage at the smoke detecting cabinet, which is situated in the wheelhouse. Each cargo compartment is individually connected to the smoke detecting cabinet by a small bore pipe through which a continuous sample of the air in each space is continually drawn by a fan. Part of the sample can be discharged into the wheelhouse to check for the smell of smoke otherwise it vents to atmosphere.

Because the ends of pipes in holds could become blocked, an indicator is fitted at the ends in the cabinet to demonstrate air flow. This telltale is usually in the form of a light propeller which is rotated by the air flow. Any smoke issuing from the pipes in the cabinet, is emphasized by the diffused lighting and should be visible, but a photo-cell detector will automatically operate an alarm.

A three-way valve is fitted on each pipeline below the smoke detector cabinet. This is normally set so that the pipe between the dry cargo space and the smoke detecting cabinet is open. When smoke is detected, the valve for the particular space is operated thus isolating the smoke detecting cabinet from the space and opening the pipe to the CO_2 cylinder battery, ready for discharge of gas to the cargo space on fire. The system of pipes for smoke detection are also used for fire extinguishing.

The CO_2 is stored in liquefied form in 45 kg solid drawn steel cylinders under pressure, or at low temperature and moderate pressure in large bulk storage tanks. In systems designed for cargo space and machinery space protection, the cylinders will be arranged for a ganged total discharge to the machinery space and limited discharge of one or more cylinders (usually by manual release) to individual cargo spaces, depending on the volume of each space. Instructions on how many cylinders are to be released for each cargo space will be displayed at the control station. The cylinders are stored vertically, with their discharge valves at the top and internal syphon tubes are arranged so that discharge is always from the bottom of the liquid. Liquid is thus discharged into the pipe system with vapour being formed only as the liquid leaves the nozzles. Without a syphon tube, CO_2 would evaporate from the liquid surface and pass through the pipe system as a gas, to give a very slow discharge.

Portable fire extinguishers

For reference purposes fires may be grouped in three classes:

1   This class covers fires in solid materials such as wood and soft furnishings. Fires in this class may be extinguished by quenching or cooling with a water.
2   This class covers fires in fluids, such as petrol, lubricating oil and grease. It is dangerous to attempt to extinguish such fires with a jet of water. A small slug of water projected beneath the surface of hot, burning oil can erupt into steam so rapidly that burning oil is thrown in all directions. A very fine water spray, used with care, can be beneficial for fighting oil fires or as a heat screen. These fires may be extinguished by smothering, i.e. being deprived of oxygen.
3   This class covers fires in electrical equipment and any extinguishing agents must be non-conductive.

The first line of defence against fire in any area of the ship, is the portable fire extinguisher. Some common portable extinguishers that have been used at sea are described.

1   soda-acid extinguisher (discharges water)
2   foam extinguisher
3   dry powder type
4   carbon dioxide (CO_2) extinguisher
5   Halon

Soda-acid extinguisher

A 9 litre soda-acid extinguisher is shown in Figure 14.29. It has a 1.63 mm thick steel shell approximately 180 mm dia. and 530 mm high. The shell ends are dished and welded or riveted to the wrapper plate. The shell must be capable of withstanding pressures of up to 14 bar in the event of a blockage occurring in the discharge nozzle. New casings are hydraulically tested to a pressure of 25 bar, for five minutes. The container is also subjected to a pressure test of 21 bar at four year intervals.

Suspended from the neck of the container is a glass phial containing sulphuric acid. The main body of the extinguisher contains sodium bicarbonate, an alkaline solution. The extinguisher must be kept upright when in use. The operator strikes the pin at the top to break the acid phial so that acid and alkali mix, to form carbon dioxide which forces the water (with the chemical remains) out of the discharge. The device must be held upright in use.

The chemical reaction is as follows:

$$H_2SO_4 + 2NaHCO_3 = Na_2SO_4 + 2H_2O + 2CO_2$$

Many portable extinguishers are discharged and refilled yearly. When refilling a soda-acid extinguisher the screwed brass cap which offers internal access

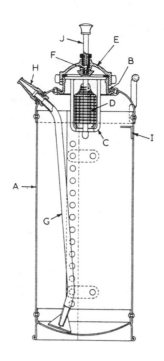

Figure 14.29   *Portable soda acid extinguisher*

A Steel container
B Neck ring
C Acid bottle cage
D Acid bottle

E Gunmental top cap
F Atmospheric valve
G Internal pipe

H Nozzle
I Alkali level indicator
J Plunger

should be carefully examined, first to see that the cap seal is in good order and secondly to check that the small holes in the sides are clear. The purpose of the holes is to release any internal pressure before the cap is fully removed. It is important to check that the ball valve F is free. This prevents liquid from rising up the internal pipe and dribbling from the nozzle in hot weather. When the appliance is operated the internal pressure closes this valve. If it is jammed open the appliance will not function correctly. Before refilling the container, the internal pipe G should be checked to see that it is firmly attached to discharge nozzle and that it is not damaged. With all extinguishers chemicals are provided in kits, complete with re-charging instructions.

Carbon dioxide and water extinguisher

The soda-acid extinguisher has largely been replaced by the CO_2/water type (Figure 14.30) which does not rely on a chemical action for is operation. The extinguishing medium in this type is water with no chemical residue. The CO_2 gas expels the water when the cartridge seal is pierced by the striker pin.

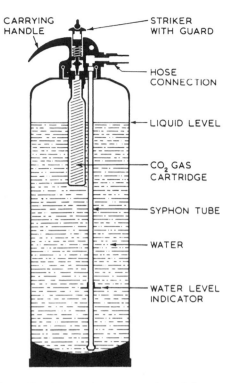

CARRYING HANDLE

STRIKER WITH GUARD

HOSE CONNECTION

LIQUID LEVEL

CO₂ GAS CARTRIDGE

SYPHON TUBE

WATER

WATER LEVEL INDICATOR

Figure 14.30 *Carbon dioxide and water extinguisher*

Chemical foam extinguisher

The outer casing of the portable (chemical) foam extinguisher (Figure 14.31) is similar in construction to the soda-acid appliance and is required to withstand the same initial test pressure of 25 bar and is retested at 21 bars every four years.

This extinguisher has a long inner container of polythene suspended from the neck ring and filled with aluminium sulphate solution. The outer container is filled to a marked level with sodium bicarbonate solution. A lead disc sits on top of the inner container and acts as a stopper. By inverting the extinguisher and shaking it, the disc is dislodged and the two solutions mix with the reaction:

$$Al_2(SO_4)_3 + 6NaHCO_3 = 2Al(OH)_3 + 3Na_2SO_4 + 6CO_2$$

The chemicals react in much the same way as those in the soda-acid extinguisher but the action is slower, giving time for bubbles to form. Foam-making substances added to the sodium carbonate determine the nature of the foam formed. The ratio of the foam produced to liquid is in the order of 8:1 to 12:1.

Because the extinguisher has to be inverted for operation, no internal pipe is fitted. When being recharged the cap seal should be examined and the pressure relief holes in the rim checked. The chemical foam extinguisher may be slow

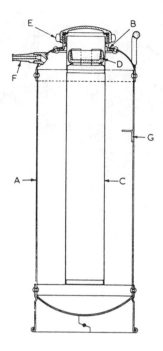

Figure 14.31 *Portable foam extinguisher*

A Steel container D Lead sealing weight F Removable nozzle
B Neck ring E Gunmetal top cap G Alkali level indicator
C Inner acid container

acting in cold conditions and unreliable if the chemical has deteriorated due to heat.

Mechanical foam extinguisher

The alternative type of portable (mechanical) foam extinguisher shown in Figure 14.32 is preferred to the chemical foam type, described previously. The appliance is filled with water and contains an inner container with a small metal bottle of liquid carbon dioxide, surrounded by a plastic bag of foam making compound. To operate this extinguisher it is necessary to strike the plunger on the cap sharply, thus piercing the seal of the CO_2 bottle. The gas then ruptures the plastic bag of foam compound and ejects the water and foam compound through a special nozzle which agitates the mixture and creates a mechanical foam. In this type of extinguisher the ratio of foam to liquid is about 8:1. The extinguisher has an internal pipe and is operated in the upright position.

Dry powder extinguisher

This type of extinguisher contains sodium bicarbonate powder with a water proofing agent such as magnesium stearate to prevent caking. The container

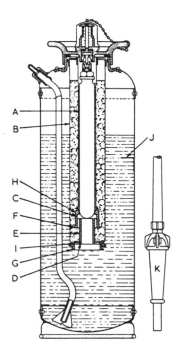

Figure 14.32 *Mechanical foam extinguisher*

A Steel phial of liquid CO_2
B Foam compound container
C-I Socket and seal for foam compound

J Water level before inserting inner container
K Foam-making nozzle

shown (Figure 14.33) holds a cartridge containing liquid carbon dioxide. On piercing its seal with the plunger pin, the gas is delivered through a tube to the bottom of the casing where it entrains the dry powder and carries it up through the discharge tube. The entrainment and CO2 pressure build up within the casing, forces all of the powder through the discharge nozzle. The powder penetrates the flame as a cloud of fine dust which appears to interfere with the chemistry of combustion. The heat of the fire will cause the sodium bicarbonate to decompose to produce carbon dioxide which helps to smother the fire. The decomposition in taking heat may give some cooling effect. The extinguisher discharges in only 15 s and it is thus important to aim the appliance accurately. The operating pressure of this container is high and it is consequently hydraulically tested to a pressure of 35 bar.

Carbon dioxide extinguisher

While carbon dioxide is used in some extinguishers as an inert propellant the gas is also used extensively as a blanketing agent. The carbon dioxide (Figure 14.34) is in liquid form and is at a pressure of 6 bar at 20°C necessitating a strong container. This type of extinguisher can only be recharged ashore. To check for leakage a record should be kept of the weight of the extinguisher.

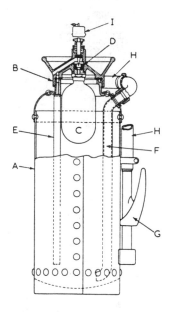

Figure 14.33   *Dry powder extinguisher*

A Steel container
B Top cap
C CO_2 capsule
D Sealing ring and disc
    pieces

E CO_2 injector tube
F Discharge tube
G Controlled discharge
H Flexible hose
I Safety cap

Alternatively the liquid level can be determined by using a special instrument with a radioactive source and a Geiger–Muller counter to detect the gas/liquid interface although this method is usually only used on large fixed CO_2 installations.

The CO_2 could be lethal if discharged accidentally in a confined space and for this reason this type of extinguisher is not permitted in the accommodation.

Portable halon extinguisher

The use of halon in portable extinguishers should be discontinued when suitable alternatives are available because of the high ozone depletion potential.

Further reading

International Conference on Safety of Life at Sea, IMCO (1974)
The Merchant Shipping (Fire Appliances) Rules.
Survey of Fire Appliances; Instructions for the guidance of surveyors, HMSO.

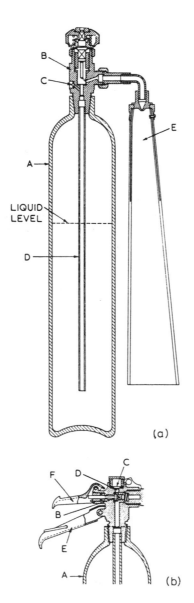

Figure 14.34 *(a) Liquid carbon dioxide extinguisher; (b) Liquid carbon dioxide extinguisher – alternative method of operation*

Key to 14.34a
A Steel cylinder
B Bronze cap with screw-down valve
C Safety bursting disc
D Liquid discharge tube
E Gas discharge horn

Key to 14.34b
A Steel cylinder
B Liquid release valve
C Safety bursting disc
D Liquid outlet
E Lifting handle
F Operating trigger

15

Safety and safety equipment

Much that is written or promulgated with respect to safety at sea is concerned with the obvious sort of risks that threaten those who are careless in any work situation. The survival of seafarers is additionally jeopardized by special factors, many of which are not obvious and not always present. Thus entry to enclosed spaces whether they are ballast or cargo tanks, pumprooms, cofferdams or even dry cargo holds may be made hazardous by the presence of dangerous liquids, toxic or flammable gas from a cargo or ballast water or by the absence of oxygen.

Closed spaces

The fore peak is an example of a tank that can remain closed and unventilated for long periods. The atmosphere in such a tank may become deficient in oxygen due to corrosion resulting from the remains of sea-water ballast or condensation. Oxygen may also be depleted by the presence in sea water of hydrogen sulphide which tends to oxidize to sulphate. Hydrogen sulphide (which is toxic) is a compound which, like ammonia, is produced by bacteria in the water, reducing sulphates and nitrogen compounds. Water containing pollutants such as hydrogen sulphide are picked up when ballasting in esturial waters. Some countries have banned the discharge into their waters of ballast from certain areas due to the presence of harmful bacteria.

Whilst oxygen is depleted, carbon dioxide may be given off by sea water due to other chemical changes. Thus a tank, apparently safe because it has been isolated by being closed, is dangerous to enter due to lack of oxygen and sometimes the presence instead of carbon dioxide and possibly other gases. The small manhole for entry at the top of the tank will not give much natural ventilation of the compartment. Fatal accidents have resulted from entry to fore peak tanks which have not been properly ventilated.

Normal oxygen content of air is about 21%. Where a test shows that there is a lower value in an enclosed space, ventilation should be continued until the correct level is reached. The air changes necessary to improve the oxygen level will have the beneficial effect of lessening the possible presence of any gas or vapour which may be harmful. A lower than normal oxygen level can cause loss of efficiency; it may cause loss of consciousness resulting in a fall, with serious or even fatal injury.

In other closed compartments oxygen may be deficient due to being absorbed by chemicals or drying paint, or because it has been excluded by the presence of other gases or vapours (e.g. vapour from cargo, refrigerant, inerting gas, smoke or fumigant). Ventilation is required before entry to any ballast tank, cargo space, pumproom or closed compartment and must be maintained while work is being carried out.

Marine safety card

The General Council of British Shipping issued a safety card with precautions and a check list, to be used by personnel intending to enter a closed compartment.

Gas analysis

Fixed oxygen analysers are fitted to ships with inert gas systems. These continuously monitor the oxygen content of the flue gas and the inert gas. Para-magnetic type analysers are frequently used. These feed an electrical signal to the oxygen indicator which is calibrated in the range 0–10% by volume. The analyser is connected to audible and visual alarms which are activated when the oxygen level rises above the desired value.

Portable oxygen content meters are also carried and used to check that individual tanks are suitably inerted. Conversely a portable meter can also be used to check whether the atmosphere in a tank will support life and thus permit personnel to enter the tank with safety. If there is any doubt that a tank atmosphere is safe, then breathing apparatus should be worn when entering.

Oxygen analyser

In order to measure the amount of oxygen in a sample from the atmosphere of a closed space or from flue gas, there must be some way of isolating it from the rest of the sample. One physical property which distinguishes oxygen from most other common gases is its para-magnetism. Faraday discovered that oxygen was para-magnetic and was, therefore, attracted by a magnetic field. The field will also induce magnetism in the oxygen, i.e. a magnetic field is intensified by the presence of oxygen and its intensity will vary with the quantity of oxygen.

Most gases are slightly diamagnetic, that is, they are repelled by a magnetic field. Thus glass spheres filled with nitrogen and mounted at the ends of a bar to form a dumb-bell (Figure 15.1) will tend to be pushed out from the strong symmetrical non-uniform, magnetic field in which they are horizontally suspended.

The dumb-bell arrangement is used in the Taylor Servomex analyser. It is suspended by a platinum ribbon in the field and, being slightly diamagnetic, it takes up a position away from the most intense part of the field. The magnets

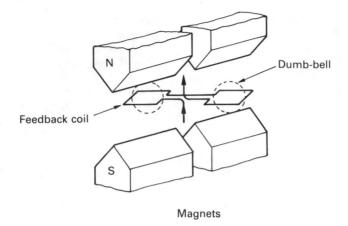

Figure 15.1   *Dumb-bell arrangement (Taylor Servomex oxygen analyser)*

and dumb-bell are housed in a chamber which has an inlet and outlet for the sample. When the surrounding gas contains oxygen, the dumb-bell spheres are pushed further out of the field due to the change produced by the para-magnetic oxygen. Torque acting on the dumb-bell is proportional to the oxygen concentration and therefore the restoring force necessary to bring the dumb-bell back to the zero position is also proportional to the oxygen concentration.

The zero position of the dumb-bell is sensed by twin photocells receiving light reflected from a mirror on the suspension. The output of the photo-cell is amplified and fed back to a coil wound on the dumb-bell (Figure 15.2) so that the torque due to oxygen in the sample is balanced by a restoring torque generated by the feedback current. Oxygen percentage is read from the meter which measures the restoring current. This is scaled to give percentage oxygen directly. Accurate calibration is obtained by using pure nitrogen for zero and normal air for setting the span at 21% oxygen.

False readings are obtained if the gas being sampled contains another para-magnetic gas. The only common gases having comparable susceptibility are nitric oxide, nitrogen dioxide and chlorine dioxide.

Oil tankers

Petroleum vapours when mixed with air can be ignited provided that the mixture limits are about 1% to 10% of hydrocarbon vapour, with the balance made up of air. The figures vary from one specific hydrocarbon to another. Below 1% the mixture is too weak and above 10% too rich. These figures may be termed the Lower and Upper Flammable limits respectively but, because of the risk with flammable vapours in an enclosed space, the terms Lower Explosive Limit (LEL) and Upper Explosive Limit (UEL) are also used. Another condition for combustion is that the oxygen content would have to be more than 11% by volume (Figure 6.9 Chapter 6).

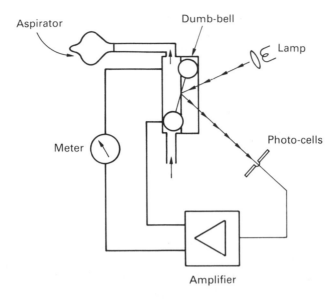

Figure 15.2 *Oxygen analyser (Taylor Servomex)*

Crude oil and the products of crude will give off vapours which are potentially dangerous depending on the volatility of the crude or product. Thus some petroleums will give off a lot of vapour at ordinary ambient temperature and tend to produce an over rich mixture. These are Class 'A' petroleums. This type of petroleum is dangerous because the over rich mixture is readily made flammable by dilution during loading, discharge or tank cleaning. A leak of vapour into the atmosphere is also dangerous.

Petroleums with moderate volatility give off less vapour but the amount can be within the flammable range. These are Class 'B' petroleums. The ullage space of a cargo tank containing Class B products may be filled with a flammable mixture.

Class 'C' petroleums give off little vapour unless heated to above the flash point.

Combustible gas indicators

A combustible gas indicator can be used to check whether there is a flammable atmosphere in a space. The most common type is the catalytic filament gas indicator (Explosimeter). This uses a heated platinum filament to catalyse the oxidation of combustible vapours. Vapour oxidation can take place even outside the limits of flammability and the heat generated will raise the temperature and hence the electrical resistance of the filament. The increase in the resistance of the filament is measured being nearly proportional to the vapour concentration in terms of flammability. The instrument can thus be calibrated terms of 0–100% of the LEL.

Precautions are necessary with these meters. The meter will read 0% for over rich mixtures and the operator needs to be quite sure that the mixture is lean (or below the LEL) and not rich (above the UEL). A visible sign of this is that, with an over-rich mixture, the meter needle will first swing over to maximum before settling on zero. Secondly the filament can be adversely affected by lead compounds or sulphur and after continual exposure to such compounds the meter can give a falsely low reading for hazardous mixtures. To ensure that this does not occur the meter should be checked against the test kits provided, when it is to be used. If the test kit indicates that the instrument is inaccurate then the filament should be changed.

Explosimeter

The atmosphere of a tank or pumproom can be tested with a combustible gas indicator which is calibrated for hydrocarbons. Frequently the scale is in terms of the lower explosive or flammable limit and marked as a percentage of the lower limit (LEL). Alternatively, the scale may be marked in parts per million (p.p.m.)

The combustible gas indicator shown diagrammatically (Figure 15.3) consists of a Wheatstone bridge with current supplied from a battery. When the bridge resistances are balanced, no current flows through the meter. One resistance is a hot filament in a combustion chamber. An aspirator bulb and

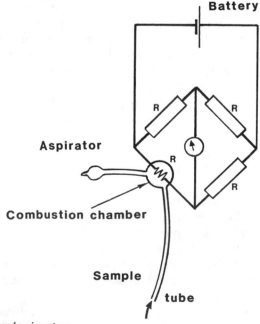

Figure 15.3   *Explosimeter*

flexible tube are used to draw a gas sample into the chamber. The gas will burn in the presence of the red hot filament causing the temperature of the filament to rise. A rise of temperature increases the resistance of the filament and this change of resistance unbalances the bridge. Current flow registers on the meter which is scaled in percentage (LEL) or p.p.m.

A lean mixture will burn in the combustion chamber, because of the filament. False readings are likely when oxygen content of the sample is low or when inert gas is present. The instrument is designed for detecting vapour in a range up to the lower flammable limit and with large percentages of gas (rich mixture) a false zero reading may also be obtained.

The instrument and batteries must be tested before use and the samples are taken from as many places as possible particularly from the tank bottom. It is possible to obtain a reading for any hydrocarbon but other combustible gases on an instrument which is scaled for hydrocarbons, false readings would be obtained. Detection of other vapours must be by devices intended for the purpose (see below).

The explosimeter is primarily a combustible gas detector but will also give guidance with regard to the safety of a space for entry by personnel. If a space has been ventilated to remove vapours, the remaining concentration can be measured by the explosimeter, provided that it is below the lower flammable or explosive range. Generally any needle deflection above zero is taken as indicating a toxic condition.

Crude oils contain all of the hydrocarbon products extracted in the refinery and many of the products are highly toxic. Benzene (C_6H_6) is an example and its low Threshold Limit Value (TLV) of 10 p.p.m. indicates this. Sour crudes carry highly toxic hydrogen sulphide (H_2S) with a TLV also of 10 p.p.m. Petrol (gasoline) has a TLV of 300 p.p.m. Entry to the cargo tanks and pumprooms of a crude oil carrier exposes personnel to these risks. There are additional hazards involved with tank entry, where inert gas has been used. The inert gas adds the risk of carbon monoxide (CO) which has a TLV of 50 p.p.m.; nitrogen dioxide (NO_2) with 3 p.p.m.; nitric oxide (NO) with 25 p.p.m. and sulphur dioxide with 2 p.p.m. Small amounts of the hydrocarbon products which are very dangerous, and any other toxic gases which may be present, require special means of detection (chemical stain tubes described below).

Threshold Limit Values are altered as proved necessary by experience. They are given in references available from health and safety authorities, which are updated annually.

Threshold limit value

Vapour concentrations are measured in terms of parts per million (p.p.m.). A guideline, based on experience, is that personnel exposed to vapour concentrations below the TLV, during normal working hours, will not be harmed but that the risk increases at concentrations above the TLV. Figures sometimes have to be revised because new factors come to light.

Cargo tanks

Cargo tanks, as the result of containing Class 'A' or 'B' cargoes, may contain toxic vapours and/or flammable mixtures. Tanks which are cathodically protected and ballasted may have accumulations of hydrogen although the hydrogen will disperse with proper ventilation, oxygen deficiency can occur due to corrosion resulting from ballast water or condensation but as mentioned previously there are other possible causes. In a tanker, the inert gas used to produce non-flammable conditions in tanks is supplied to reduce the relative oxygen content. A gas-free tank can become dangerous again if there is sludge or scale remaining or if a pipeline containing liquid or gas is opened.

Pumprooms

Pumprooms are subject to leakage from pump shaft seals and pipeline valve glands. The liquid accumulates in bilges (which should be kept clear). Volatiles will form vapour which from sour crude or benzene are toxic. Oxygen can be lacking because it has been displaced by other vapours or inert gas used in an emergency. A pumproom which has remained closed for a long period will, like a ballast tank, lose its oxygen from corrosion.

Cofferdams

Cofferdams and other spaces adjacent to those with cargo can become contaminated due to leakage. They may also be short of oxygen as the result of corrosion, etc.

Chemical tankers

Cargo tanks, pumprooms and other closed spaces, on chemical as on other tankers, may be deficient in oxygen or may contain flammable and/or toxic vapour. Additionally, many liquid cargoes are corrosive and toxic, others are toxic substances which are absorbed through the skin or by ingestion (swallowing). Chemical tanker problems are multiplied by the number and variety of chemicals carried and by the range of risks. Reference books are necessary and available; they contain data on the different hazards and methods of combating them.

Liquid residues in tanks and pumprooms must be considered as potentially dangerous. They may not be easily identifiable (many corrosive liquids are difficult to distinguish from water) and the content can only be guessed from knowledge of previous cargo. Such liquid may be corrosive, or a poison which can be absorbed through the skin. If volatile, the vapour may be toxic or flammable.

Contamination by gases and vapours can be checked provided that the presence of a particular substance is suspected.

Chemical stain tubes

Testing for contaminants in spaces where the variety precludes the provision of special detecting instruments, is made possible with the use of tubes packed at one end with chemical granules that change colour on contact with a particular gas or vapour. The glass stain tubes are sealed to protect the detecting chemicals, the end being broken off before use. There is a different chemical tube for each substance.

The pump or aspirator (Figure 15.4) for taking the test has a long sample tube on its suction side and the stain tube is fitted on the discharge side after first purging to clear air and to fill the tube with a sample from the space. The result, obtained when using Dräger type instruments, can be read directly from the tube after the prescribed number of pumping strokes or found from a chart based on the number of strokes to completely change the colour of the chemical. Chemicals have a two-year shelf life.

General precautions

Any closed space requires ventilation before entry and the ventilation must be maintained during the time that work is being carried out. If there is reason to suspect lack of oxygen or the presence of toxic vapours then ventilation or gas freeing is started some time before entry and the atmosphere is checked before going in.

Liquids in bilges or in the bottoms of tanks should be drained as far as possible. The liquid may itself be dangerous, or dangerous vapour may be produced if it is volatile. Scales and sludges, particularly when heated, tend to give off vapours. These should be removed before entry. Valves and pipelines

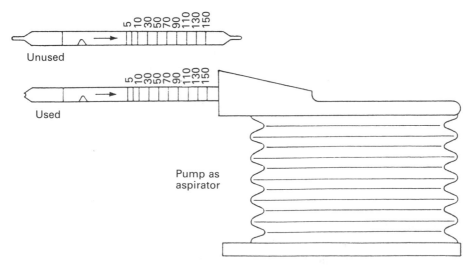

Figure 15.4 *Dräger testing device*

should not be opened as contamination can occur from liquid or vapour in the pipe.

Before working on oil or chemical cargo pumps or pipelines, they should be washed through. After pumping certain chemicals, it may be necessary to use a solvent to wash the pump. As an extra precaution, protective clothing and breathing apparatus may be necessary. Work on pumps should only be carried out when tanks are in a safe condition. Hydraulic or steam lines to pump motors must be closed securely and the power system shut down.

Any portable lights used must be gas tight and safe. When a closed space has to be entered, there must be a second person in attendance at the entrance or, if it is a routine, other personnel should be made aware that entry is intended. The duty of the watcher at the entrance or anybody else involved, is to go for assistance in the event of trouble.

When the situation requires the wearing of breathing apparatus and lifeline the set must be thoroughly checked and signals arranged with those in attendance.

The location of rescue equipment, and the method of using it, must be generally known.

Gas free certificate

When work has to be carried out in port a certificate may be required stating that a space is gas free or that it is safe for hot work (welding, burning). The certificate is obtained from an authorized industrial chemist after tests to prove that any gas present is below the lower explosive limit and below the TLV in quantity. As well as being gas free the compartment must also be free from oil or other residues and from scale or sludge.

Breathing apparatus

For entry to enclosed spaces or when fighting a fire it may be necessary to use breathing apparatus either because of a dangerous atmosphere (toxic gas or smoke) or because a space has been closed to deprive a fire of oxygen. In some ships the equipment comprises a helmet attached by a length of hose to a bellows or a hand operated rotary air pump; others are equipped with self-contained units with one or two compressed air cylinders.

The units with an air hose and bellows or hand operated pump are usually equipped with a helmet which covers the man's head. Operation is quite simple but it is important to memorize a system of signals to indicate to the man operating the pump whether more air is required or if the person wearing the helmet is in trouble. The signals are displayed on a tag attached to the apparatus and should be carefully noted.

The self-contained units vary in detail but consist essentially of one or two compressed air cylinders fitted to a harness which the man wears on his back. Each cylinder is charged to a pressure of 200 bar and contains enough air to sustain a man for approximately 20 minutes at a hard-working rate or 40 minutes if he is at rest. A reducing valve, set at a pressure of about 5.5 bar, is fitted on the cylinder outlet pipe together with a pressure gauge. A bypass valve is usually incorporated. The bottles are connected to a close-fitting face mask by a length of low pressure hose. Interposed between the reducing valve and the mask, is the demand valve which adjusts to the rate of breathing. This valve may be mounted on the front of the harness or on the back of the harness waist band. A non-return valve in the face mask permits the expulsion of air. It is essential when putting the face mask on to be sure that it is correctly adjusted and fits the face properly. This can be checked by stopping the supply of air at the demand valve and breathing in. Any leakage around the edge of the mask should then be obvious.

Emergency generating sets

Diesel-driven emergency generator sets are normally fitted above the bulkhead deck. These are arranged to provide power to essential circuits in the event of the main power plant being out of action because of fire, flooding or any other reason. They must be capable of starting reliably from cold and must have their own independent supply of fuel. In passenger ships they are arranged to start automatically if the main power supply fails. Machines of up to 60 kW may be started by a motor drawing its power from a battery. This will normally be charged from the mains via a suitable transformer/rectifier. Larger machines may be started by compressed air but this must be from a separate reservoir located above the bulkhead deck.

Some emergency generator prime movers are air-cooled diesel engines. These have good cold start characteristics and there is no liquid cooling system requiring attention.

Emergency bilge pumps for passenger ships

To meet the contingency of a passenger ship machinery space being flooded and the power-plant in consequence being put out of action, an emergency bilge pump is required unless other conventional bilge pumps are distributed through several separate compartments. Power is derived from an emergency generating-set. The function of the emergency bilge pump is not necessarily to pump out a compartment which has been badly holed, but to deal with leakage in adjacent compartments, which may result from straining of the bulkhead structure, small cracks which have developed or fractured pipes. As the compartment in which a pump is placed may be badly holed, the pump should be capable of working when completely submerged and having to pump out adjoining compartments.

The pump must be self-priming and capable of overcoming a reasonable amount of air leakage into the bilge-piping system. The emergency bilge pump

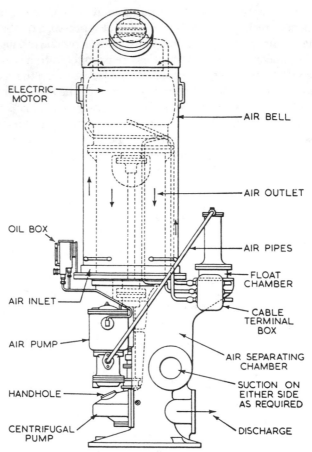

Figure 15.5 *Patent emergency bilge pump*

should be capable of working continuously as an ordinary bilge pump, so as to avoid the necessity of fitting an additional pump.

The emergency bilge pump (Figure 15.5) motor is placed in an air-bell of proportions such that water cannot reach the motor. A quick-running two-throw air-pump driven by worm gearing runs all the time the pump is being used. This pump takes either air or water or both, and not only primes the centrifugal pump but also deals with air leakage.

Power operated water tight doors

The adequate watertight subdivision of ships is effected by steel watertight bulkheads or transverse divisions extending from the tank top of the double bottom to above the margin line, bulkhead deck or freeboard deck of the ship.

It may be necessary to provide doors in some of the bulkheads. These doors must be properly watertight when closed and capable of being opened and shut locally as well as remotely in the event of an emergency. In some vessels it is considered adequate to provide a simple hand-operated rack-and-pinion system to open the doors. In such cases an extended spindle to a point above the waterline is all that is required to provide remote operation.

In vessels which have a large number of watertight bulkheads pierced by access doors below the waterline some sort of powered system is necessary. A number of systems are in use, both electrical and hydraulic. A circuit diagram for a hydraulically operated system is shown in Figure 15.6. At each door a motor driven pump supplies oil, at a pressure of 48 bar to a direct acting cylinder through a control valve which is actuated electrically by solenoids when on bridge control. The d.c. supply for the pump motors is obtained under normal conditions, from an a.c./d.c. transformer rectifier unit common to all

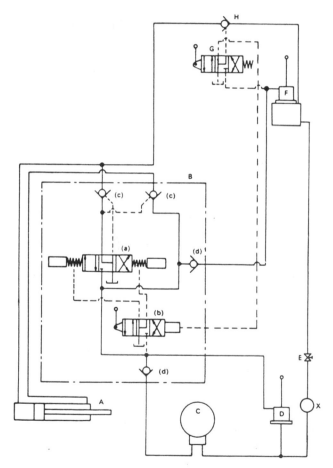

Figure 15.6 *Hydraulic watertight door (Stone Manganese Marine Ltd)*

A Door cylinder
B Door control valve unit, comprising (a) main valve (solenoid); (b) pilot valve; (c) pilot operated check valves; (d) check valves
C Pump unit
D Hand pump (local)

E Stop valve
F Hand pump (remote) and tank
G Control valve (remote)
H Pilot operated check valve (remote)
X Strainer

doors in the installation. In an emergency the supply is taken from the ship's batteries.

Local control

A control is located adjacent to each door for shut/open or for intermediate positions. The control lever is operated manually from either side of the bulkhead and movement of the lever actuates a pilot valve controlling the pump motor circuit under all conditions of control. Advantage is taken of the area differential of the piston in the door operating cylinder for closing and opening the door. A slightly greater force is available in unsealing (opening) the door.

The control lever is spring loaded to the mid-position to ensure a hydraulic lock within the cylinder, thus preventing any possible movement of the door due to the motion of the ship.

Bridge control

The bridge controller is designed to close (in sequence) a maximum of twenty doors within a specified time of 60 s. This period includes a 10 s audible alarm period at each door before the closing movement starts and the alarm continues to sound until each door is fully closed.

Close and re-open controls energize solenoids at the door control valve to start the pump motor. Limit switches, actuated by the movement of each door, control the electric circuit.

Any door which is re-opened locally while the system is under the closed condition from the bridge controller will automatically reclose when the local control lever is released. Indication that power is available to close the doors is given by a 'power on' amber light, with a test button provided.

Emergency control

Under conditions when no power is available, the doors may be closed and opened by a manually operated pump and control valves at either of two positions:

(a)   Adjacent to each door from either side of the bulkhead, and
(b)   From a position on the bulkhead deck.

Light indication

A coloured light indicator on the bridge shows the position of each door: green for shut and red for open. The appropriate lights are duplicated alongside the manually operated emergency pumps on the bulkhead deck. To ensure that power is always available all the indicator lights and alarms are connected direct to the d.c. supply and to the ship's batteries.

Lifeboat davits

Boat davits (Figure 15.7) vary in design to suit the load to be handled and the layout of a ship, but the principles are common to most davits.

When lowering a lifeboat no mechanical assistance apart from gravity is applied. The only manual function required of the operator is to release the winch handbrake and hold it at the off position during the lowering sequence. If the operator loses control of the brake lever in difficult conditions, the attached

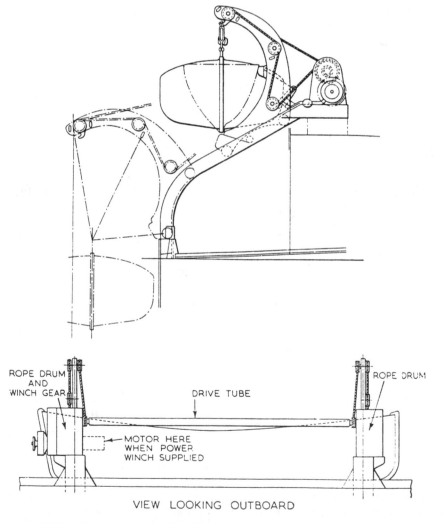

VIEW LOOKING OUTBOARD

Figure 15.7 *Overhead gravity davits (Welin Davit & Engineering Co Ltd)*

weight will provide a positive means of application and the boat will be held at any intermediate position. This condition applies throughout the outboard movement of the boat, from its stowed position until it is waterborne. Figure 15.9 shows the hand brake arrangement. The main brake (on the left in Figure 15.8 is fitted with two shoes, pivoted at one end and coupled at the other with the weighted lever, by a link. The lever projects from the casing through a watertight seal. The shoes are Ferodo-lined and have a normal useful life of five years or more.

Figure 15.8 shows in section the main brake described above and the centrifugal brake (shown on the right-hand side of the drawing). The centrifugal brake limits the rate of descent of the boat when the handbrake is not engaged. Shoes of calculated weight act on the inner surface of a stationary drum, being thrown out by centrifugal effect against the restraining springs. The lowering speed of the boat can be kept within the predesigned limit of 36 m/min. A ratchet arrangement ensures that the drums will not reverse and the boat drop back towards the water in the event of a power failure when a boat is being hoisted.

The brakes require regular inspection for wear and after replacement must be properly tested. The handbrake must obviously be able to hold the boat and to arrest downward movement after a limited free run to test the limiting effect of the centrifugal brake.

Lifeboat engines

The SOLAS 1974 Convention requires that lifeboat engines, where fitted, should be compression ignition engines. Both water-cooled and air-cooled engines are installed in lifeboats. Because of their cold starting characteristics, simplicity and low maintenance requirements, air-cooled engines might be favoured for open lifeboats, but water cooled engines are usual. For totally enclosed lifeboats with water-cooled engines a small single-pass heat exchanger (usually just a large bore tube) may be arranged on the outside of the lifeboat bottom, for cooling of the freshwater circuit by the sea.

Hydraulic cranking system

Some lifeboat engines and the larger diesel-driven emergency generator sets may be fitted with a hydraulic starter motor. The Startorque system (Figure 15.10) is such a device which uses an automatically charged accumulator to provide power to the hydraulic cranking motor. The accumulator is precharged with nitrogen to a pressure of 83 bar.

The system accumulator can be re-charged by a hand pump B or by an engine driven pump I. The stored energy in the accumulator is released by a hand-operated valve F, assisted by a check valve which allows a small quantity of oil to pass, enabling full engagement of the starter motor pinion with the flywheel. The valve then opens fully allowing full flow to the hydraulic

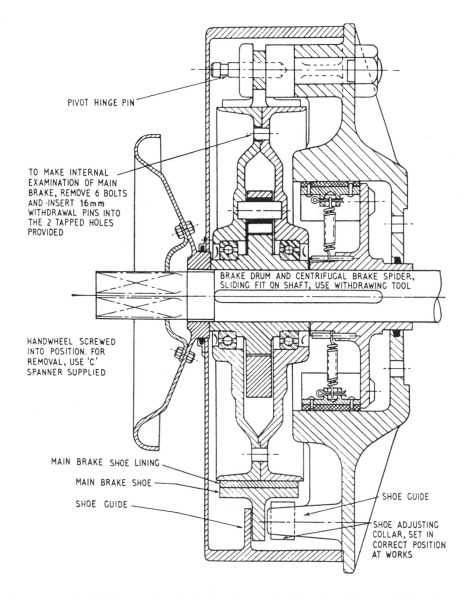

PIVOT HINGE PIN

TO MAKE INTERNAL
EXAMINATION OF MAIN
BRAKE, REMOVE 6 BOLTS
AND ·INSERT 16mm
WITHDRAWAL PINS INTO
THE 2 TAPPED HOLES
PROVIDED

BRAKE DRUM AND CENTRIFUGAL BRAKE SPIDER,
SLIDING FIT ON SHAFT, USE WITHDRAWING TOOL

HANDWHEEL SCREWED
INTO POSITION. FOR
REMOVAL, USE 'C'
SPANNER SUPPLIED

MAIN BRAKE SHOE LINING

MAIN BRAKE SHOE

SHOE GUIDE

SHOE GUIDE

SHOE ADJUSTING
COLLAR, SET IN
CORRECT POSITION
AT WORKS

Figure 15.8 *Welin davit winch. Section of brakes showing centrifugal
brake (Welin Davit & Engineering Co Ltd)*

cranking motor which generates enough torque to start the engine. The oil
returns to the reservoir where it is pumped back to the accumulator by either
the hand pump or the engine driven pump.

An off-loading valve H protects the system from being overcharged and
spills at 20.7 bars back to the reservoir on an open circuit maintaining flow
through the re-charging pump I at all times. The unit cranks the engine at about
375 rev/min for nine revolutions although these figures may be varied to suit
particular engines by modifying the size of the accumulator.

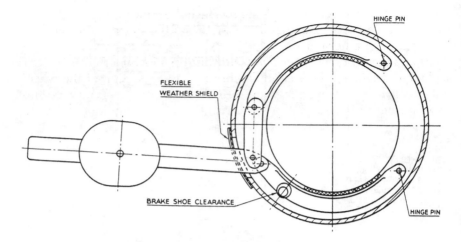

Figure 15.9   *Section of main brake (Welin Davit & Engineering Co Ltd*

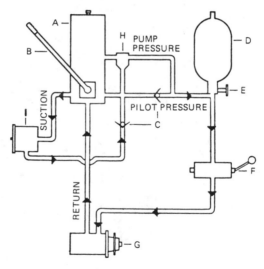

Figure 15.10   *Schematic of Startorque system showing principles of operation*

A Oil reservoir and filter
B Hand pump
C Non-return valves
D Hydraulic accumulator
E Hand shut-off valve

F Starter operating valve
G Hydraulic cranking motor
H Off loading value
I Mechanical recharging pump

Whistles and sirens

Audible signals, to indicate the presence of a ship in poor visibility or to inform other vessels of the ship's intended movements, have long been used at sea. Steam, air and electric whistles have all been fitted for this duty. Some have audible ranges of as much 9 nautical miles.

The air and steam whistles operate on much the same principle, namely the working fluid causes a diaphragm to vibrate and the sound waves generated are amplified in a horn.

The arrangement of a Super Tyfon air whistle is shown in Figure 15.11. The diaphragm details can be seen in Figure 15.12 and a section of the whistle's control valve is shown in Figure 15.13. Units of this type may be found working from air pressures of 6 to 42 bar with air consumptions in the range

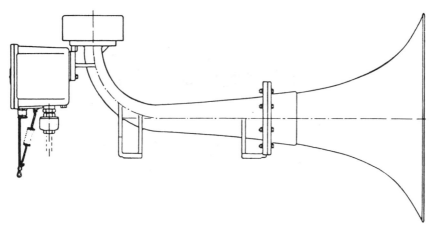

Figure 15.11 *Super Tyfon air whistle (Kockums, Sweden)*

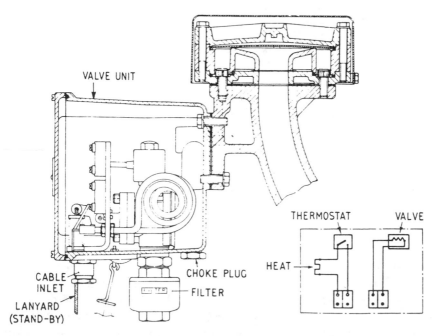

Figure 15.12 *Detailed arrangement of diaphragm (Kockums, Sweden)*

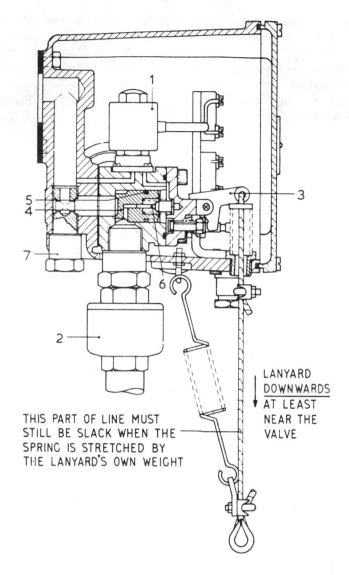

THIS PART OF LINE MUST
STILL BE SLACK WHEN THE
SPRING IS STRETCHED BY
THE LANYARD'S OWN WEIGHT

LANYARD
DOWNWARDS
↓ AT LEAST
NEAR THE
VALVE

Figure 15.13   *Control valve (Kockums, Sweden)*

1. Pilot valve	3. Lever	5. Housing	7. Choke plug
2. Filter	4. Piston	6. Spindle	

25–35 litres/s. Variously sized choke plugs (7) are fitted depending on the supply pressure. Alteratively an adjustable choke may be provided instead of the plug. It is important that the correct choke setting is selected to match the maximum supply pressure. If this is too high for the setting the diaphragm might break; if the pressure is too low the volume will be inadequate.

Primary control of the whistle is afforded by one or more push buttons located at strategic points on the bridge. By pushing the button an electric circuit activates the solenoid pilot valve (1) which then causes piston (4) to move, allowing air to pass to the diaphragm. An automatic device is usually fitted which permits the selection of one or more automatic periodic signals. The Tyfon auto-control (Figure 15.14) allows automatic selection of either a 1 or 2 minute cycle. The normal duration of signal is 5 s. The electric circuit for this equipment is so arranged that, when on automatic signal, depression of the manual button overrides the preselected sequence.

Secondary control of the whistle is by a lanyard which directly operates lever (3) (Figure 15.13) allowing air on to the diaphragm.

Air enters the whistle valve unit via a filter (2) which requires cleaning occasionally. An additional filter is frequently fitted at the lowest point of the air supply line and this will require draining periodically. A routine inspection of this filter at monthly intervals is recommended.

Should the diaphragm require changing, the dirt cover should be removed and the 12 retaining bolts should be unscrewed. it is not necessary to remove the bottom flange. The O ring, on which the diaphragm seats, should always be renewed and it is important to tighten the 12 retaining screws evenly.

Whenever work is to be done on the whistle the air and the electrical supply to the unit must be isolated.

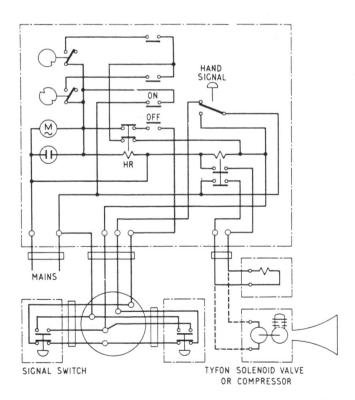

Figure 15.14 *Wiring diagram for Tyfon auto-control whistle (Kockums, Sweden)*

Electric whistles

An Electro-Tyfon whistle is shown in Figure 15.15 from which it can be seen that the electric motor drives a reciprocating piston through a gear train and crank. This generates an air pressure which vibrates the diaphragm.

Further reading

International Conference on Safety of Life at Sea, IMCO (1974).
Fire Prevention and Detection, *Marine Engineers' Review*, 1980.
The Merchant Shipping (Fire Appliances) Rules.
Survey of Fire Appliances; Instructions for the Guidance of Surveyors. HMSO.
Abbott, H. (1987) *Safer by Design*, The Design Council.

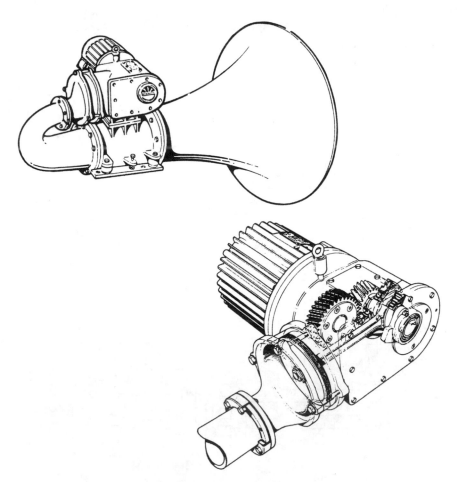

Figure 15.15   *Sectional arrangement of electro-Tyfon whistle (Kockums Ltd)*

Bignell, V., Peters, G. and Pym, C. (1977) *Catastrophic Failures*, The Open University Press.

Kletz, T. *HAZOP & HAZAN*, The Institution of Chemical Engineers.

Kletz, T. (1988) *Learning from Accidents in Industry*, Butterworth-Heinemann.

16

Control and instrumentation

The periodically unmanned machinery space was made possible first by an evolutionary process which took place over a number of years and finally by the introduction of bridge control, which for diesel engines, had long been possible. During the years of progress and refinement, many improvements were made to various types of machinery to enhance reliability. Some types of equipment such as steam reciprocating pumps for engine room services and cargo work, were discarded in favour of centrifugal pumps which are simple and reliable. Automatic operation had been introduced for refrigeration equipment as the then modern CFC type refrigerants, took over from CO_2 systems. Control equipment for auxiliary boilers and engine cooling circuits, followed, so eliminating other routine duties. Instrumentation and alarms had been improved and then fitted more extensively to give more complete monitoring with shut down as appropriate. Long before the advent of the UMS certificate, main machinery was operating with little more than routine attention, as the engine room staff carried out maintenance.

Monitoring and control equipment together with the various alarms, must continue to operate when any or all machinery and systems being monitored, have failed. The power supply for electrical alarms and control equipment is therefore independent of the main supply and may be based on a 24 V direct current (d.c.) system. The application of a moderate voltage, direct current supply to control means that the equipment is simple and safe, with batteries incorporated for independent operation should there be a main electrical power failure. The 24 V d.c. supply may be provided through a transformer and rectifier arrangement with batteries in float or kept fully charged for emergencies or power can be supplied from batteries which are alternately on load or on charge.

Low pressure d.c. system

A ship's 440 V alternating current (a.c.) electrical power system is, of course, highly dangerous and not easily adaptable for control purposes. Alternating current supplies operating at 55 V or more are considered as potentially lethal partly because of the frequency of 60 Hz (or 50 Hz) which can cause muscular contraction in shock victims. Additionally the alternating nature of the supply causes generation of nuisance stray currents and problems with solenoids.

Direct current electrical systems operating at battery voltage, when used for control and emergency systems, are safe, ideal for operation of solenoids, require minimum size of both wire and insulation, can be driven from mains through transformers and rectifiers with stand-by batteries to provide emergency power.

The mains a.c., transformed down to 30 V (necessarily greater than the 24 V batteries for charging) and rectified, supplies the control system and maintains the emergency battery charge. In the event of mains failure, the de-energized solenoid activates a connection between batteries and the system while also isolating it from the mains.

Instrument and control air

The derivation of quality air for control systems is dealt with in Chapter 2.

Control system

The simple control loop has three elements, the measuring element, the comparator element and the controlling element. The loop may be effected pneumatically, electronically or hydraulically. In some instances the control loop will be a hybrid system perhaps utilizing electronic sensors, a pneumatic relay system and hydraulic or electric valve actuators. Each system has its strengths and weaknesses:

Pneumatics – require a source of clean dry air – can freeze in low temperature, exposed conditions, but equipment is well-proven and widely used.

Electronics – good response speeds with little or no transmission losses over long distances, easily integrated with data logging system, required to be intrinsically safe in hazardous zones.

Hydraulics – require a power pack, may require accumulator for fail-safe action – compact and powerful and particularly beneficial in exposed conditions.

Measurement of process conditions

The range of parameters to be measured in merchant ships includes temperatures, pressures, liquid levels, speed of rotation, flow, electrical quantities and chemical qualities. Instrumentation used for remote information gathering purposes invariably converts the measured parameter to an electrical signal which may be used to indicate the measured value on a suitably calibrated scale, provide input information to a data logger or computer, initiate an alarm or provide a signal for a process controller.

As stated earlier, however, the more favoured means of providing process control information (as opposed to information display only) is to use a pneumatic system.

Electrical transducers

Any device used to measure one parameter in terms of another, such as change of temperature by change in electrical resistance, is called a transducer. The following are examples of electrical transducers used in shipboard instrumentation systems.

Temperature measurement

Liquid in glass, mercury in steel and vapour pressure type thermometers have been used on ships for many years. The three main types employing electrical properties are the resistance thermometer, the thermocouple and the thermistor. These are ideal for the provision of input to local or panel mounted displays, data loggers or control systems.

Essentially, a resistance thermometer is a precision resistor with a known temperature coefficient of resistance (i.e. change of resistance with temperature). The majority of resistance thermometers used in marine systems have as their active element a coil of fine platinum wire mounted on a ceramic former. The common standard of calibration is 100 Ω at 0°C, increasing by approx. 0.385 Ω per °C up to 100°C. For more accurate calibration, the manufacturer's temperature/resistance tables should be consulted. Resistance thermometer elements may be housed in many configurations to suit particular applications. The most widely used housing is a stainless steel tube surmounted by a threaded portion and connecting head for mounting in pipelines or tanks. Another type comprises a length of mineral-insulated, stainless steel or copper covered cable, with the resistance thermometer element built into one end. This design is useful to measure temperatures in difficult locations such as the sterntube outboard bearings.

Thermocouples are formed by the junction of two wires of dissimilar metal. When the free ends of these wires are connected to a measuring circuit, a voltage will appear across the instrument terminals which is a function of the difference in temperature between the junction of the two thermocouple wires (hot junction) and the instrument terminals (cold junction). This voltage is known as a thermo-electric e.m.f., and is different for the various thermo-couple materials used. Typical thermocouple combinations using specially developed alloys are copper/constantan, iron/constantan and chromel/constantan, the latter producing the largest signal (approx. 53 mV at 700°C). In many cases, it is not practicable to run the thermocouple leads back to the measuring instrument without some break point. This may be achieved by:

(a) Extension leads. These are simply cables of the same materials as the thermocouple element.
(b) Compensating cables. These cables may be made of a cheaper alloy having the same thermo-electric properties as the thermocouple, but being incapable of withstanding the same environmental conditions.
(c) Copper cables. If copper wires are introduced at an intermediate point in

the cable run, then the thermocouple will measure the difference in temperature between the hot junction and the point at which the change to copper wires is introduced.

The main advantage of thermocouples over resistance thermometers, are mechanical strength and when necessary small dimensions. The disadvantages are a small working signal, the problem of controlling or compensating for the cold junction temperature and lower accuracy.

The thermistor has many of the advantages of both thermocouples and resistance thermometers. Common types take the form of a small bead of semi-conducting material, from which two measuring leads are led away to a terminal arrangement, with mechanical protection in the various forms suggested for resistance thermometers. The thermistor element of one type, exhibits an extremely large negative temperature coefficient in some cases thousands of ohms for a temperature shift of 100°C. These devices can be made very small, very rugged and with extremely accurate resistance/temperature characteristics.

Pressures

The majority of pressure transducers operate by first producing a mechanical motion proportional to applied pressure, from which is derived an electrical signal by some secondary mechanism.

The types most common are the bourdon tube/potentiometer mechanism, in which the motion of the free end of tube is used to move the slide of a potentiometer, and the diaphragm/strain gauge type. There are two generic types of strain gauge, known as bonded and unbonded gauges. Bonded gauges are cemented to the diaphragm, and consist essentially of a grid of conducting material which exhibits a small change in electrical resistance when its shape is changed due to the flexing of the diaphragm. In unbonded types the active element is generally a wire stretched between rigid supports which are displaced by the motion of the diaphragm (Figure 16.1).

As strain gauges only produce a very small change in resistance, they are normally used in Wheatstone bridge circuits.

Measurement of liquid levels

The majority of engine room service tanks need only be monitored between fixed high and low limits, so that float type level switches are adequate. For bunker tanks, a widely used system is the 'bubbler' type level gauge, in which tubes are led to the tanks. The pressure necessary to pass a small volume of air through the tubes is reflected by manometer tubes calibrated in terms of tank level.

Capacitative type sensors may also be used, in which the level in a tank is measured by the change in capacitance in a circuit comprising two concentric

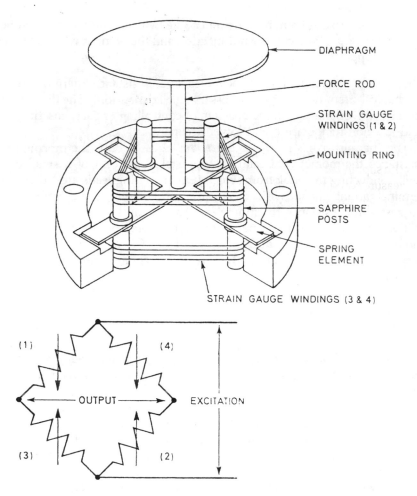

Figure 16.1   *Pressure transducer (Bell & Howell Consolidated Electrodynamics)*

elements, the tank fluid being the dielectric. Pressure transducers may be used for measuring levels in vented tanks.

For more difficult tasks such as boiler drum or condenser hot-well levels, pneumatic level transmitters are generally used.

Flow measurement

Flows of fuel oil, feed water and steam are often required for control purposes and efficiency calculations. Fuel flow is usually measured by a positive displacement meter, in which the process fluid drives a rotor with a known discharge per revolution, the rate of revolution being an analogue of fuel flow.

For feedwater and steam, orifice plates or flow nozzles are used in conjunction with differential pressure transducers, flow being proportional to the square root of the pressure drop across the constrictive element.

Torque measurement

Many ships have torque-meters and power-meters fitted to their propeller shafts. A number of types have been used, some incorporating optical devices to measure the angular displacement of a section of the shaft; others used electric slip rings for the same purpose. Transducers are now being widely used to measure torque, since they are robust and do not require slip rings which are prone to misreading because of dirt or bad electrical contact.

The Torductor torque transducer

The ring type Torductor torque transducer consists of three identical pole-rings with poles mounted on them. The number of poles fitted is chosen as a multiple of four to reduce the influence of the ring joints necessary for easy mounting of the torque transducer around the shaft. The poles are fitted with coils having alternately reversed winding directions. The middle ring is displaced half the pole-pitch relative to the outer rings and the distance between the rings is approximately equivalent to half the pole-pitch. The middle ring is normally used as primary and excited with 50 or 60 Hz. The two outer rings are used as secondaries and connected in series, with mutually reversed winding directions, as indicated by the letters A and B in Figure 16.2, which shows the development of the shaft surface under the ring type Torductor torque transducer and the projection of the poles. The primary poles are marked N and S depicting a certain instant in the magnetizing cycle.

If the shaft is unloaded and without internal stresses, the magnetic fields between the different N- and S-poles will be symmetrical so that the zero equipotential lines will be situated symmetrically under the secondary poles A and B. The secondary flux and hence the secondary voltage is thus zero at zero stress.

When torque is applied to the shaft, the principal stresses ± 0 indicated in Figure 16.2 are obtained. The permeability in the direction of tension, i.e. between the poles B and S and between A and N, is then increased, while the permeability in the direction of compression, i.e. between the poles B and N and between A and S is decreased. Thus all A-poles come magnetically nearer to the N-poles and all the B-poles magnetically nearer to the S-poles. The result is magnetically the same as if the secondary rings had been tangentially displaced in mutually opposite directions and opposite to the torsion of the shaft. The resulting fluxes through the poles co-operate in inducing an output voltage in the series-coupled windings. The output is normally of the order of 10 V and 1 mA, i.e. large enough for an instrument without any amplification.

As the ring type Torductor torque transducer measures almost uniformly around the shaft, the 45° stresses are virtually integrated around the

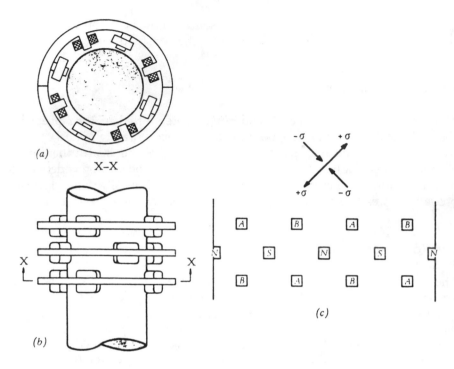

Figure 16.2   *Principle of ring type Torductor torque transducer*

(a) and (b) Physical arrangement   ±Principal stresses
(c) Development of shaft surface under the   N, S. Primary poles
    Torductor poles   A, B. Secondary poles

circumference and the modulation of the output voltage is thus reduced to a very small value.

The effective response time of the Torductor is mainly determined by the exciting frequency and the desired degree of filtering of the output signal. For 50–60 Hz excitation it can be of the order of 10–30 ms, dependent on the circuitry chosen.

The ring type Torductor torque transducer can be statically calibrated with good accuracy. The calibration has to be performed with the torque transducer mounted around the shaft as the sensitivity is dependent on its composition and heat treatment. This is however, a drawback which the Torductor torque transducer in principle shares with all other torque transducers.

The Torductor tachometer

The tachometer (Figure 16.3) is a three-phase a.c. generator. The frequency, and not the voltage, is used as a measure of the rate of rotation. This avoids the problem of brushes, non-linearity in the amplitude of the output voltage and conductor resistances. The electronic equipment consists of both analogue and digital circuits. It should be mentioned that a very stable oscillator, common to

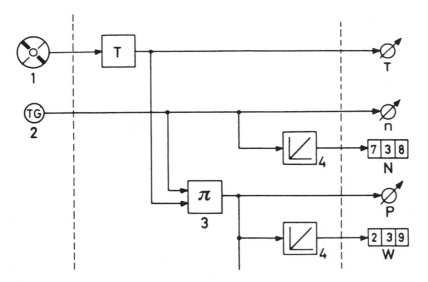

Figure 16.3 *Torductor power meter*

1. Torque transducer
2. Tacho-generator
3. Multiplier
4. Integrator
T. Torque
n. Rate of rotation
N. Total number of revolutions
P. Shaft horsepower
W. Total power output

the entire equipment, is used as a reference for generating pulses having a predetermined duration. The operation of the electronic equipment is briefly as follows.

The tachometer signal is converted into a square wave having a constant pulse length and with a frequency proportional to the tachometer frequency. This square wave controls an electronic contact with a constant input voltage. The mean current through this will then be proportional to the tachometer frequency. The current is fed into an a.c. amplifier, whose output voltage is thus directly proportional to the rate of rotation. The direction of rotation and thus also the polarity of the output signal, is determined through sensing of the phase sequence. The total number of revolutions is indicated on a counter on the engine-room console. A digital circuit imparts to the counter one pulse for every tenth revolution of the shaft.

The shaft (delivered) horsepower is obtained according to the same principle as the rate of rotation. If the constant input voltage of the electronic contact is replaced by a voltage proportional to the torque, the mean current through the contact will be proportional to the shaft horsepower. The current is fed as previously to an a.c. amplifier, which in its turn drives a pointer instrument.

The total power output is obtained through integration of the power signal, which is therefore fed into an amplifier with capacitance feedback serving as an electronic integrator. A level discriminator senses when the output voltage from the amplifier exceeds a certain value. The integrator then receives a resetting pulse via an electronic contact. Each of these pulses represents a certain amount of energy and the total number of pulses the total output of the machinery. Indication is accomplished with an electro-mechanical counter.

Pneumatic control systems

For this discussion, measurement is the value of a process quantity or quality obtained by some suitable measuring device. It may be recorded or indicated by a pen or pointer on a scale. The terms 'pen' and 'measurement' will sometimes be used synonymously. For example a 'change in pen position upscale' means an increase in measurement.

A simple control loop is shown in Figure 16.4. The measuring element determines the pressure downstream of the control valve and transmits this to the comparator element, which is a pneumatic controller. The pneumatic controller compares the measured value with a manually set desired value and generates a correcting signal which causes a control valve to open or close. A Foxboro Model 40 Controller is shown schematically in Figure 16.5 which serves to illustrate the way in which pneumatic controllers work. The desired value sought in the controlled system is set by turning the control setting index knob, thus positioning the index. A system of levers and links is arranged so that whatever position the index is placed in any deviation by the pen from the pre-set index will result in a proportional movement of the horizontal link attached to the proportioning lever.

The particular controller shown in Figure 16.5 is known as a proportional or single-term controller. This type of instrument produces a correcting pneumatic signal in the range 0.2–1 bar which is proportional to the deviation of the measured value (pen position) from the desired value (index position). Like most proportional controllers the Model 40 incorporates a device for altering the ratio of actuating signal change to feedback position change; thus the relationship between pen change and output change, called proportional band, is adjustable to suit the process. The effects of proportional band adjustment are illustrated in Figure 16.6.

The basic device used to modify the signal pressure is the flapper nozzle unit. A regulated air supply is passed through a reducing valve of fine capillary bore

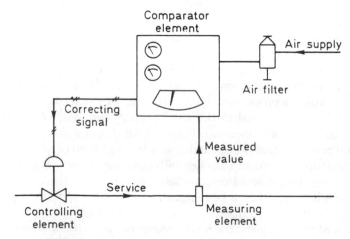

Figure 16.4 *Simple pressure control loop*

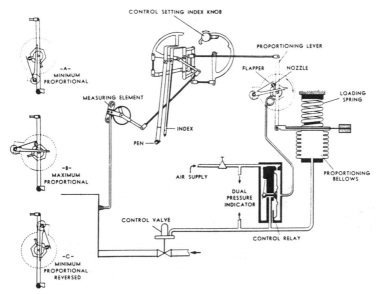

Figure 16.5   *Model 40 controller – proportional (Foxboro Yoxall)*

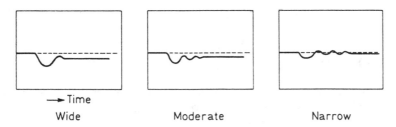

→ Time

Wide   Moderate   Narrow

Figure 16.6   *Effects of successive increments of proportional bandwidth*

and thence to a nozzle of much greater diameter. If the nozzle is unobstructed the pressure in the intervening tubing will be low. If the flapper is re-positioned closer to the nozzle, the pressure will rise in the section between the restriction and nozzle. By connecting the input of an amplifying relay (the control relay of Figure 16.5) between the restriction and the nozzle, a flapper movement in the region of 0.015 mm can be made to generate a signal change of 0.83 bar. The three graphs in Figure 16.7 show the behaviour of the process variable after a sudden load change. A wide proportional band brings the variable back to a stable value quickly but at a new value substantially below the original set point. This effect is known as offset. The moderate and narrow proportional band widths show successively longer stabilization times and smaller offsets.

The proportional band adjustment is therefore very important, as it enables the control system to be matched to the characteristics of the process under control. It is obvious from Figure 16.6 that the essence of band width control is finding the best compromise to prevent excessive cycling after a load change, which is an undesirable effect.

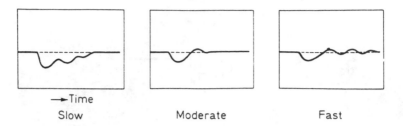

→ Time

Slow   Moderate   Fast

Figure 16.7   *Effects of successive increments of reset rate*

Two-term controller

In services such as piston and cylinder cooling water, there will be large and frequent thermal load changes during manoeuvring. In such cases the offset produced by single-term control may be unacceptable, in which case a two-term controller may be used.

The two-term controller has both proportional and integral actions. Integral action consists of adding to the proportional correcting signal a further increment which will counteract the offset which would otherwise result. The combined effect is to reset the control valve at a rate proportional to the deviation of the measured value from the setpoint value. Thus, for a large load change, the speed of reset is rapid initially, slowing down as the deviation reduces.

The ratio between process deviation and speed of reset is adjustable and the effects of different reset rates are shown in Figure 16.7, the proportional band remaining at a fixed value. The slow reset rate restores the variable to its initial value after a substantial drop following a sudden load change. The moderate rate brings the condition to normal more quickly and with less deviation. With fast reset, the initial deviation is even less, but this further increase in reset rate causes prolonged hunting of the variable about the set point. In this illustration the moderate reset rate is therefore the optimum setting.

Reset action creates serious instability during system start up, when the value of the controlled variable is well below the set point. Integral action is therefore only brought into effect when the system has reached normal operating conditions.

Three-term controllers

Three-term controllers employing derivative action may be encountered occasionally. In three-term controllers an excess correction is made which is proportional to the rate of change of the controlled variable. This has the effect of reducing the deviation and stabilization time, and may be met in more advanced combustion control systems.

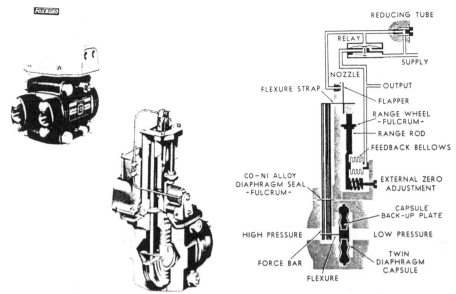

Figure 16.8 *Model 13A d/p cell differential pressure transmitter (Foxboro Yoxall)*

The transmitter

The most common method of generating the pneumatic signal feeding the controller is to use a force–balance device. An example of a force–balance device used to transmit a pneumatic signal proportional to a process pressure change can be found in Foxboro's Model 13A d/p differential pressure transmitter shown in Figure 16.8.

The basic components – the body, force–balance transmitter system and air connection block – are in separate units. Process piping stresses are thus confined to the body and air piping stresses to the air connection block, so that the force–balance transmitter system is completely isolated from exterior stresses.

The replaceable twin-diaphragm capsule is clamped between separable sections of the body and separates high- and low-pressure chambers in the body. The interior of the capsule is filled with a silicone fluid having a viscosity of 500 centistokes which in combination with restrictions in the capsule backup plate provides a frequency cut-off point at approximately 1 Hz.

A Co–Ni alloy diaphragm, located at the top of the high-pressure chamber, seals off the measuring cavity and also acts as a fulcrum for the force bar. A 'C' flexure connects the twin diaphragms of this capsule and the force bar. The force bar, because of its fixed fulcrum has a specific mechanical advantage (d_1/D_1) and is connected to the range rod by the flexure strap. A feedback bellows and zero adjustment spring are located at the other end of the range rod. The range wheel, which acts as a fulcrum, will give a different mechanical advantage (d_2/D_2) to the range rod for each different position on the rod.

Pressures are applied to the high- and low-pressure sides of the twin-diaphragm capsule. The difference between these pressures creates tension in the 'C' flexure, and a pulling force (F_1) is exerted on the lower end of the force bar. The effective area of the diaphragm times pressure equals force. The resulting moment $(F_x D_1)$ at the other end of the force bar causes the flapper to tend to move toward the nozzle. The output of the relay is increased, and the force (F_2) of the feedback bellows on the range rod creates a moment that will balance the force (F_1) of the differential pressure across the capsule and its resultant moment on the force bar. The balance of forces results in an output pressure change which is in some proportion to the change in input differential force.

The flapper-nozzle assembly is located at the highest point in the transmitter mechanism. The position of the flapper is controlled by the minute motion of the force bar/range rod combination.

For each value of measurement, the capsule will assume a definite position as will the force bar and flapper and thus produce the same output when the flapper-nozzle is in the throttling band. Total movement at the capsule for full output is less than 0.25 mm.

Figure 16.9 shows the 13A d/p cell transmitter as a double force-bar mechanism – two force bars connected by a flexure strap. The value of force F_x is important in actual calculation but is better represented as an unknown F_1, the maximum force of the differential pressure which may have a value between 0.5 to 21.6 m of water is variable; d_1/D_1, the mechanical advantage of the force bar, is fixed. F_x, the unknown force, balances F_x. F_2, being the maximum feedback force at 1 bar, is constant. d_2/D_2, the mechanical advantage of the range rod, is variable. In order to increase the range of the transmitter, the range wheel is moved up, increasing d_2, decreasing D_2 – thus increasing the

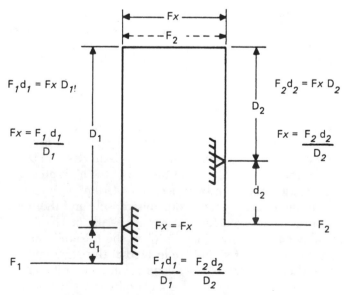

Figure 16.9   *d/p cell force-balance moments (Foxboro Yoxall)*

ratio d_2/D_2. Now F_1 must be larger to produce the same balance, requiring an increased range of differential force. The range wheel has a 10 to 1 (maximum to minimum span) adjustment capability.

The externally adjustable zero spring creates a force against the range rod. Its purpose is to preload the range rod to provide transmitter output of 0.2 bar with no differential across the twin-diaphragm capsule.

In applying this transmitter to some measurements, such as liquid level, density, or reversing flow it becomes desirable to preload force F_x in a direction which will provide span elevation or suppression. The adjustable spring exerts a force through the force bar to the diaphragm capsule, setting minimum output 0.2 bar at the minimum range value desired.

Temperature transmitter

Figure 16.10 shows a pneumatic force–balance device used to transmit a 0.2–1 bar pneumatic signal proportional to measured temperature. A gas filled thermal element, which consists of bulb, capillary and bellows capsule, provides an output signal which is linear with temperature. In the Foxboro Model 12A instrument shown, accuracy is claimed to be $\pm 1/2\%$ of span for any span within ranges from $-73°C$ to $+540°C$. Cryogenic systems down to $-270°C$ are available.

Pneumatic transmitters are also available for various functions such as rate of flow measurement and tank level gauging.

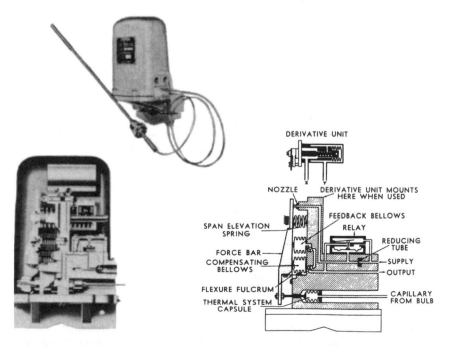

Figure 16.10 *Model 12A temperature transmitter (Foxboro Yoxall)*

Electronic transmitters

Although pneumatic instruments are the most common signal transmitters used in ships there is a growing number of electronic transmitters found at sea. These are of varying types but it is interesting to note that Foxboro has adopted the force balance systems used in its pneumatic transmitters, for Foxboro electronic systems. Figure 16.11 is a schematic drawing of a Foxboro E13 series electronic differential pressure transmitter suitable for flow, liquid level and low pressure measurement applications.

This instrument measures differential pressure from 0–127 mm water to 0–21.5 m water and transmits a proportional 10–50 mA d.c. signal. The similarities in this instrument and that shown in Figure 16.8 are obvious. Instead of the flexure strap and flapper nozzle found in the pneumatic transmitter, however, the force bar pivots about the Co–Ni alloy seal (4) and transfers a force to the vector mechanism (5).

The force transmitted by the vector mechanism to the lever system (11) is dependent on the adjustable angle. Changing this angle adjusts the scan of the instrument. At point (6) the lever system pivots and moves a ferrite disk, part of a differential transformer (7) which serves as a detector. Any position change of the ferrite disk changes the output of the differential transformer determining

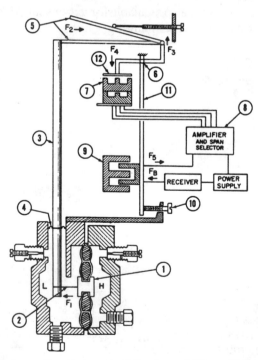

Figure 16.11   *Schematic diagram of Foxboro E13 series electronic differential pressure transmitters (Foxboro Yoxall). For key to numbers, see text*

the amplitude output of an oscillator (8). The oscillator output is rectified to a d.c. signal and amplified, resulting in a 10–50 mA d.c. transmitter output signal. A feedback motor (9) in series with the output signal, exerts a force proportional to the error signal generated by the differential transformer. This force rebalances the lever system. Accordingly, the output signal of the transmitter is directly proportional to the applied differential pressure at the capsule.

For any given applied differential pressure, within the calibrated measurement range, the ferrite disk of the detector is continuously throttling, maintaining an output signal from the amplifier proportional to the measurement and retaining the force balance system in equilibrium.

Control valves

The reputation of the diaphragm operated pneumatic control valve as a final element in marine control systems has been established over many years. The control valve consists of four basic parts: the actuator, packing box, valve body and valve trim.

The diaphragm actuator receives pneumatic signals from the controller, which usually has an output in the range 0.20–1.03 bar. Control valves for such services are designed so that a change of signal from 0.20–1.03 bar will cause the valve to travel from the fully open to fully closed position. Actuators may be direct acting, in which an increasing air signal causes the valve stem to move downwards, or reverse acting. They are usually arranged so that failure of the control signals causes the valve to open, and handwheels are provided for manual operation in this event. Sometimes direct-action control valves are used for pressure control applications, in which the process fluid itself is applied to the diaphragm.

Diaphragm motion is transmitted to the valve plug by the stem which passes through the packing box. In most standard valves the packing consists of several V-rings of p.t.f.e. (polytetrafluorethylene), loaded by a spring which provides sealing pressure at low service pressures, and arranged so that as the service pressure rises further sealing pressure is applied see Figure 16.13.

Valve types

The flow characteristic of a valve is determined by the contours of its trim, i.e. its plug and its cage (or its seat). The characteristic may be defined as the relationship between valve lift and the resulting low. Designs for quick opening, equal percentage, linear and parabolic characteristics are available. The trim design also governs the range ability of a valve, which is a measure of maximum and minimum controlled flow.

Single-seated valves, guided by the valve plug stem or by vanes on the underside of the valve plug will be found. These offer a tight shut-off but have the disadvantage of having to work against the unbalanced force created by

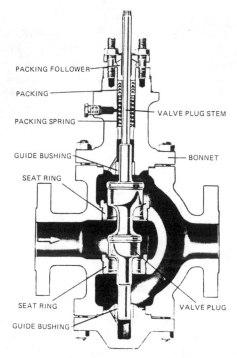

PACKING FOLLOWER

PACKING

PACKING SPRING

VALVE PLUG STEM

GUIDE BUSHING

BONNET

SEAT RING

SEAT RING

VALVE PLUG

GUIDE BUSHING

Figure 16.12   *Double-ported valve with V-pup valve plug giving equal percentage flow characteristics (Fisher Control Valves Ltd)*

the pressure of the process fluid passing between the valve plug and seat. To overcome this unbalanced force some valves will be found in which there are two plugs on the same stem (Figure 16.12). Known as double-ported valves these are arranged so that the two forces act in opposition.

The valve has the disadvantage however that tight shut-off cannot be obtained unless a soft seat is employed (because of difficulties in getting the valve plugs to touch both seats simultaneously).

It is more usual to find control valve trims of the types shown in Figure 16.13 in which a ported cage is utilized as a valve guide. These can be arranged as shown in Figure 16.13d to give tight shut-off and, in the case of the type shown in Figure 16.13c without the problems of an unbalanced force acting on the plug. The valve characteristic is governed by the contour of the ports in the valve cage.

Valve selection and sizing

Control valves are selected to suit each particular application, and the process characteristics to be taken into account are pressure, temperature, allowable pressure drop, stability, stream velocity, nature of the process fluid, etc. The effects of these conditions on the individual components must be carefully assessed to ensure that the valve will give many years of trouble free service.

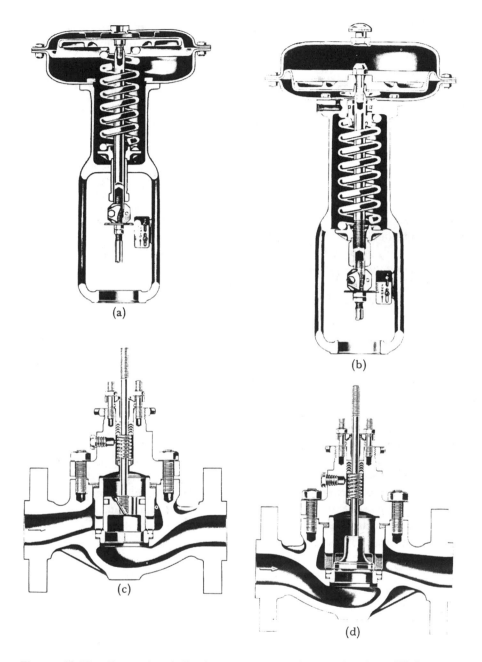

Figure 16.13   *Example of diaphragm operated control valves (Fisher Control Valves Ltd.)*

(a) direct acting;   (c) balanced trim. Design ED
(b) reverse acting;   (d) Design EC with full-capacity standard valve plug for viscous fluids

A control valve must be correctly sized in order to control the desired quantity of fluid correctly. If the size is chosen simply to suit pipeline sizes it is most likely that poor control will result due to the valve capacity being too large. The most sophisticated controller will not be able to correct the condition, and in addition the seating surfaces of the valve are likely to wear rapidly. Control valve manufacturers will accept responsibility for the specification and sizing of control valves, providing they are given complete information regarding the service conditions to be met.

Alarm systems and data loggers

The first step towards centralized control of marine machinery was to extend the conventional control and instrumentation facilities to a central control console, usually housed in a special air conditioned control room. Some of the consoles were very large having a great deal of information on individual large gauges. For later installations and with the widespread use of micro-processors it became the practice to integrate the three basic instrumentation functions — alarm monitoring, display of data and recording — within one electronic system. These systems have been fitted in a large number of ships, including refrigerated cargo vessels where the data logging facility is ideal for recording cargo space temperature. Alarm scanning is the most important function performed by this type of equipment.

Scanning speeds vary between one and 400 channels per second for analogue parameters, and the accuracy of alarm comparison is generally within 1% of the measurement range. An extremely complex machinery arrangement can be checked for malfunction twice every second, and alarm thresholds can be set very close to normal operating conditions, so giving practically instantaneous response to potentially dangerous situations. The development of high-reliability alarm-scanning systems is an important accompaniment to the increasing use of multi-engined propulsion systems, higher engine powers and operation with unattended engine rooms.

Several types of equipment are employed in ships, and whilst the details of operation vary, the basic arrangements are similar. Such equipment may be regarded as comprising four sections: primary measurement, signal selection, signal processing and control of output units. Figure 16.14 shows a typical arrangement.

Primary measurements

The plant conditions to be recorded or checked are measured by transducers of various kinds. The term transducer applies to a range of devices which produce electrical signals proportional to various measurements such as temperature, pressure, etc. Most transducers develop a change in resistance corresponding to the primary measurement, which can be evaluated by the signal processing

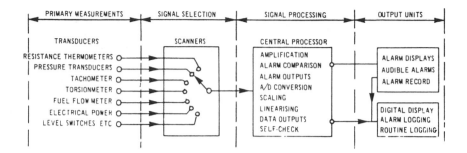

Figure 16.14 *Four basic sections of alarm scanning/data logging system.*

system. A typical example is the resistance thermometer. Other transducers such as tachometers and some ship speed logs, generate a voltage.

An analogue transducer gives an output which varies with the magnitude of the measured condition. The term transducer is also more loosely applied to switching devices which only give a change of output at one or more fixed points.

Signal selection

The basic principle of scanning systems is that a number of measurements are evaluated sequentially by one high-quality signal processing system. The transducer signals are selected singly for evaluation by means of relays or solid-state switching networks. The relays or transistor switches are operated by signals from a scan control unit, which is generally regulated by an electronic clock. Scanning speeds vary according to the type of signal selection system used, and on the speed of response of the signal processing equipment.

Relay scanners are usually limited to about 10 channels per second because of limited relay life and the delay required for the signal to stabilize after switching. Solid state scanners do not suffer from these limitations to any great extent and analogue scanning speeds of 400 channels per second are thus made possible.

A high scan rate gives the system a very short response time to alarm conditions, which is very important with modern high-rated machinery. It also enables the logging system to tabulate a cascade of faults in proper chronological order, enabling the operator to identify the source of trouble when presented with a complex fault situation.

Signal processing

The first stage in signal processing is amplification of the low voltage transducer signal, which is typically in the range 0–100 mV and signal amplifiers will raise this level to, say, 0–5 V. The most important part of signal evaluation is comparison of the signal level with upper and/or lower alarm limits. It therefore follows that this operation should be carried out as early as possible in the signal processing chain, so that failures in other parts of the signal processor do not affect the alarm monitoring function. Alarm comparison is therefore ideally performed whilst the signal is in analogue form, by comparing the transducer voltage level with alarm limits set on potentiometers. It should be noted that in some systems this most important function is carried out in the digital part of the system. In this case the alarm limits are set by means of pins inserted in matrix boards.

When an alarm limit is exceeded, sub-routines are set in action which control the output units such as alarm display windows, audible alarms and alarm printers. The digital section of the system converts the analogue data into digital form for presentation on the multi-point indicator or printer.

Each transducer signal is referred to an analogue-to-digital converter (ADC), which reproduces the voltage signal in numerical form. Digital systems use the binary system of numbers by means of which a quantity can be expressed on 1's and 0's, and quantities are transmitted and operated upon while in this form. Binary to decimal conversion occurs at the output units.

The digitized transducer signal is referred to a scaling unit which multiplies the signal by an appropriate constant so that its numerical value corresponds to engineering units such as °C. The scale unit will contain multipliers for each of the transducer measurement ranges in the system's repertoire. For those transducers which produce non-linear signals, the scale unit must apply a multiplier which varies with the magnitude of the transducer signal. Some transducers have offset characteristics, where their reading at, say, 0°C is some value other than zero volts. The scale unit or computer may therefore be called upon to perform addition, multiplication, linearization and combinations of all three, changing its routine as each transducer signal is processed.

Control of output units

When alarm conditions occur, they are announced by a klaxon and identified by flashing windows bearing the identities of the alarm channels. When an alarm acknowledge button is pressed, the klaxon is silenced and the window remains illuminated until the fault is cleared.

The digital display may be used to indicate the value of any channel, by setting three decade switches to the corresponding channel numbers. Alarm limit settings for each channel may be shown on the digital display. Where a typewriter is not employed, a handwritten log may be taken by simply switching in the channels required and noting the data presented. Typewriters can be used to print routine logs of all or part of the data processed by the system. Each log entry is time-identified, and means are provide to adjust the system time to match ship's time.

The full value of a printer can be realized by incorporating alarm history printing. All alarm conditions, together with value and time, are recorded in red on occurrence and again in black on clearance.

Parallel entry instrumentation systems

One disadvantage of alarm scanning systems is that a failure in the central sections of the equipment can cause loss of facilities on all channels. For this reason they must be made to a very high standard of reliability which makes them relatively expensive. Also it is advisable for the ship to carry a fully comprehensive spares kit if it is engaged in deep sea trading.

Generally, the more channels that are monitored by such equipment, the more cost-effective they become, as the cost per channel is reduced. For smaller ships, where the data logging facility is not important, parallel entry instrumentation systems may be employed. The essential difference is that alarm comparators and alarm lamp drive circuits are provided for each channel, instead of being shared by all inputs. The transducer signals may then be manually selected by means of rotary switches, for digitization, scaling and display at a central facility. Alarm printing may also be incorporated in parallel entry systems.

Special instruments

In addition to the conventional measurements, small sub-systems are employed for special functions such as fuel viscosity control, turbocharger vibration alarms, bearing wear-down alarms, flame failure devices for boilers, fire alarm systems, and others. Many of these devices are dealt with in relation to the machinery they serve, in this book and elsewhere.

Bridge control for diesel engines

Bridge control for diesel-driven vessels available through the installation of controllable-pitch propellers (CPPs) or reverse/reduction gear boxes, does not require a facility for engine starting from the bridge. It is normal with various ships, ferries, tugs and other workboats which manoeuvre with CPPs or reversing gearboxes, for the engines to be started locally in the machinery space with the use of a manually operated air start system. For starting with direct coupled controllable-pitch propellers, the hydraulic system must be operating and the control lever must be set at zero pitch. With reversing gearboxes, the control must be at the neutral position so that there is no drive to the propeller. For engines driving CPPs through clutches and gearboxes, clutches must be disengaged.

When the engine running is satisfactorily, control for the CPP or reversing gearbox is from the bridge. Manoeuvring may be carried out for a CPP, through a simple pitch control with the engine running at constant speed or through a combinator which adjusts pitch and engine speed. With a reversing gearbox, the ahead or astern clutch is engaged at idling revolutions and speed is then adjusted as required.

Remote control of a direct reversing diesel engine which is directly coupled to a fixed pitch propeller, requires that the functions of starting and reversing must be able to be carried out from the bridge. The controls must accommodate critical running speeds by avoiding other than brief operation in any barred speed range to minimize excessive vibration. General temperatures and pressures are maintained correct by the automatic controls for ancillary engine systems, for all types of diesel installations.

Conventional local control systems for direct reversing engines were easily adapted to basic remote operation from the bridge, because hydraulic or pneumatic servo-systems were already installed to assist manual operation. The setting of ahead or astern running direction through a servo-motor system, can be achieved with a lever or an actuator. The air start system is normally activated by a lever operated pilot valve but an actuator can be used to produce the same result. Engine speed can be set by using a lever, a remotely operated positioner or a governor with a split-field motor adjustment.

It is essential, with bridge control, that all operations demanded by the engine should take place automatically without intervention by bridge personnel who should manipulate only a simple handle. Indication is required before use of bridge control, that the turning gear is disengaged, starting air is available at the correct pressure, cooling water, lubricating oil, fuel oil and other services are satisfactory. Speed and direction indicators demonstrate that the movement order has been obeyed.

The following sequence is initiated when starting:

Camshaft correctly positioned.
Starting air admitted.
Fuel admitted.
Starting air shut off.
Speed adjusted to the value required.

Failure of an engine to start, results in the cycle being repeated, still under automatic control. Usually after three false starts the operation ceases and an alarm functions.

Boiler control systems

Automatic control of boilers and turbines is a much more complex problem than that of diesels and does not lend itself to any precise directions. Nevertheless it offers the greatest scope for efficiency and economy in both manpower and fuel consumption.

Dealing first with steam raising, the efficiency of modern complicated high-efficiency steam and feed-water systems depends on the correct relationship and operation of a large number of independent controls. They are all inter-related and each variation of main engine load, sea temperature, etc. requires a different combination of values; optimum efficiency is rarely achieved without some form of automatic control. Automated boiler control is therefore highly cost-effective.

The two basic considerations are:

(a) Maintenance of steam pressure by control of fuel oil.
(b) Optimum combustion control by regulation of combustion air flow and maintenance of correct ratio of fuel and air at all combustion rates.

Fuel consumption is influenced by flue-gas temperatures and excessive oxygen, fan load, low superheat, low boiler pressure, temperature of condensate, feed-water temperature, circulating-water-pump load and the temperature of lubricating oil to the turbines.

Methods and systems adopted for automatic control vary considerably, and leading firms in this field have developed and installed them with good results. An analysis of their methods would entail an extensive outline of systems and components and is beyond the scope of this chapter.

The principal items requiring control in an automatic system may be grouped as follows:

Boiler system	*Turbine and reduction gear*
Steam pressure	Speed
Steam temperature (superheat)	Bleeder valve control
Water level	Lubricating-oil temperature
Feed pump	Overspeed
Feed-water temperature	Condensate system
Fuel oil	Temperature of astern turb
Forced draught fan	
Air heater	
Smoke density	

Note that many subsidiary services not included in this list need to be included in a complete system. The subject is dealt with in greater detail in *Marine Steam Boilers* (Milton and Leach), another publication in the Butterworth Marine Series.

Semi-automatic systems

Semi-automatic systems operating over a narrow range have been in use for many years. They are capable of maintaining automatic control during manoeuvring, because the turn-down range of burners has been extended by a wide-range of steam-assisted pressure jet burners which effectively atomize the fuel even at low pressures. The insertion and withdrawal of burners whilst manoeuvring is now unnecessary.

Controls for generators

In unattended machinery installations it is necessary to provide certain control facilities for the electrical generating plant. These may vary from simple load sharing and automatic starting of the emergency generator, to a fully comprehensive system in which generators are started and stopped in accordance with variations in load demand.

Medium-speed propulsion plants normally use all-diesel generating plant. Turbine ships obviously use some of the high quality steam generated in the main boilers in condensing or back-pressure turbo-generators, with a diesel generator for harbour use. The usual arrangement on large-bore diesel propulsion systems is a turbo-generator employing steam generated in a waste-heat boiler, plus a diesel generator for use while manoeuvring, for port duty, and for periods of high electrical demand.

Diesel generators

The extent of automation can range from simple fault protection with automatic shut-down for lubricating oil failure, to fully automatic operation. For the latter, the functions to be carried out are:

Preparations for engine starting.
Starting and stopping engines according to load demand.
Synchronization of incoming sets with supply.
Circuit breaker closure.
Load sharing between alternators.
Maintenance of supply frequency and voltage.
Engine/alternator fault protection.
Preferential tripping of non-essential loads.

When diesel generators are arranged for automatic operation, it is good policy to arrange for off-duty sets to be circulated with main engine cooling water so that they are in a state of readiness when required. Pre-starting preparations are then simply limited to lubricating oil priming.

It is necessary to provide fault protection for lubricating oil and cooling services, and in a fully automatic system these fault signals can be employed to start a stand-by machine, place it on load and stop the defective set.

Turbo-generators

The starting and shut-down sequences for a turbo-generator are more complex than those needed for a diesel-driven set, and full automatic control is therefore less frequently encountered. However, control facilities are often centralized in the control room, together with sequence indicator lights to enable the operator to verify each step before proceeding to the next. Interlocks may also be employed to guard against error.

The start up sequence given below is necessarily general, but it illustrates the principle and may be applied to remote manual or automatic control:

Reset governor trip lever.
Reset emergency stop valve.
Start auxiliary L.O. pump.
Start circulating pump.
Apply gland steam.
Start extraction pump.
Start air ejectors.
Open steam valve to run-up turbine.

Computers

A number of ships have been fitted with computers programmed to carry out a great variety of tasks embracing navigation, ships' housekeeping, crew wages, machinery survey and overhaul, spares and stock control, weather routing, cargo discharge and loading. General purpose industrial computers have also been employed for the task of machinery alarm scanning and data logging.

Further reading

BS. 1523. Glossary of terms used in automatic control regulating systems; Section 2 – Process control; Section 3 – Kinetic control, British Standards Institution.
Automation in Merchant Ships, Fishing News Books.
Gray, D. *Centralised and Automatic Control in Ships*, Pergamon Press.
Guide for Shipboard Centralised Control and Automation, American Bureau of Shipping.
Guidance Note NI 134. CN3 Automated Ships, Bureau Veritas.
Jones, E. B. (1971) *Instrument Technology; Vol. 3 Telemetering and Automatic Control*, Butterworths.
Roy, G. J. (1978) *Notes on Instrumentation and Control*, 2nd edn, Butterworth-Heinemann.
'Rules and Regulations for the Construction and Classification of Steel Ships', reprint of Chapter 1, Lloyds Register of Shipping.
'Rules for the Classification and Construction of Steel Ships', reprint of Vol IV, Chapter 9, Germanischer Lloyd.
'Rules for the Construction and Classification of Steel Ships', reprint of Chapters VII and XV, Det norske Veritas.
Wilkinson, P. T. C., Fundamentals of Marine Control Engineering, Whitehall Press.

Index